Elementary Algebra

Elementary Algebra

RICHARD E. JOHNSON
University of New Hampshire

MARGARET ULLMANN MILLER
Skyline College

STEVEN D. KERR
Weber State College

The Benjamin/Cummings Publishing Company, Inc.
Menlo Park, California • Reading, Massachusetts
London • Amsterdam • Don Mills, Ontario • Sydney

Sponsoring Editor: Susan A. Newman
Production Editor: Greg Hubit
Coordinator: Margaret Moore
Cover Designer: Henry Breuer
Cover Photographer: George Fry
Book Designer: Janet Bollow
Artist: Michael Fornalski

Copyright © 1981 by The Benjamin/Cummings Publishing Company, Inc.
Philippines copyright 1981 by The Benjamin/Cummings Publishing Company, Inc.

All rights reserved. No part of this publication may be reproduced, stored in a retrieval system, or transmitted, in any form or by any means, electronic, mechanical, photocopying, recording, or otherwise, without the prior written permission of the publisher. Printed in the United States of America. Published simultaneously in Canada.

Library of Congress Cataloging in Publication Data

Johnson, Richard E   1913–
 Elementary algebra.

 Includes index.
 1. Algebra. I. Miller, Margaret Ullmann, joint
author. II. Kerr, Steven D., joint author.
III. Title.
QA152.2.J62   512.9   80-26297
ISBN 0-8053-5052-7

ABCDEFGHIJ-DO-8321

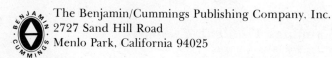

The Benjamin/Cummings Publishing Company, Inc.
2727 Sand Hill Road
Menlo Park, California 94025

Preface

OUR GOAL

Our goal as authors of this text is to provide a clear, understandable, and solid foundation in those fundamentals of algebra which are necessary to pursue other mathematics courses and to solve problems in other disciplines. Since algebra is a generalization of arithmetic, we begin with the operations and properties of natural and positive rational numbers and then introduce the arithmetic of integers. These are followed by a comprehensive coverage of the usual topics of an "elementary" or "introductory" algebra course presented in what we have found to be a logical and useful order. We begin by thoroughly covering linear equations in one and two variables. We then cover polynomial expressions, factoring, and other more challenging topics, in the belief that students are more likely to experience success with these latter topics after mastering the simpler algebraic concepts and skills.

THE FEATURES

The features of our text can be described under two categories: *style of presentation* and *exercises, word problems, and applications*. We have written and designed our text with students and instructors in mind and we invite you to familiarize yourself with these features, since they will assist you in using our book to its best advantage.

Style of Presentation

Writing style We have made every effort to present the mathematics in a straightforward, clear, and concise manner.

Worked examples in the text are numerous and include step-by-step explanations.

Running commentary enhances the text by explaining each step. The commentary is in italic type and is adjacent to the appropriate lines of the worked examples. A ■ symbol is used between a step and its running commentary whenever there might be confusion between them, such as when both are equations.

Quick reinforcement boxes conclude each section. This feature gives students an opportunity to work exercises that pertain to the section and to get immediate feedback on their understanding of the material. Answers to the quick reinforcements are printed upside down in the boxes.

Flowcharts are used as graphical aids to assist in the presentation and understanding of more complex ideas.

Checks appear after students have read through an example problem. Students are shown how to replace variables with solutions in a step-by-step procedure to verify these solutions and to encourage good problem-solving habits.

Exercises, Word Problems, and Applications

Chapter-opening word problems and related photographs begin each chapter. Each previews the content that will be presented in that chapter. Answers to these word problems are given at the back of the book.

Exercise sets conclude each section and each chapter. The chapter review exercises can serve as a chapter test. The exercise sets include a generous number of drill exercises and word problems that are paired ("a" and "b" types) to twice test similar concepts and are graded. Answers to the "a" exercises are given at the back of the book.

Applications are presented throughout the text in worked examples, specially designated sections ("Applied Problems"), and exercise sets to motivate new concepts as well as applications of these concepts.

Illuminating problems are a feature unique to our elementary algebra book. Each time a new concept is tested in an exercise set, it is designated by a light-bulb symbol, ♡. *Complete solutions* for these illuminating problems are included at the back of the book.

𝕹𝖔𝖘𝖙𝖆𝖑𝖌𝖎𝖆 𝕻𝖗𝖔𝖇𝖑𝖊𝖒𝖘 appear in many exercise sets. These are word problems selected from *New Elementary Algebra . . . for Schools & Academies* (Ivison, Blakeman, Taylor & Co.) by Horacio N. Robinson, L.L.D., published in 1875. By their inclusion we hope to entertain and to demonstrate the continuity of the study of algebra.

Calculator activities are included in most of the section and chapter exercise sets. These activities involve the manipulation of numbers that are best handled by using the calculator.

SUPPLEMENTS

The supplements for this textbook have been prepared with care and attention to detail.

The *Instructors' Manual*, prepared by Margaret Ullmann Miller, contains two sample tests for each chapter, a sample final examination, and answers to "(b)" exercises.

The *Study Guide*, prepared by Lucille Groenke (Mesa Community College, Arizona), contains additional worked examples and exercises. Particular attention is paid to those topics for which students usually need extra drill and review.

ACKNOWLEDGMENTS

We wish to acknowledge and thank the many reviewers of our book (a list appears below) whose guidance has been invaluable. We also wish to extend thanks to Lucille Groenke for her dedication and to Margaret Moore and Janet Greenblatt for their technical and editorial expertise. We all believe that without the drive and encouragement of our sponsoring editor, Susan Newman, this book would not have been published. Finally, a special thanks goes out to the family support of Dan, Jeremy, and Joshua.

Manuscript Reviewers

Raymond McNamee, Triton College
Michael Detlefsen, Slippery Rock State College
T. N. Bhargava, Kent State University
Una Bray, Marymount Manhattan College
Susan Forman, Bronx City College
Maxine Goldberg, San Francisco State University
Lucille Groenke, Mesa Community College

James Hilton, Grossmont College
Jeff Needle, Southwestern College
Eric Nelson, Chabot College
Ruth Anne Fish, Foothill College
Dwann F. Veroda, El Camino College
Robert P. Balles, Gavilan College
Robert G. Russell, West Valley College
Douglas W. Peterson, Green River Community College
Paul Eldersveld, College of Dupage
Edgar M. Chandler Jr., Phoenix College
Roy M. Honda, West Valley College
Mary E. Graham, Skyline College

R. E. J.
M. U. M.
S. D. K.

Brief Contents

1 NUMBERS — 1

2 ORDER — 43

3 EQUATIONS IN ONE VARIABLE — 77

4 INEQUALITIES IN ONE VARIABLE — 117

5 EQUATIONS IN TWO VARIABLES AND THEIR GRAPHS — 137

6 SYSTEMS OF LINEAR EQUATIONS — 187

7 INEQUALITIES IN TWO VARIABLES — 219

8 POLYNOMIALS — 245

9 DIVIDING AND FACTORING POLYNOMIALS — 277

10 RATIONAL EXPRESSIONS — 315

11 RADICALS — 361

12 QUADRATIC EQUATIONS — 395

Answers to "(a)" Exercises — 429
Index — 493

Detailed Contents

Each *section* ends with a list of key terms and an exercise set. Most of the exercise sets also include calculator activities.

1 NUMBERS 1
- **1-1** Types of Numbers 1
- **1-2** Sets, Symbols, and Numerical Expressions 4
- **1-3** Equations and Formulas 13
- **1-4** Basic Properties of Numbers 23
- **1-5** Additional Properties 27
- **1-6** Review of Rational-Number Arithmetic 31
- **1-7** Review of Geometrical Formulas 36
- Review Exercises and Calculator Activities 38

2 ORDER 43
- **2-1** Order Among Numbers 43
- **2-2** Order Properties 48
- **2-3** Addition and Subtraction of Real Numbers 50
- **2-4** Multiplication and Division of Real Numbers 56
- **2-5** Powers and Scientific Notation 62
- **2-6** Square Roots and Approximations 67
- Review Exercises and Calculator Activities 73

3 EQUATIONS IN ONE VARIABLE 77
- **3-1** Equations and Their Solutions 77
- **3-2** Methods of Solving Equations 82
- **3-3** Equations in the Real World 87
- **3-4** More Solving 96
- **3-5** More Applied Problems 102
- **3-6** More Formulas 110
- Review Exercises and Calculator Activities 114

4 INEQUALITIES IN ONE VARIABLE 115
- **4-1** Inequalities and Their Solutions 117
- **4-2** Methods of Solving Inequalities 121
- **4-3** Graphing Inequalities 127
- Review Exercises and Calculator Activities 134

5 EQUATIONS IN TWO VARIABLES AND THEIR GRAPHS — 137

- 5-1 Equations in Two Variables — 137
- 5-2 Coordinate Planes — 142
- 5-3 Linear Equations in Two Variables — 151
- 5-4 The Slope of a Line — 161
- 5-5 Equations of Lines — 169
- 5-6 Nonlinear Graphs — 176
- Review Exercises and Calculator Activities — 181

6 SYSTEMS OF LINEAR EQUATIONS — 187

- 6-1 Systems of Two Linear Equations in Two Variables — 187
- 6-2 Substitution Method of Solving a System — 193
- 6-3 Addition Method of Solving a System — 197
- 6-4 Applied Problems — 206
- Review Exercises and Nostalgia Problems — 217

7 INEQUALITIES IN TWO VARIABLES — 219

- 7-1 Linear Inequalities in Two Variables — 219
- 7-2 Graphs of Inequalities in Two Variables — 223
- 7-3 Systems of Linear Inequalities — 230
- Review Exercises and Calculator Activities — 242

8 POLYNOMIALS — 245

- 8-1 The Language of Polynomials — 245
- 8-2 Sums and Differences of Polynomials — 253
- 8-3 Products of Monomials — 258
- 8-4 Products of Polynomials — 262
- 8-5 Some Special Products — 268
- Review Exercises and Calculator Activities — 273

9 DIVIDING AND FACTORING POLYNOMIALS — 277

- 9-1 Division by a Monomial — 277
- 9-2 Division of Polynomials — 284
- 9-3 Monomial Factors — 287
- 9-4 Special Binomial Factors — 290
- 9-5 Factoring Trinomials — 297
- 9-6 Quadratic Equations — 304
- Review Exercises and Calculator Activities — 312

10 RATIONAL EXPRESSIONS — 315

- **10-1** Simplifying Rational Expressions — 315
- **10-2** Multiplying and Dividing Rational Numbers — 319
- **10-3** Adding and Subtracting Rational Expressions — 324
- **10-4** Complex Fractions — 332
- **10-5** Properties of Exponents — 336
- **10-6** Evaluating Rational Expressions — 340
- **10-7** Rational Equations and Formulas — 345
- **10-8** Applied Problems — 351
- Review Exercises and Calculator Activities — 357

11 RADICALS — 361

- **11-1** Real Numbers — 361
- **11-2** Simplifying Roots — 367
- **11-3** Operations with Square Roots — 375
- **11-4** Algebraic Expressions — 381
- **11-5** Radical Equations, Formulas, and Applications — 385
- Review Exercises and Calculator Activities — 392

12 QUADRATIC EQUATIONS — 395

- **12-1** Solving Quadratic Equations by Factoring — 395
- **12-2** Solving Quadratic Equations by Taking Square Roots — 399
- **12-3** Completing the Square — 402
- **12-4** Solving Quadratic Equations by Completing the Square — 404
- **12-5** The Quadratic Formula — 407
- **12-6** Applied Problems — 411
- **12-7** Graphs of Quadratic Equations — 417
- Review Exercises, Nostalgia Problems, and Calculator Activities — 426

Answers to "(a)" Exercises — 429
Index — 493

Elementary Algebra

1 Numbers

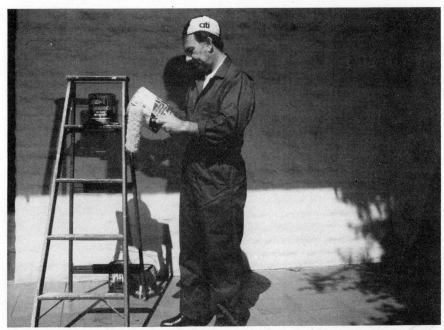

Adrian Perenon is planning to paint his house. The house has six walls, each with a length of 12 meters and a height of 3.5 meters. If a can of paint covers 65 square meters, how much area will Adrian be painting, and how many cans of paint will he need?

1-1 TYPES OF NUMBERS

Numbers allow you to describe how many apples are in a box, how far an earth satellite travels in one day, how cold it is outside, and how long the diagonal of a square is.

There are 27 apples in the box.	27 *is an integer*
The satellite travels 700,000 kilometers in one day.	700,000 *is an integer*
The temperature outside is $-8.4°$ Celsius (°C).	$-8.4 = -\frac{84}{10}$ *is a rational number*
The length of the diagonal of the square shown in Figure 1-1 is $\sqrt{2}$.	$\sqrt{2}$ (*read "the square root of 2"*) *is an irrational number*

2 CHAPTER 1: NUMBERS

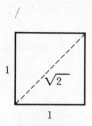

FIGURE 1-1

As these examples show, there are various kinds of numbers. The most common numbers are the **integers**:

$$\ldots, -5, -4, -3, -2, -1, 0, 1, 2, 3, 4, 5, \ldots$$

The list goes on forever in both directions, as implied by the dots (. . .)

There are negative integers such as -4 and -79, the integer zero (0), and positive integers such as 3 and 146.

Numbers other than integers are often needed to record everyday events. For example, a sprinter runs the hundred-meter dash in 10.7 seconds, a bag of potatoes costs $2.83, and a gold nugget weighs $4\frac{2}{3}$ grams. Each of the numbers

$$10.7 = \frac{107}{100}, \quad 2.83 = \frac{283}{100}, \quad 4\frac{2}{3} = \frac{14}{3}$$

is an example of a rational number. A **rational number** is any number that can be expressed as a quotient of two integers. Observe that every integer is also a rational number; for example, the integer 3 equals the rational number $\frac{3}{1}$.

While the rational numbers are sufficient for carrying out most business transactions, they are not sufficient in other applications of mathematics. For example, we might need to know the exact length of the diagonal of a given square or the exact circumference of a given circle in order to solve a problem in geometry. The length of the diagonal of the square in Figure 1-2 is exactly $\sqrt{2}$, and the circumference of the circle in Figure 1-3 is exactly π (read "pi"). The numbers $\sqrt{2}$ and π cannot be expressed as quotients of integers.

FIGURE 1-2

Circumference = π

FIGURE 1-3

1-1: TYPES OF NUMBERS 3

However, each can be *approximated* by a rational number:

$\sqrt{2} \approx 1.41$ $\approx$ *means "is approximately equal to"*

$\pi \approx \dfrac{22}{7}$

Numbers such as $\sqrt{2}$ and π, which cannot be expressed as quotients of integers, are called **irrational numbers.** (The prefix "ir" means "not.")

The numbers used in this book are called **real numbers.** Every real number is either a rational number or an irrational number. Real numbers can be combined with the operations of addition (+), subtraction (−), multiplication (×), and division (÷). For example,

$8 + 5 = 13$

$\dfrac{17}{3} - \dfrac{7}{3} = \dfrac{10}{3}$ 3 = common denomenator

$7 \times 9 = 63$ *Also written* $7 \cdot 9$ *or* $(7)(9)$

$12 \div 6 = 2$ *Also written* $12/6$ *or* $\dfrac{12}{6}$

These are the four operations of arithmetic. They are also the operations you can perform on any hand-held calculator. Because these operations combine numbers *two* at a time, they are called *binary* operations.

Quick Reinforcement

Label each of the following numbers as either an integer, a rational number, or an irrational number, using the most precise label.

(a) $\tfrac{3}{4}$ R   (b) $\sqrt{3}$ IR   (c) -4 I   (d) 2 R   (e) 0 I

Answers (a) Rational (b) Irrational (c) Integer (d) Rational (e) Integer

---

EXERCISES 1-1

Identify each number with the most precise label, choosing from the following list: integer, rational number, irrational number.

💡 1. (a) $\tfrac{7}{8}$ R   (b) 6 I
   2. (a) $\sqrt{7}$ IR   (b) $\tfrac{1}{3}$ R

4 CHAPTER 1: NUMBERS

3. (a) $\sqrt{81}$ (b) 0
4. (a) 1.01 (b) $\sqrt{2}$
5. (a) 77 (b) 0.6
6. (a) $\sqrt{5}$ (b) 14
7. (a) -3 (b) $\frac{1}{100}$
8. (a) $\sqrt{101}$ (b) 2
9. (a) -0.54 (b) $\sqrt{100}$
10. (a) $\frac{9}{10}$ (b) -103

Identify the numbers in each sentence with the most precise label, choosing from the following list: integer, rational number, irrational number.

11. (a) A class had 37 men and 29 women.
    (b) A circle had a circumference of π centimeters.
12. (a) The hypotenuse of a triangle was $\sqrt{3}$ meters.
    (b) The distance around the equator is approximately 40,000 kilometers.
13. (a) A player's batting average was 0.389.
    (b) A jogger ran $\frac{7}{8}$ of a kilometer.
14. (a) The diameter of the sun is 149,600,000 kilometers.
    (b) The temperature yesterday was $21\frac{1}{2}$°C.
15. (a) A pay raise of $2.50 an hour was negotiated.
    (b) A tuning fork vibrating at 264 cycles per second produces the note C.

1-2 SETS, SYMBOLS, AND NUMERICAL EXPRESSIONS

A collection of objects, such as a rock collection, a stamp collection, or a collection of numbers, is called a **set**. The objects in a set are called **elements**. In algebra, you are primarily interested in sets whose elements are numbers.

Braces { } are frequently used to indicate sets:

{2, 4, 6, 8, 10}   The set consisting of the first five positive even integers

{−1, −2, −3}   The set made up of the negative integers −1, −2, −3

$\left\{\frac{1}{5}, \frac{2}{5}, \frac{3}{5}, \frac{4}{5}\right\}$ The set whose elements are the four rational numbers shown

1-2: SETS, SYMBOLS, AND NUMERICAL EXPRESSIONS

Clearly, 4 is an element of the set $\{2, 4, 6, 8, 10\}$, whereas $\frac{6}{5}$ is not an element of the set $\{\frac{1}{5}, \frac{2}{5}, \frac{3}{5}, \frac{4}{5}\}$. We symbolize these two facts by

$4 \in \{2, 4, 6, 8, 10\}$ $\quad \in$ *means "is an element of"*

$\frac{6}{5} \notin \{\frac{1}{5}, \frac{2}{5}, \frac{3}{5}, \frac{4}{5}\}$ $\quad \notin$ *means "is not an element of"*

A set can have no elements. Such a set is denoted by

ϕ or $\{\ \}$

and is called the **empty set** or **null set**. At the other extreme, a set can have an infinite number of elements. For example, the set of all positive integers is an infinite set. Because it is impossible to list all its elements, it is convenient to use the notation

$\{1, 2, 3, 4, 5, \ldots\}$ *The three dots (...) indicate the list continues forever: next comes 6, then 7, and so on*

Sets are often designated by letters of the alphabet. For example, S and T designate sets as follows:

$S = \{1, 3, 5, 7, 9\}$ *S is the set whose elements are the first five positive odd integers*

$T = \{3, 6, 9, 12\}$ *T is the set whose elements are the first four positive multiples of 3*

Given sets S and T, their **union** is designated by $S \cup T$ and consists of all elements in S or T. For example, if

$S = \{1, 3, 5, 7, 9\}, \quad T = \{3, 6, 9, 12\}$

then

$S \cup T = \{1, 3, 5, 6, 7, 9, 12\}$ *1 is in S, 3 in S and T, 5 in S, 6 in T, 7 in S, 9 in S and T, 12 in T. No other number is in S or T.*

Given sets S and T, their **intersection** is designated by $S \cap T$ and consists of all elements in both S and T. If S and T are as above, then

$S \cap T = \{3, 9\}$ *3 and 9 are the only numbers in both $S = \{1, 3, 5, 7, 9\}$ and $T = \{3, 6, 9, 12\}$.*

As a further illustration of union and intersection, let F designate the football team and B the basketball team at State College. Then the set

$\quad F \cup B \quad$ The union of F and B

consists of all State College students that are on the football team or the basketball team. The set

$\quad F \cap B \quad$ The intersection of F and B

consists of all State College students that are on both teams. If no student is on both teams, then

$\quad F \cap B = \phi$

Two sets are **equal** if they contain exactly the same elements. For example, if $A = \{1, 2, 3\}$ and $B = \{3, 2, 1\}$, then $A = B$. Consider the sets $C = \{2, 4, 6\}$ and $D = \{2, 4\}$. Every element of D is an element of C. You can say that D is a **subset** of C and write

$\quad D \subseteq C \quad \subseteq$ means "is a subset of"

Note that C is *not* a subset of D,

$\quad C \not\subseteq D, \quad \not\subseteq$ means "is not a subset of"

since set C contains the element 6, which is not in D.

Problem 1 Determine which statements are true for the sets

$\quad A = \{5, 9, 7\}, B = \{2, 4, 7\}, C = \{5, 9, 8, 7, 4\}$

(a) $A = B$
(b) $B \cap C = \{4, 7\}$
(c) $A \subseteq C$
(d) $A \cup B = \{2, 9, 8, 7, 4\}$

SOLUTION:

(a) False, 5 is an element of A but not of B.
(b) True, 4 and 7 are in both B and C. No other element is in both.
(c) True, every element of A is also an element of C.
(d) False, 8 is in neither A nor B, and 5 is in $A \cup B$.

One feature of mathematics is its extensive use of symbols. You may be surprised to learn that the numerals themselves are symbols used to identify numbers. Thus, the integer six is identified by the numeral 6, the rational number two-thirds by the numeral $\frac{2}{3}$, and the square root of two, an irrational number, by $\sqrt{2}$. Other symbols designate operations with numbers and sets:

$+, -, \times, \div, \cup, \cap$ [union, intersection]

The letters of the alphabet are also symbols, and they are used in mathematics to represent objects such as sets and numbers. In the same way that a TV announcer uses Brand X to stand for any of the competing products, so does a mathematician use a **variable** x as a symbol standing for any number in some set of numbers. Of course, other letters, such as y, t, n, or T, are also used for variables. Given a variable x, the numbers it stands for are called **values** of x.

Using a variable, you can write

$$\{x \mid x \text{ is a positive integer}\}$$

of all x — such that — x is a positive integer [the set]

for the set of all positive integers. If **I** designates the set of all integers,

$$\mathbf{I} = \{\ldots, -3, -2, -1, 0, 1, 2, 3, \ldots\}$$

then

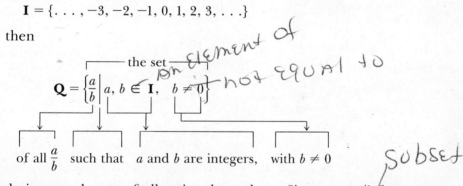

$$\mathbf{Q} = \left\{ \frac{a}{b} \,\middle|\, a, b \in \mathbf{I},\ b \neq 0 \right\}$$

of all $\frac{a}{b}$ — such that — a and b are integers, with $b \neq 0$ [the set; an element of; not equal to; subset]

designates the set of all rational numbers. Since $n = n/1$ for every integer n, every integer is a rational number, that is, $\mathbf{I} \subseteq \mathbf{Q}$. If $\mathbf{I}_r$ designates the set of all irrational numbers, and **R** designates the set of all real numbers, then

$$\mathbf{R} = \mathbf{Q} \cup \mathbf{I}_r \qquad \mathbf{Q} \cap \mathbf{I}_r = \phi$$

A collection of numerals and operation symbols such as

$3 + (4 \times 6) - 5$ *Parentheses, ( ), are called grouping symbols, other grouping symbols are brackets, [ ], and braces, { }.*

is called a **numerical expression.** Each numerical expression describes a number, its value. To find the value of the expression above,

8 CHAPTER 1: NUMBERS

carry out the indicated operations. The parentheses around 4×6 indicate that this operation is to be performed first. Thus, you evaluate $3 + (4 \times 6) - 5$ as follows.

$3 + (4 \times 6) - 5$
$3 + 24 - 5$
$27 - 5$
22 Its value is twenty-two

To evaluate a numerical expression that does not contain parentheses or other grouping symbols, it is conventional to *work from left to right and to do the multiplications and divisions first, then the additions and subtractions*. Some examples follow.

Evaluate $3 \times 5 + 4 \times 7$ Do multiplications first
$15 + 28$ Then do additions
43 Its value

Evaluate $12 - 4 \div 2 + 7 \times 3$ Do divisions and multiplications first
$12 - 2 + 21$ Then do subtractions and additions
$10 + 21$
31 Its value is 31

If more than one pair of grouping symbols is used in a numerical expression, carry out the indicated operations in the innermost groupings first. Then do the indicated operations in the next innermost groupings, and so on, until the grouping symbols are all gone. You finish the evaluation as shown in the two earlier examples.

Evaluate $[7 \times (2 + 8) - 9] + 11$ The innermost grouping is (2×8)
$[7 \times 10 - 9] + 11$ The next grouping is $[7 \times 10 - 9]$
$[70 - 9] + 11$
$61 + 11$
72 Its value is seventy-two

1-2: SETS, SYMBOLS, AND NUMERICAL EXPRESSIONS

Evaluate $[3 \times (5 \times 2 + 6) - 8] \times (4 - 2)$ $(5 \times 2 + 6)$ *is innermost grouping*

$[3 \times (10 + 6) - 8] \times (4 - 2)$ $(10 + 6)$ *is innermost grouping*

$[3 \times 16 - 8] \times (4 - 2)$ *You can work with either remaining grouping at this stage*

$[48 - 8] \times (4 - 2)$

40×2

80 *Its value is eighty*

It is common practice to write

$7 \cdot 2$ or $(7)(2)$

in place of

7×2 *Similarly, $a \cdot b$ is written as ab and $x \cdot (a + b)$ as $x(a + b)$*

and to write

$7(2 + 8)$

in place of

$7 \times (2 + 8)$ or $7 \cdot (2 + 8)$

Also,

$7 \div 2$

is often written as

$\frac{7}{2}$

A collection of numerals, variables, and other symbols that becomes a numerical expression when values are assigned to the variables is called a **mathematical expression.** Thus

$3(7 + x)$ *Variable x*

is a mathematical expression. If you give x the value 5, the resulting numerical expression is

3(7 + 5)   *Replace x by 5*

3 · 12

36   *Its value is thirty-six*

If you give x a different value, say 13, the resulting numerical expression has a different value:

3(7 + 13)   *Replace x by 13*

3 · 20

60   *Its value is sixty*

As a practical example of a mathematical expression, consider a company that produces smoke alarms. Suppose it costs the company $12.50 for each smoke alarm it produces. Then it costs $25 to produce two smoke alarms, $62.50 to produce five, and $500 to produce forty. Evidently, x smoke alarms cost $12.50x$ dollars to produce. For this mathematical expression, the variable x was given values 2, 5, and 40. The mathematical expression $12.50x$ had corresponding values 25, 62.50, and 500.

Another example of a mathematical expression is

$x[3 + 2(y - 4)]$ *Variables x and y*

If you give x the value 4 and y the value 7, the resulting numerical expression is

$4[3 + 2(7 - 4)]$ *Replace x by 4, y by 7*

$4[3 + 2(3)]$

$4[3 + 6]$

$4 \cdot 9$

36   *Its value is thirty-six*

Quick Reinforcement

Evaluate each of the following expressions.

(a) $13 - 2(6 \div 2 + 1) + 1$

(b) $(4 + x)(x + 1)$ with x given the value 6

If $A = \{1, 3, -\frac{1}{2}\}$ and $B = \{-3, 0.1, 1\}$, then determine each of the following.

(c) $A \cup B$ (d) $A \cap B$

Answers (a) 6 (b) 70 (c) $\{1, 3, -\frac{1}{2}, -3, 0.1\}$ (d) $\{1\}$

EXERCISES 1-2

Given sets $A = \{2, 4, 6, 8, 10\}$, $B = \{0, 3, 6, 9, 12\}$, and $C = \{0, 4, 8, 12, 16\}$, list the elements of each of the following.

1. (a) $A \cup B$ (b) $A \cap B$
2. (a) $B \cap C$ (b) $B \cup C$
3. (a) $A \cap (B \cup C)$ (b) $A \cap (B \cap C)$
4. (a) $(B \cap C) \cup A$ (b) $(B \cup C) \cup A$
5. (a) $(B \cup A) \cap (B \cup C)$ (b) $(A \cap C) \cup (A \cap B)$
6. (a) $A \cup \{3, 4, 5, 6\}$ (b) $B \cap \{6, 8, 10, 12\}$

Determine which statements are true for the following sets:

$M = \{4, 6, 8, 10\}$ $N = \{4, 5, 6, 7\}$ $O = \{4, 6\}$ $P = \{4, 6, 8, 10\}$

7. (a) $O \subseteq M$ (b) $M \subseteq N$
8. (a) $N = O$ (b) $N = P$
9. (a) $M \cap N = O$ (b) $O \cup P = P$
10. (a) $N \cup M = \{4, 5, 6, 7, 8, 9, 10\}$ (b) $N \cap P = \{4, 5, 6, 7, 8, 10\}$
11. (a) $M = P$ (b) $O \subseteq P$
12. (a) $N \subseteq M$ (b) $P \subseteq O$

Evaluate each of the following expressions.

13. (a) $(6 + 5) \cdot 3 + 9$ (b) $(1 + 2 \cdot 4) \cdot 2 + 3$
14. (a) $5 \cdot 3 + 2 \cdot (-7)$ (b) $(6 \cdot 9) \cdot 0 + 3$
15. (a) $(15 + 2 \cdot 1) \cdot 2 + (-1)$ (b) $5(6 + 4) + 5 \cdot 2$
16. (a) $(4 \cdot 3) \cdot 3 + 2(2 \cdot 5)$ (b) $4 \cdot 5 + 4 \cdot 6 + 4 \cdot 3$
17. (a) $2(7 + 4) + 6 \cdot 3 + 6 \cdot 4$ (b) $4 \cdot (8 + 5 + 7 + 12) + (-88)$
18. (a) $6 \div 3 \cdot 2$ (b) $6 + 4 \div 2$
19. (a) $(2 \cdot 2 + 2) \div 6$ (b) $12 \div (7 - 2 \cdot 3)$
20. (a) $3(3 \cdot 4 - 24 \div 2)$ (b) $2[(4 + 1) - 5]$
21. (a) $13 - 2[1 + (4 - 1)]$ (b) $(15 + 1) - 6[4 \div (3 - 1)]$
22. (a) $[4 - 3(2 - 1) + 3 \div (6 - 3)]$ (b) $15 - 2[13 - 4(12 - 10)]$
23. (a) $[3 + 6(4 - 2)] - [4(10 - 5) \div 2]$ (b) $6(4 - 2) - 3[12 - (4 - 1)] \div 27$

12 CHAPTER 1: NUMBERS

Evaluate each of the following expressions for the given values of x and y.

24. (a) $x(3 + 7)$ for x given the value 2
 (b) $3(x - 1)$ for x given the value 7
25. (a) $5(7 - y)$ for y given the value 5
 (b) $y(9 + 2)$ for y given the value 3
26. (a) $5(x - y)$ for x given the value 7, y given the value 5
 (b) $2(y + x)$ for x given the value 3, y given the value 2
27. (a) $x(3 + y)$ for x given the value 2, y given the value 3
 (b) $y(7 - x)$ for x given the value 4, y given the value 2
28. (a) $2[x + y(8 - 3)]$ for x given the value 2, y given the value 5
 (b) $x[7 - y(6 - 4)]$ for x given the value 3, y given the value 2
29. (a) $y[3 + 2(x + 1)]$ for x given the value 3, y given the value 4
 (b) $x[12 - 3(y - 2)]$ for x given the value 2, y given the value 6

Insert grouping symbols in each of the following to give the desired answer.

30. (a) $4 + 2 \cdot 3 = 18$ (b) $15 - 3 + 1 = 11$
31. (a) $9 \div 5 - 2 = 3$ (b) $6 \cdot 5 + 1 = 36$
32. (a) $14 - 3 \cdot 2 - 2 = 14$ (b) $14 - 3 \cdot 2 - 2 = 0$
33. (a) $15 \div 3 \cdot 2 = 10$ (b) $15 \div 3 + 2 = 3$
34. (a) $8 \cdot 4 - 2 \cdot 3 = 48$ (b) $8 \div 4 + 10 \div 5 = 4$

Develop as many different answers for the given expressions as possible. For each, show the proper grouping.

35. (a) $2 \cdot 3 + 5$ (b) $17 - 5 \cdot 2$
36. (a) $18 - 6 \div 2 + 1$ (b) $9 \cdot 5 - 2 \cdot 3 \div 13$

Calculator Activities

Evaluate each of the following expressions. Give your answer to best accuracy for each given set of numbers. (Note: Best accuracy is usually defined as an accuracy no better than the least accurate measure in the given set of numbers. For example, 2.1×4.37 would have a best accuracy of 9.2.)

1. (a) $[(843 - 78) \div 15 + 4] \div 11$ (b) $23 \times (138 + 52) \div 10$
2. (a) $(731.97 - 44.03) \cdot 1.235 - 66.9$ (b) $44.29 \times (1.06 + 44.21 + 93.7)$
3. (a) $5.093 - 0.92(2.81 + 2.06)$ (b) $12{,}305(85{,}031 - 77{,}772) \div 5$
4. (a) $[7.2 - (4.002 + 6.7 \div 2.11)] + 3.9(8 - 2.04)$
 (b) $27(\tfrac{1}{4} - 0.143) + 5.3(6 - 2\tfrac{1}{2})$

Evaluate each of the following expressions for the given values of x and y to the best accuracy, as defined above.

5. (a) $x(4.3 - 3.9)$ for x given the value 1.7
   (b) $7.31(y - 2.09)$ for y given the value 9.66
6. (a) $3.991(x + y)$ for x given the value 0.093, y given the value 1.225
   (b) $3.19[4.275 + x(1.994 - y)]$ for x given the value 5.22, y given the value 1.032
7. (a) $y[9.2 + 3.1(x + 4.7)]$ for x given the value 1.1, y given the value 8.7
   (b) $x[1095 - 63(y - 47)]$ for x given the value 83, y given the value 72

1-3 EQUATIONS AND FORMULAS

A statement consisting of two mathematical expressions connected by an equals sign (=) is called an **equation.** The two expressions are called the *sides* of the equation. For example,

$$3 + 7 = 5 \times 2$$

| left side | | right side |

An equation is a statement about the equality of two numerical expressions. It is a *true* statement if the two sides have the same value and *false* if they do not. Since $3 + 7$ and 5×2 have the same value, ten, the equation above is true.

Another example of an equation is

$$3 + (11 + 6) = (4 + 5) + 10$$

| left side | | right side |

Since the left side has the value 20:

$3 + (11 + 6)$

$3 + 17$

20

whereas the right side has the value 19:

$(4 + 5) + 10$

$9 + 10$

19

the equation above is false.

An equation involving variables, such as

$12 + x = 3x + 6$

is neither true nor false until values are given to the variable x. For example, if x is given the value 3, the equation is true:

$12 + x = 3x + 6$
$12 + 3 = 3 \cdot 3 + 6$
$12 + 3 = 9 + 6$
$15 = 15$ *True*

On the other hand, the equation is false if x is given the value 7:

$12 + x = 3x + 6$
$12 + 7 = 3 \cdot 7 + 6$
$12 + 7 = 21 + 6$
$19 = 27$ *False*

An equation involving two or more real-life variables is called a **formula.** For example,

$P = 4x$

is a formula for the perimeter P of a square (distance around the square) in terms of the length x of one side (see Figure 1-4). This formula becomes either true or false when values are given to the variables x and P. It is true when x is given the value 12 and P the value 48:

$P = 4x$
$48 = 4 \cdot 12$
$48 = 48$ *True*

It is false when x is given the value 10 and P the value 32:

$P = 4x$
$32 = 4 \cdot 10$
$32 = 40$ *False*

1-3: EQUATIONS AND FORMULAS   15

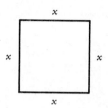

FIGURE 1-4   Perimeter = 4x

The variables P and x are in the same units, such as centimeters, inches, or miles.

Given any value of x, you can easily find a value of P such that the equation $P = 4x$ is true:

If x has the value 5 centimeters, the formula is true if P has the value $4 \cdot 5$, or 20, centimeters.

If x has the value 10 meters, the formula is true if P has the value $4 \cdot 10$, or 40, meters.

Thus, a square garden having a length of 10 meters on each side requires 40 meters of fencing to enclose it.

Quick Reinforcement

Tell whether each equation is true or false for the given values of the variables.

(a) $P = 2l + 2w$, given $P = 8$, $l = 1$, $w = 3$

(b) $5 + x = 2 - x$, given $x = 2$

Answers (a) True (b) False

Many statements about numbers can be expressed by equations.

STATEMENT	EQUATION
1. The sum of two numbers is seven.	$x + y = 7$ x and y are variables
2. The coin box in the coke machine contains nickels, dimes, and quarters, 146 coins altogether.	$w + x + y = 146$ w is the number of nickels x is the number of dimes y is the number of quarters

3. John is twice as old as his son Nick.

$a = 2b$

a is John's age in years
b is Nick's age in years

4. Marcus has ten fewer stocks in his portfolio than Pete.

$N = M - 10$

N is the number of stocks Marcus has
M is the number of stocks Pete has

You can easily assign values to the variables that will make each equation true. To make equation 1 true, you might let x have the value 2 and y the value 5:

$x + y = 7$
$2 + 5 = 7$ True

Of course, there are many other values that make the equation true. For example,

$x + y = 7$ x has value $\frac{5}{2}$, y has value $\frac{9}{2}$
$\frac{5}{2} + \frac{9}{2} = 7$ True

For equation 2, you might let w have the value 100, x the value 40, and y the value 6:

$w + x + y = 146$
$100 + 40 + 6 = 146$ True

For equation 3, you could let b have the value 22, in which case a must be given the value $2 \cdot 22$ to make the equation true:

$a = 2b$
$44 = 2 \cdot 22$ True

For equation 4, you could let M have the value 150, in which case N must be given the value 140 to make the equation true:

$N = M - 10$
$140 = 150 - 10$ True

Quick Reinforcement

Translate the following statements into mathematical expressions.
(a) Two more than half a number
(b) Five minus the sum of one and a number

Answers: (a) $\frac{1}{2}x + 2$ (b) $5 - (1 + x)$

Here are some well-known formulas with examples of how they are used.

$$\boxed{A = lw}\quad \text{A is the area of a rectangle, l is length, w is width}$$

For example, if a rectangular garden is 15 meters long and 8 meters wide, then its area is

$$A = 15 \cdot 8 = 120 \text{ square meters}$$

$$\boxed{C = 2\pi r}\quad \text{C is the circumference of a circle with radius r}$$

$$\boxed{A = \pi r^2}\quad \text{A is the area of a circle with radius r ($r^2 = r \cdot r$)}$$

For example, the bottom of a tepee is a circle with a radius of 3 meters. Therefore, the circumference of the circle is

$$C = 2\pi \cdot 3 \approx 19 \text{ meters} \qquad \text{Using } \pi \approx \tfrac{22}{7}$$

The area of the circle is

$$A = \pi 3^2 = 9\pi \approx 28 \text{ square meters}$$

$$\boxed{C = \frac{5}{9}(F - 32)} \qquad \text{F stands for Fahrenheit, C for Celsius (temperature)}$$

$$\boxed{F = \frac{9}{5}C + 32}$$

For example, if the Fahrenheit temperature is 50°, then the Celsius temperature is

$$C = \frac{5}{9}(50 - 32) = \frac{5}{9}(18) = 10°C$$

If the Celsius temperature is 25°, then the Fahrenheit temperature is

$$F = \frac{9}{5}(25) + 32 = 45 + 32 = 77°F$$

$$\boxed{S = 4\pi r^2} \qquad \text{S is the surface area of a sphere of radius r}$$

$$\boxed{V = \frac{4}{3}\pi r^3} \qquad \text{V is the volume of a sphere with radius r ($r^3 = r \cdot r \cdot r$)}$$

For example, if a balloon has a radius of 10 meters, then its surface area is

$$S = 4\pi \cdot 10^2 = 400\pi \approx 1257 \text{ square meters}$$

The volume of gas in the balloon is

$$V = \frac{4}{3}\pi \cdot 10^3 = \frac{4000\pi}{3} \approx 4189 \text{ cubic meters}$$

$$\boxed{A = \frac{h}{2}(a + b)} \qquad \begin{array}{l}\text{A is the area of the trapezoid in Figure 1-5, where} \\ \text{h is the height and a and b the two parallel sides (bases)}\end{array}$$

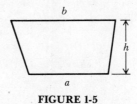

FIGURE 1-5

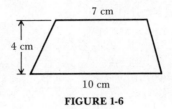

FIGURE 1-6

For example, a trapezoid with a height of 4 centimeters and parallel sides of 10 and 7 centimeters, as shown in Figure 1-6, has an area

$$A = \frac{4}{2}(10 + 7) = 2 \cdot 17 = 34 \text{ square centimeters}$$

$\boxed{I = PRT}$ I is simple interest, P is principal, R is rate, T is time

For example, if you put $160 in a bank that pays 6% per year simple interest, and you leave the money in the bank for 3 years, your interest will be

$$I = 160 \times 0.06 \times 3 = \$28.80 \qquad P = 160, R = 0.06, T = 3$$

$\boxed{v = -9.7t + k}$ v is velocity of a freely falling object, k is initial velocity in meters per second, t is time in seconds

For example, a ball is thrown straight up with an initial velocity of 40 meters per second. Its velocity 3 seconds later is

$$v = -9.7 \cdot 3 + 40 \qquad\qquad t = 3, k = 40$$
$$= -29.1 + 40 = 10.9 \text{ meters per second}$$

Quick Reinforcement

The area of a triangle is $A = \frac{1}{2}bh$, where h = height and b = base.

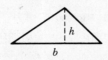

Determine the area of the triangle shown in the figure, where $h = 3$ and $b = 10$ millimeters.

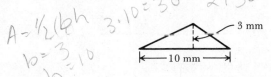

Answer: 15 square millimeters

EXERCISES 1-3

Tell whether each equation is true or false for the given values of the variables.

1. (a) $P = 2l + 2w$, given $P = 10$, $l = 2$, $w = 3$
   (b) $P = 4x$, given $P = 20$, $x = 5$
2. (a) $y + 3 = 5 - y$, given $y = 1$
   (b) $x - 5 = x + 3$, given $x = 8$
3. (a) $P = 2l + 2w$, given $P = 30$, $l = 6$, $w = 7$
   (b) $m - 3 = 2m - 6$, given $m = 3$
4. (a) $m - 3 = 2m - 6$, given $m = 4$
   (b) $P = 4x$, given $P = 8$, $x = 1$

Translate each of the following statements into mathematical expressions.

5. (a) Six less than a number
   (b) Twelve more than three times a number
6. (a) The sum of a number and twice the number
   (b) The quotient of a number and four
7. (a) The difference between a number and seven
   (b) The product of three times a number and six
8. (a) Ten minus the product of a number and 3
   (b) The sum of the product of three and a number, and four
9. (a) The quotient of the sum of a number and three, and the number
   (b) The product of the sum of two and a number, and the sum of three and the number

Evaluate each of the following formulas.

10. (a) $P = 4x$ for x given the value of 6
    (b) $P = 4x$ for x given the value of 11

11. (a) $A = l \cdot w$ for $l = 3, w = 2$
    (b) $A = l \cdot w$ for $l = 14, w = 9$
12. (a) $C = \frac{5}{9}(F - 32)$ for $F = 27°$ (b) $C = \frac{5}{9}(F - 32)$ for $F = 76°$
    (This formula converts Fahrenheit temperatures to Celsius temperatures.)
13. (a) $F = \frac{9}{5}C + 32$ for $C = 16°$ (b) $F = \frac{9}{5}C + 32$ for $C = 65°$
    (This formula converts Celsius temperatures to Fahrenheit temperatures.)
14. (a) $V = \frac{4}{3}\pi r^3$ for $\pi = 3.14, r = 3$ (b) $V = \frac{4}{3}\pi r^3$ for $\pi = 3.14, r = 15$
    (This formula finds the volume of a sphere.)
15. (a) $I = PRT$ for $P = \$6000, R = 0.05, T = 2$
    (b) $I = PRT$ for $P = \$2000, R = 0.06, T = \frac{1}{2}$
    (This formula finds the simple interest of a given amount of money at a rate R for a time T.)
16. (a) $A = \frac{1}{2}(a + b)h$ for $a = 2, b = 3, h = 8$
    (b) $A = \frac{1}{2}(a + b)h$ for $a = \frac{2}{3}, b = \frac{1}{8}, h = 24$
    (This formula finds the area of a trapezoid.)
17. (a) $v = -9.7t + k$ for $t = 0.2, k = 8$
    (b) $v = -9.7t + k$ for $t = 2.8, k = 200$
    (This formula finds the velocity of a moving object at a given time t and starting speed k.)

Use variables and mathematical symbols to abbreviate these statements.

18. (a) John has three more mystery books than biographies.
    (b) Sandra has five fewer silver dollars than her brother.
19. (a) Martha is four times as old as her son.
    (b) The library budget is one-third of the total city budget.
20. (a) The large box of paper clips has two and one-half times as many clips as the small box.
    (b) Seven times Matt's age is his uncle's age.
21. (a) The small light bulb has one-third the wattage of the large one.
    (b) Herbert put in five more hours of study for the biology final than his friend Irwin.
22. (a) The sum of Juan's pay and Lucia's pay is $121.
    (b) The number of coins in the two banks in Lee's room is 265.
23. (a) The recipe has 750 milliliters more flour than sugar.
    (b) The lecture class is half as long as the laboratory class.
24. (a) The new computer has three times the memory capacity of the old one.
    (b) The interest rate is twice what it was seven years ago.

22   CHAPTER 1: NUMBERS

25. (a) The laboratory has six more bunsen burners than it had last year.
    (b) The first petrie dish of bacteria culture has three and one-half times as much as the second petrie dish of the same culture.
26. (a) The sum of the family's two mortgage payments was $835.
    (b) The speed of the sailboat was 3 knots less than the motorboat.

Translate each mathematical expression into an equivalent verbal statement.

27. (a) $5 - \frac{1}{4}$ (b) $7 + x$
28. (a) $2 \cdot x$ (b) $8 \div 3$
29. (a) $(3 + x) - 4$ (b) $5 - 4 \cdot y$
30. (a) $\dfrac{(x - 1)}{3}$ (b) $\dfrac{5 + p}{q}$
31. (a) $2 + x = 5$ (b) $y - 3 = 7$
32. (a) $4x + 1 = 9$ (b) $\frac{1}{2}(3 + p) = 2$
33. (a) $4 - (x + 2) = 2x$ (b) $\frac{1}{3}x = x - 4$

Calculator Activities

Evaluate each of the following formulas. Give your answer to best accuracy for each given set of numbers. (Note: Best accuracy is usually defined as an accuracy no better than the least accurate measure in the given set of numbers.)

1. (a) $P = 2l + 2w$ for $l = 27.34$ cm and $w = 6.2$ cm
   (b) $P = 2l + 2w$ for $l = 929.8$ m and $w = 330.07$ m
   (This formula finds the perimeter of a rectangle.)
2. (a) $V = l \cdot w \cdot h$ for $l = 2.1$ cm, $w = 3.02$ cm, $h = 0.95$ cm
   (b) $V = l \cdot w \cdot h$ for $l = 0.902$ m, $w = 0.0951$ m, $h = 0.271$ m
   (This formula finds the volume of a rectangular solid.)
3. (a) $A = P(1 + r)$ for $P = \$32{,}050$, $r = 5\frac{1}{4}\%$
   (Don't forget to convert the rate percent into a decimal!)
   (b) $A = P(1 + r)$ for $P = \$2470$, $r = 9.85\%$
   (This formula computes the current amount of money (A) after one year with an initial investment of P dollars at an interest rate r; the compounding of the interest is done once in the year.)
4. (a) $L = \frac{3}{20}G + 1$ for $G = 63$ (a badger)
   (b) $L = \frac{3}{20}G + 1$ for $G = 112$ (a pig)
   (This formula computes a prediction of the longevity of a mammal (in years), given its gestation period G in days.)

5. For any triangle, the sum of the three angles is 180°:

   $\angle A + \angle B + \angle C = 180°$

   (a) Find $\angle C$, given $\angle A = 23.87°$, $\angle B = 92.4°$
   (b) Find $\angle A$, given $\angle C = \angle B = 3.043°$

1-4 BASIC PROPERTIES OF NUMBERS

There are rules for algebra, just as there are rules for the game of chess. Before you can begin to solve problems in algebra, you must become familiar with the rules. These rules take the form of **properties** of equality, addition, and multiplication of real numbers.

The properties of equality are given below.

Reflexive Property

For every real value of a, the equation $a = a$ is true.

For example, $4 = 4$ and $\frac{2}{3} = \frac{2}{3}$ are true.

Symmetric Property

For all real values of a and b, if $a = b$ is true, then $b = a$ is also true.

For example, $\frac{1}{2} = \frac{3}{6}$ is true; therefore, $\frac{3}{6} = \frac{1}{2}$ is also true.

Transitive Property

For all real values of a, b, and c, if $a = b$ and $b = c$ are true, then $a = c$ is also true.

For example, $\frac{4}{10} = \frac{2}{5}$ and $\frac{2}{5} = \frac{6}{15}$ are true; therefore, $\frac{4}{10} = \frac{6}{15}$ is also true.

Additive Property

For all real values of a, b, and c, if $a = b$ is true, then $a + c = b + c$ is also true.

For example, $7 = 5 + 2$ is true; therefore, $7 + 8 = (5 + 2) + 8$ is also true.

Multiplicative Property

For all real values of a, b, and c, if $a = b$ is true, then $a \cdot c = b \cdot c$ is also true.

For example, $9 = 6 + 3$ is true; therefore, $9 \cdot 5 = (6 + 3) \cdot 5$ is also true.

24 CHAPTER 1: NUMBERS

The six basic properties of addition and multiplication of real numbers are given below.

1. Closure Property

For all real numbers a and b, there exist a unique real number $a + b$ and a unique real number $a \cdot b$. (*unique* means "one and only one.")

2. Commutative Property

For all real values of a and b, the equations

$a + b = b + a$ *Addition is commutative*

$ab = ba$ *Multiplication is commutative*

are true.

For example,

$7 + 5 = 5 + 7$ *Each side is equal to* 12

$7 \cdot 5 = 5 \cdot 7$ *Each side is equal to* 35

3. Associative Property

For all real values of a, b, and c, the equations

$a + (b + c) = (a + b) + c$ *Addition is associative*

$a(bc) = (ab)c$ *Multiplication is associative*

are true.

For example,

$$7 + (9 + 14) \qquad (7 + 9) + 14$$
$$7 + 23 \qquad\qquad 16 + 14$$
$$30 \qquad\qquad\qquad 30$$

Thus, as predicted by the associative property for addition, $7 + (9 + 14) = (7 + 9) + 14$.

Similarly,

$$10 \cdot (8 \cdot 7) \qquad (10 \cdot 8) \cdot 7$$
$$10 \cdot 56 \qquad\qquad 80 \cdot 7$$
$$560 \qquad\qquad\qquad 560$$

Thus, as predicted by the associative property for multiplication,
$10 \cdot (8 \cdot 7) = (10 \cdot 8) \cdot 7$

4. Distributive Property

For all real values of a, b, and c, the equation $a(b+c) = ab + ac$ is true.

For example,

$6(17 + 3)$ $\quad$ $6 \cdot 17 + 6 \cdot 3$
$\quad\downarrow$ $\quad\quad\quad\quad\quad\downarrow$
$6 \cdot 20$ $\quad\quad$ $102 + 18$
$\quad\downarrow$ $\quad\quad\quad\quad\quad\downarrow$
$\quad 120$ $\quad\quad\quad\quad 120$

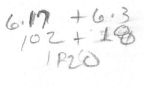

Thus, as predicted by the distributive property,
$6(17 + 3) = 6 \cdot 17 + 6 \cdot 3$

The numbers 0 and 1 have special properties as follows.

5. Identity Property

For every real value of a, the equations

$a + 0 = a \quad$ 0 is the additive identity element
$a \cdot 1 = a \quad$ 1 is the multiplicative identity element

are true.

For example,

$24 + 0 = 24$
$17 \cdot 1 = 17$

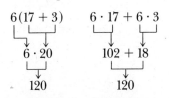

6. Inversive Property

For every real number a, there exists a unique real number $-a$ such that $a + (-a) = 0$ is true.

For every nonzero real number a, there exists a unique real number $\dfrac{1}{a}$ such that $a \cdot \dfrac{1}{a} = 1$ is true.

The number $-a$ is called the **additive inverse**, or **negative**, of a. The number $1/a$ is called the **multiplicative inverse**, or **reciprocal**, of a. For example:

The additive inverse of 21 is -21, and $21 + (-21) = 0$.

The additive inverse of -21, written $-(-21)$, is 21; that is, $-(-21) = 21$, since $(-21) + 21 = 0$.

The additive inverse of $\frac{2}{3}$ is $-\frac{2}{3}$, and $\frac{2}{3} + (-\frac{2}{3}) = 0$.

The multiplicative inverse of 8 is $\frac{1}{8}$, and $8 \cdot \frac{1}{8} = 1$.

Quick Reinforcement

Identify the properties used to make each equation true.
(a) $3 + (7 + 5) = (3 + 5) + 7$
(b) $4 \cdot (2 \cdot 3) = (4 \cdot 3) \cdot 2$

Answers (a) Commutative, associative of addition (b) Commutative, associative of multiplication

EXERCISES 1-4

Name the property that makes each of the following equations true.

1. (a) $5 + (7 + 4) = (5 + 7) + 4$ (b) $(2 \cdot 6) \cdot 3 = 2 \cdot (6 \cdot 3)$
2. (a) $9 + 0 = 9$ (b) $8 \cdot 1 = 8$
3. (a) $1 \cdot 12 = 12 \cdot 1$ (b) $15 + 0 = 0 + 15$
4. (a) $17 = 17$ (b) $9 \cdot 3 = 3 \cdot 9$
5. (a) $5 \cdot (6 \cdot 7) = (5 \cdot 6) \cdot 7$ (b) $(17 + 2) + 1 = 17 + (2 + 1)$
6. (a) $12 \cdot 6 + 8 \cdot 6 = (12 + 8) \cdot 6$ (b) $5(2 + 3) = 5 \cdot 2 + 5 \cdot 3$
7. (a) $4 \cdot \frac{1}{4} = 1$ (b) $7 + (-7) = 0$
8. (a) $6 \cdot (8 + 3) = 6 \cdot 8 + 6 \cdot 3$ (b) $2 \cdot (1 \cdot 3) = (2 \cdot 1) \cdot 3$
9. (a) $5 + (-5) = 0$ (b) $4 \cdot 8 + 4 \cdot 3 = 4(8 + 3)$
10. (a) $7 + 0 = 7$ (b) $\frac{1}{6} \cdot 6 = 1$
11. (a) Since $\frac{1}{2} = \frac{2}{4}$, then $\frac{2}{4} = \frac{1}{2}$ (b) $-3 = -3$
12. (a) Since $8 = 6 + 2$, then $8 + 3 = 6 + 2 + 3$
    (b) Since $\frac{2}{3} = \frac{4}{6}$ and $\frac{4}{6} = \frac{8}{12}$, then $\frac{2}{3} = \frac{8}{12}$
13. (a) Since $3 = 2 + 1$, then $5 \cdot 3 = 5 \cdot (2 + 1)$
    (b) Since $9 = \frac{18}{2}$, then $\frac{18}{2} = 9$
14. (a) Since $5 = \frac{10}{2}$ and $\frac{10}{2} = \frac{20}{4}$, then $5 = \frac{20}{4}$
    (b) Since $4 + 5 = 9$, then $2 + 4 + 5 = 2 + 9$

15. (a) $8 = 8$
    (b) Since $3 \cdot 4 = 12$, then $2 \cdot 3 \cdot 4 = 2 \cdot 12$
16. (a) $\frac{1}{2} \cdot 2 = 1$
    (b) $6 + (-6) = 0$
17. (a) $10 + 5 = 5 + 10$ ~Commutive Addition~
    (b) Since $7 + 3 = 10$, then $6(7 + 3) = 6 \cdot 10$
18. (a) Since $\frac{6}{2} = 3$, then $3 = \frac{6}{2}$
    (b) $19 \cdot 1 = 1 \cdot 19$

1-5 ADDITIONAL PROPERTIES

Addition and multiplication are binary operations, that is, they combine numbers two at a time to give another number. Thus, the sum or product of three or more numbers, such as

$3 + 5 + 10 + 2$ $\quad$ 3, 5, 10, *and* 2 *are called the* **terms** *of the sum*

or

$3 \times 5 \times 10 \times 2$ $\quad$ 3, 5, 10, *and* 2 *are called the* **factors** *of the product*

is defined as the number obtained by combining the terms or factors two at a time from left to right.

$$
\begin{array}{cc}
3 + 5 + 10 + 2 & 3 \times 5 \times 10 \times 2 \\
8 + 10 + 2 & 15 \times 10 \times 2 \\
18 + 2 & 150 \times 2 \\
20 & 300
\end{array}
$$

Thus

$3 + 5 + 10 + 2 = 20 \quad 3 \times 5 \times 10 \times 2 = 300$

Using the associative and commutative properties of addition, you can add the terms of a sum in any order and still get the same answer. For example, let us prove that $3 + 5 + 10 + 2 = (5 + 2) + (3 + 10)$.

$3 + 5 + 10 + 2 = (3 + 5) + 10 + 2$ *By definition*

$= (5 + 3) + 10 + 2$ *By commutative property*

$= [(5 + 3) + 10] + 2$ *By definition*

$= [5 + (3 + 10)] + 2$ *By associative property*

$= 5 + [(3 + 10) + 2]$ *By associative property*

$= 5 + [2 + (3 + 10)]$ *By commutative property*

$= (5 + 2) + (3 + 10)$ *By associative property*

Similarly, you can multiply the factors of a product in any order and still get the same answer. For example,

$6 \times 5 \times 11 = (6 \times 5) \times 11$ *By definition*

$= (5 \times 6) \times 11$ *By commutative property*

$= 5 \times (6 \times 11)$ *By associative property*

$= 5 \times (11 \times 6)$ *By commutative property*

$= (5 \times 11) \times 6$ *By associative property*

$= 5 \times 11 \times 6$ *By definition*

Thus,

$6 \times 5 \times 11 = 5 \times 11 \times 6 = 330$

$30 \times 11 \quad\quad 55 \times 6$

The number 0 is the additive identity element; thus, $a + 0 = a$. Under multiplication, however, 0 acts in a completely different way; for all real values of a,

$a \times 0 = 0$

is true.

1-5: ADDITIONAL PROPERTIES

Let us show that $3 \times 0 = 0$ as an indication of the truth of this property. You know that

$$1 \times 0 = 0 \qquad 1 \times a = a \text{ for every number } a$$

Therefore,

$$\begin{aligned} 2 \times 0 &= (1+1) \times 0 \\ &= 1 \times 0 + 1 \times 0 \qquad \textit{Distributive property} \\ &= 0 + 0, \text{ or } 0 \end{aligned}$$

Continuing,

$$\begin{aligned} 3 \times 0 &= (1+2) \times 0 \\ &= 1 \times 0 + 2 \times 0 \qquad \textit{Distributive property} \\ &= 0 + 0, \text{ or } 0 \end{aligned}$$

Thus, $3 \times 0 = 0$.

The equation

$$a \cdot b = 0$$

is true if either a or b has the value 0 and false otherwise. For example, the equation

$$7x = 0$$

is true if x has the value 0 and false otherwise.

Quick Reinforcement

Indicate the property used at each step.

$(5 + 6) + (-5)$
(a) $= (6 + 5) + (-5)$
(b) $= 6 + [5 + (-5)]$
(c) $= 6 + 0$
(d) $= 6$

Answers (a) Commutative of addition (b) Associative of addition (c) Additive inverse (d) Additive identity

EXERCISES 1-5

Indicate the correct property used in each step: commutative of addition, commutative of multiplication, associative of addition, associative of multiplication, distributive property, additive identity, additive inverse, multiplicative identity, or multiplicative inverse.

1. (a) $7 + (3 + 4)$
    $(7 + 3) + 4$
    $10 + 4$
    14

    (b) $(9 + 3) + (-9)$
    $(3 + 9) + (-9)$
    $3 + [9 + (-9)]$
    $3 + 0$
    3

2. (a) $(3 \times 2) \times 5$
    $3 \times (2 \times 5)$
    3×10
    30

    (b) $(4 \times 6) + (4 \times 5)$
    $4 \times (6 + 5)$
    4×11
    44

3. (a) $7 \times (3 + 2)$
    $(7 \times 3) + (7 \times 2)$
    $21 + 14$
    $14 + 21$
    35

    (b) $(5 \times 15) \times (2 \times 2)$
    $5 \times (15 \times 2) \times 2$
    $5 \times (2 \times 15) \times 2$
    $(5 \times 2) \times (15 \times 2)$
    10×30
    300

4. (a) $(3 \times 4) \times (\frac{1}{3} \times \frac{1}{4})$
    $3 \times (4 \times \frac{1}{3}) \times \frac{1}{4}$
    $3 \times (\frac{1}{3} \times 4) \times \frac{1}{4}$
    $(3 \times \frac{1}{3}) \times (4 \times \frac{1}{4})$
    1×1
    1

    (b) $[5 \times (-2)] + (5 \times 2)$
    $5 \times [(-2) + 2]$
    5×0
    0

5. (a) $(7.2 \times 4.3) + [7.2 \times (-4.3)]$
    $7.2 \times [4.3 + (-4.3)]$
    7.2×0
    0

    (b) $4 \times (0.25 + 0)$
    $(4 \times 0.25) + (4 \times 0)$
    $1 + 0$
    1

6. (a) $8 + [3 + (-8)] + (-3)$
    $8 + [(-8) + 3] + (-3)$
    $[8 + (-8)] + [3 + (-3)]$
    $0 + 0$
    0

    (b) $5 \times (7 \times \frac{1}{5}) \times \frac{1}{7}$
    $5 \times (\frac{1}{5} \times 7) \times \frac{1}{7}$
    $(5 \times \frac{1}{5}) \times (7 \times \frac{1}{7})$
    1×1
    1

7. (a) $(2x + 3y) + (3x + y)$
    $2x + (3y + 3x) + y$
    $2x + (3x + 3y) + y$
    $(2x + 3x) + (3y + y)$
    $(2 + 3)x + (3 + 1)y$
    $5x + 4y$

    (b) $5p + [(-6m) + 3p] + 6m$
    $5p + [3p + (-6m)] + 6m$
    $(5p + 3p) + [(-6m) + 6m]$
    $(5 + 3)p + [(-6) + 6]m$
    $8p + [(-6) + 6]m$
    $8p + (0 \times m)$
    $8p + 0$
    $8p$

8. (a) $3 + [5(x + 2)]$
$3 + [5x + 10]$
$3 + [10 + 5x]$
$[3 + 10] + 5x$
$13 + 5x$

(b) $\frac{2}{3} + [4(y + \frac{1}{3})]$
$\frac{2}{3} + [4y + \frac{4}{3}]$
$\frac{2}{3} + [\frac{4}{3} + 4y]$
$[\frac{2}{3} + \frac{4}{3}] + 4y$
$\frac{6}{3} + 4y$
$2 + 4y$

9. (a) $(1.3 \times -4.1y) + (1.3 \times 4.1y)$
$1.3 \times (-4.1y + 4.1y)$
$1.3 \times (-4.1 + 4.1)y$
$1.3 \times (0)y$
1.3×0
0

(b) $(5.0 \times 0.1) \times (0.2 \times 10.0)$
$5.0 \times (0.1 \times 0.2) \times 10.0$
$5.0 \times (0.2 \times 0.1) \times 10.0$
$(5.0 \times 0.2) \times (0.1 \times 10.0)$
1×1
1

10. (a) $8 \times (3w + 5y)$
$(8 \times 3w) + (8 \times 5y)$
$24w + 40y$

(b) $[5(t + 0.4)] + 2.1$
$[5t + 2.0] + 2.1$
$5t + [2.0 + 2.1]$
$5t + 4.1$

1-6 REVIEW OF RATIONAL-NUMBER ARITHMETIC

Every rational number has the form $\frac{a}{b}$ (or a/b or $a \div b$), where a and b are integers, with $b \neq 0$. Each integer a is also the rational number $a/1$.

Equality of rational numbers

$$\frac{a}{b} = \frac{c}{d} \quad \text{if and only if } a \cdot d = b \cdot c$$

For example,

$\frac{3}{6} = \frac{1}{2}$ because $3 \cdot 2 = 6 \cdot 1$

$\frac{4}{9} \neq \frac{2}{3}$ because $4 \cdot 3 \neq 9 \cdot 2$

Multiplication of rational numbers

$$\frac{a}{b} \cdot \frac{c}{d} = \frac{a \cdot c}{b \cdot d}$$

For example,

$\frac{5}{9} \cdot \frac{2}{3} = \frac{5 \cdot 2}{9 \cdot 3} = \frac{10}{27}$

$1 \cdot \frac{3}{11} = \frac{1}{1} \cdot \frac{3}{11} = \frac{1 \cdot 3}{1 \cdot 11} = \frac{3}{11}$ Always, $1 \cdot \frac{a}{b} = \frac{a}{b}$

32 CHAPTER 1: NUMBERS

$$\frac{3}{5} \cdot \frac{5}{3} = \frac{3 \cdot 5}{5 \cdot 3} = \frac{15}{15} = 1 \qquad \text{Always, } \frac{a}{b} \cdot \frac{b}{a} = 1; \frac{a}{b} \text{ and } \frac{b}{a} \text{ are } \underline{\text{reciprocals}} \text{ of each other}$$

Reducing rational numbers

$$\frac{a \cdot b}{a \cdot c} = \frac{b}{c} \quad \text{because} \quad (ab)c = (ac)b$$

For example,

$$\frac{10}{15} = \frac{5 \cdot 2}{5 \cdot 3} = \frac{2}{3}$$

$$\frac{32}{56} = \frac{4 \cdot 8}{7 \cdot 8} = \frac{4}{7}$$

Division of rational numbers

$$\frac{a}{b} \div \frac{c}{d} = \frac{a}{b} \cdot \frac{d}{c} = \frac{ad}{bc} \qquad \text{You can write } \frac{\frac{a}{b}}{\frac{c}{d}} \text{ for } \frac{a}{b} \div \frac{c}{d}$$

For example,

$$\frac{5}{9} \div \frac{2}{7} = \frac{5}{9} \cdot \frac{7}{2} = \frac{35}{18}$$

$$\frac{4}{\frac{1}{2}} = \frac{4}{1} \cdot \frac{2}{1} = \frac{8}{1} = 8$$

$$\frac{\frac{2}{3}}{5} = \frac{2}{3} \div \frac{5}{1} = \frac{2}{3} \cdot \frac{1}{5} = \frac{2}{15}$$

Factors and multiples of integers. If a, b, and c are integers such that $c = ab$, then a or b is called a **factor** of c, and c is called a **multiple** of a or b. For example, 6 is a factor of 54 because $54 = 6 \cdot 9$, and 28 is a multiple of 2 because $2 \cdot 14 = 28$.

An integer p greater than 1 is called a **prime number** if 1 and p are the only positive factors of p. If an integer a greater than 1 has positive factors other than 1 and a, then a is called a **composite number**. For example, 7 is a prime number because 1 and 7 are the only positive factors of 7. Other prime numbers are 3, 5, 41, and 113. The number 2 is a factor of 4, so 4 is composite. Other composite numbers are 15, 33, and 91.

1-6: REVIEW OF RATIONAL-NUMBER ARITHMETIC

An important property of integers is that every integer greater than 1 can be expressed as a product of prime numbers. For example,

$$60 = \underbrace{2 \cdot 2 \cdot 3 \cdot 5}_{primes} \qquad 147 = \underbrace{3 \cdot 7 \cdot 7}_{primes}$$

Least common multiple (LCM). Given integers a and b, the smallest positive integer that is a multiple of both a and b is called the **least common multiple** (LCM) of a and b. For example, 12 is the LCM of 4 and 6; it is the smallest integer that is a multiple of both 4 and 6. Also, 40 is the LCM of 8 and 10.

You can find the LCM of two integers by looking at their prime factors. For example, to find the LCM of 252 and 270, first factor them into primes:

$$252 = 2 \cdot 2 \cdot 3 \cdot 3 \cdot 7 \qquad 270 = 2 \cdot 3 \cdot 3 \cdot 3 \cdot 5$$

The LCM must contain the prime factors of both integers. Thus, the LCM must contain as factors two 2s, three 3s, one 5, and one 7.

$$\text{LCM of 252 and 270} = 2 \cdot 2 \cdot 3 \cdot 3 \cdot 3 \cdot 5 \cdot 7 = 3780 \qquad \textit{Not easily guessed!}$$

Addition and subtraction of rational numbers

To add or subtract two rational numbers with the same denominator,

$$\frac{a}{b} + \frac{c}{b} = \frac{a+c}{b}, \qquad \frac{a}{b} - \frac{c}{b} = \frac{a-c}{b}$$

For example,

$$\frac{2}{3} + \frac{5}{3} = \frac{7}{3} \qquad \frac{5}{2} - \frac{1}{2} = \frac{4}{2} = 2$$

To add or subtract two rational numbers with different denominators, find the LCM of the denominators and proceed as above.

For example, find $\frac{2}{3} + \frac{1}{5}$. The LCM of 3 and 5 is 15.

$$\frac{2}{3} = \frac{2}{3} \cdot \frac{5}{5} = \frac{10}{15}, \qquad \frac{1}{5} = \frac{1}{5} \cdot \frac{3}{3} = \frac{3}{15}$$

$$\frac{2}{3} + \frac{1}{5} = \frac{10}{15} + \frac{3}{15} = \frac{13}{15}$$

Now find $\frac{17}{18} - \frac{2}{33}$. First find the LCM of 18 and 33.

$$18 = 2 \cdot 3 \cdot 3, \qquad 33 = 3 \cdot 11$$

The LCM is $2 \cdot 3 \cdot 3 \cdot 11$, or 198. Thus,

$$\frac{17}{18} = \frac{17 \cdot 11}{18 \cdot 11} = \frac{187}{198}, \qquad \frac{2}{33} = \frac{2 \cdot 6}{33 \cdot 6} = \frac{12}{198}$$

$$\frac{17}{18} - \frac{2}{33} = \frac{187}{198} - \frac{12}{198} = \frac{175}{198}$$

Decimals. You can convert fractions into decimal notation by the process of long division. For example, convert $\frac{1}{8}$ into a decimal.

```
     0.125
 8)1.000
   8
   ‾‾
    20
    16
    ‾‾
     40
     40
     ‾‾
      0
```

Thus, $\frac{1}{8} = 0.125$

You can also convert decimals into fractions. For example, express the decimals 0.15 and 2.015 as quotients of integers.

$$0.15 = \frac{15}{100} = \frac{5 \cdot 3}{5 \cdot 20} = \frac{3}{20}$$

$$2.015 = \frac{2015}{1000} = \frac{5 \cdot 403}{5 \cdot 200} = \frac{403}{200}$$

Percent. Percent means per one hundred. For example,

5% means $\frac{5}{100} = 0.05$

$8\frac{1}{4}\% = 8.25\%$ means $\frac{8.25}{100} = 0.0825$ *Dividing by* 100 *moves the decimal point*

0.1% means $\frac{0.1}{100} = 0.001$ *two places to the left*

Just as you can convert from a percent to a decimal, so can you convert from a decimal to a percent. For example,

$0.05 = \frac{5}{100}$ or 5%

$10.5 = 10.50 = \frac{1050}{100}$ or 1050% *Multiplying by* 100 *moves the decimal point two places to the right*

$0.0825 = \frac{8.25}{100}$ or 8.25%

1-6: REVIEW OF RATIONAL-NUMBER ARITHMETIC

We will be concerned with several applications of percent in later chapters. The key thing to remember is

$n\%$ of x is y

means

$$\frac{n}{100} \cdot x = y$$

EXERCISES 1-6

Simplify each of the following as completely as possible. Reduce all rational numbers when possible.

1. (a) $\frac{5}{15}$ (b) $\frac{5}{8} \cdot \frac{7}{9}$
2. (a) $1 \cdot \frac{3}{5}$ (b) $\frac{24}{32}$
3. (a) $\frac{3}{8} \cdot \frac{16}{9}$ (b) $\frac{2}{3} \div \frac{4}{7}$
4. (a) $\frac{225}{25}$ (b) $\frac{4}{5} \cdot \frac{5}{32}$
5. (a) $\frac{8}{5} \div \frac{30}{5}$ (b) $\frac{7}{11} \cdot 11$
6. (a) $\frac{5}{\frac{1}{10}}$ (b) $\frac{\frac{3}{4}}{3}$
7. (a) $\frac{128}{32}$ (b) $\frac{12}{35} \cdot \frac{14}{9}$
8. (a) $\frac{90}{49} \div \frac{45}{14}$ (b) $\frac{\frac{1}{4}}{\frac{1}{2}}$

State whether each of the following numbers is prime or composite. If it is composite, express it as a product of prime numbers.

9. (a) 11 (b) 21
10. (a) 9 (b) 5
11. (a) 32 (b) 24
12. (a) 100 (b) 23
13. (a) 4 (b) 40
14. (a) 17 (b) 72
15. (a) 20 (b) 31

Find the least common multiple (LCM) of each of the following sets of numbers.

16. (a) 3, 5, 9 (b) 2, 5, 8
17. (a) 9, 15, 20 (b) 3, 5, 7

18. (a) 8, 12   (b) 2, 10
19. (a) 7, 9, 15   (b) 4, 16, 32
20. (a) 5, 30, 45   (b) 21, 35, 14

Simplify each of the following as completely as possible, reducing all rational numbers when possible.

21. (a) $\frac{1}{2} + \frac{3}{4}$ (b) $\frac{2}{3} - \frac{1}{6}$
22. (a) $\frac{7}{8} - \frac{1}{4}$ (b) $\frac{3}{10} + \frac{2}{15}$
23. (a) $\frac{5}{6} + \frac{4}{15}$ (b) $\frac{3}{8} - \frac{7}{24}$
24. (a) $\frac{5}{16} + \frac{5}{24} + \frac{5}{15}$ (b) $\frac{4}{9} + \frac{11}{12} + \frac{13}{15}$

1-7 REVIEW OF GEOMETRICAL FORMULAS

Perimeter. The perimeter P of a plane geometrical figure is the distance around the figure.

Rectangle   Square   Triangle   Circle

$P = 2l + 2w$ $P = 4s$ $P = a + b + c$ $P = 2\pi r$

FIGURE 1-7   **FIGURE 1-8**   **FIGURE 1-9**   **FIGURE 1-10**

Area. The area A of a plane geometrical figure is the number of square units it contains.

Rectangle   Square   Triangle   Circle

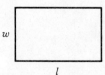

$A = l \cdot w$ $A = s \cdot s = s^2$ $A = \frac{1}{2} b \cdot h$ $A = \pi r^2$

FIGURE 1-11   **FIGURE 1-12**   **FIGURE 1-13**   **FIGURE 1-14**

Volume. The volume of a geometrical solid is the number of cubic units it contains.

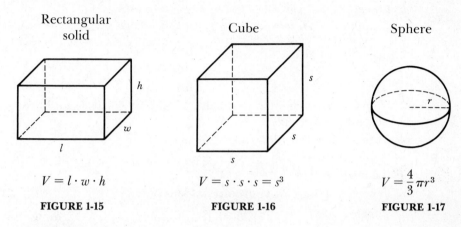

Rectangular solid	Cube	Sphere
$V = l \cdot w \cdot h$	$V = s \cdot s \cdot s = s^3$	$V = \dfrac{4}{3}\pi r^3$
FIGURE 1-15	**FIGURE 1-16**	**FIGURE 1-17**

An additional formula states that the angles of a triangle add up to 180°.

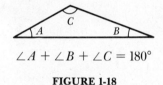

$$\angle A + \angle B + \angle C = 180°$$

FIGURE 1-18

Two angles A and B are complementary if $\angle A + \angle B = 90°$.

KEY TERMS

Integers	Equation
Rational numbers	Formula
Irrational numbers	Properties
Real numbers	Additive identity
Set	Multiplicative identity
Elements	Additive inverse (negative)
Empty set (null set)	Multiplicative inverse (reciprocal)
Union	Terms
Intersection	Factors
Equal	Factor
Subset	Multiple
Variable	Prime number
Numerical expression	Composite number
Mathematical expression	Least common multiple

REVIEW EXERCISES

Identify each of the following numbers by the most precise label, choosing from the following list: integer, rational number, irrational number.

1. (a) -2.3 (b) $\sqrt{5}$
2. (a) $\frac{3}{8}$ (b) $\sqrt{100}$
3. (a) 1200   (b) 1.031
4. (a) The speed of light in a vacuum is 30,000,000,000 centimeters per second.
   (b) A team won $\frac{4}{5}$ of the games they played.
5. (a) The distance from Mr. Everyman's house to his office was 21.3 kilometers.
   (b) The hypotenuse of a given triangle is $\sqrt{6}$.

Given the sets $A = \{0, 2, 5, 7\}$, $B = \{1, 2, 3\}$, and $C = \{3, 5, 7\}$, list the elements of each of the following.

6. (a) $(A \cap B) \cup C$ (b) $A \cup B \cap C$
7. (a) $A \cup B \cap C$ (b) $(A \cup B) \cap (A \cup C)$

Determine which statements are true for the sets

$$W = \{1, 5, 9\} \quad X = \{1, 5, 9, 11\} \quad Y = \{1, 5, 9, 13\} \quad Z = \{1, 5, 9\}$$

8. (a) $Z \subseteq Y$ (b) $Y \subseteq X$
9. (a) $X = Y$ (b) $W = Z$
10. (a) $X \cap Y = W$ (b) $W \cup X = Y$

Evaluate each of the following expressions.

11. (a) $3 \cdot (7 + 5) + 2 \cdot 4 + (-1)$ (b) $(4 \cdot 3 \cdot 2 \cdot 8) \cdot 2 + 6 \cdot 0$
12. (a) $6 \cdot 4 + 8 \cdot 3 - 4$ (b) $5 \cdot 3 + 5 \cdot 5 - 5 \cdot 4$

Evaluate each of the following expressions for the given values of x and y.

13. (a) $x(12 + y)$ for x given the value 2, y given the value 3
    (b) $y(14 - x)$ for x given the value 10, y given the value 5
14. (a) $3[x + 2(y + 1)]$ for x given the value 2, y given the value 4
    (b) $x[30 - 4(y - 3)]$ for x given the value 3, y given the value 5

Insert grouping symbols in each of the following to give the desired answer.

15. (a) $9 + 7 \div 2 = 8$ (b) $6 \cdot 3 - 2 = 6$
16. (a) $5 \cdot 2 - 3 \cdot 3 = 1$ (b) $5 \cdot 3 - 2 \cdot 3 = 15$

Tell whether each equation is true or false for the given values of the variables.

17. (a) $x + 2 = 2x - 1$, given $x = 3$
    (b) $y - 7 = y + 3$, given $y = 1$

18. (a) $p = 2l + 2w$, given $l = 2, w = 4, p = 14$
    (b) $p = 4x$, given $p = 32, x = 8$

Translate each of the following statements into mathematical expressions.

19. (a) Four more than three times a number
    (b) The product of a number and one-half the number
20. (a) The quotient of the sum of a number and three, and four
    (b) The difference between the sum of a number and $\frac{1}{2}$, and 6

Evaluate each of the following formulas.

21. (a) $C = \frac{5}{9}(F - 32)$ for $F = 33°$
    (b) $C = \frac{5}{9}(F - 32)$ for $F = 59°$
22. (a) $V = l \cdot w \cdot h$ for $l = 0.2, w = 0.7, h = 10$
    (b) $V = l \cdot w \cdot h$ for $l = 62, w = 9, h = \frac{1}{2}$
23. (a) $I = PRT$ for $P = \$2100, R = 0.08, T = \frac{1}{3}$
    (b) $I = PRT$ for $P = \$33{,}000, R = 0.095, T = 3$

Use variables and mathematical symbols to abbreviate these statements.

24. (a) The sum of Miyo Lee's salary and Kazo Lee's salary is $530 each week.
    (b) Francine Singh sold seven more cars this month than the next best salesperson at the dealership.
25. (a) The orchard has seven times as many apple trees as peach trees.
    (b) The computer room has five fewer terminals than the maximum number the computer can accommodate.

Translate each mathematical expression into an equivalent verbal statement.

26. (a) $5 \div x$ (b) $(3 + y) \cdot 7$
27. (a) $5x - 2 = 8$ (b) $\frac{1}{2}x = x + 4$

Name the property that makes each of the following equations true.

28. (a) $5 \cdot (8 + 2) = 5 \cdot 8 + 5 \cdot 2$ (b) $\frac{1}{3} \cdot 3 = 1$
29. (a) $2 \cdot (3 + 2) = (3 + 2) \cdot 2$
    (b) Since $7 = \frac{14}{2}$ and $\frac{14}{2} = \frac{49}{7}$, then $7 = \frac{49}{7}$

Indicate the correct property used in each step, choosing from the following list: commutative property of addition or multiplication, associative property of addition or multiplication, distributive property, additive identity or inverse, multiplicative identity or inverse.

30. (a) $[7 + (-5)] + 5$ (b) $7 \times (2 + \frac{1}{7})$
    $7 + [(-5) + 5]$ $7 \times (\frac{1}{7} + 2)$
    $7 + 0$ $(7 \times \frac{1}{7}) + (7 \times 2)$
    7 $1 + 14$
                               15

40 CHAPTER 1: NUMBERS

31. (a) $4a + (3b + 6a) + 9b$
     $4a + (6a + 3b) + 9b$
     $(4a + 6a) + (3b + 9b)$
     $(4 + 6)a + (3 + 9)b$
     $10a + 12b$

    (b) $[3.2(m + 5)] + (-16)$
     $[3.2m + 16] + (-16)$
     $3.2m + [16 + (-16)]$
     $3.2m + 0$
     $3.2m$

Simplify each of the following as completely as possible, reducing all rational numbers when possible.

32. (a) $\frac{26}{13}$ (b) $\frac{12}{25} \cdot \frac{5}{6}$
33. (a) $\frac{2}{3} \div \frac{5}{12}$ (b) $\frac{2/5}{8}$

State whether each of the following numbers is prime or composite. If it is composite, express it as a product of prime numbers.

34. (a) 27 (b) 43
35. (a) 13 (b) 55

Find the LCM of each of the following sets of numbers.

36. (a) 4, 12, 9 (b) 6, 15, 11
37. a) 2, 7, 9 (b) 5, 15, 45

Simplify each of the following as completely as possible, reducing all rational numbers when possible.

38. (a) $\frac{4}{5} - \frac{7}{15}$ (b) $\frac{5}{12} + \frac{3}{16}$
39. (a) $\frac{8}{33} + \frac{5}{22}$ (b) $\frac{7}{10} + \frac{9}{20} + \frac{1}{2}$

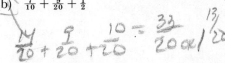 **Calculator Activities**

Evaluate each of the following expressions.

1. (a) $252(741 - 291) + 289 \div 17$
   (b) $1542 \div (1241 - 2 \cdot 235) + 21(41 + 150)$
2. (a) $[1410 - (698 \div 4) + 1] - 3(4 + 95.21)$
   (b) $5 - 1949 \div (1745 + 204) \cdot 2$
3. (a) $0.021 + (5 \div 0.41 - 2)(1510 - 49.28)$
   (b) $[9.2 - (4.01 + 2.53 \div 2)] + 5.1(4 - 3.201)$
4. (a) $\frac{3}{8} \cdot 385.03 \div 4.12 - 9.2$
   (b) $0.15(\frac{1}{2} - 0.31) + 109 - \frac{2}{5} \cdot 0.81 + 1$

Evaluate each of the following expressions for the given values of x and y.

5. (a) $x(7.2 - x)$ for x given the value 3.82
   (b) $y(y + 2.713)$ for y given the value 0.901

6. (a) $x[4.1 - y(x + 0.013)]$ for x given the value 8.7, y given the value 2.05
   (b) $3.2[x + y(x - 0.042)]$ for x given the value 1.3, y given the value 9.74

Evaluate each of the following formulas.

7. (a) $I = PRT$ for $P = \$5500$, $R = 8\frac{1}{2}\%$, $T = \frac{1}{12}$
   (b) $I = PRT$ for $P = \$45,300$, $R = 9\frac{3}{8}\%$, $T = 21\frac{1}{2}$
8. (a) $v = -9.7t + k$ for $t = 3.78$ seconds, $k = 9.02$ meters per second
   (b) $v = -9.7t + k$ for $t = 5.021$ hours, $k = 88.9$ kilometers per hour
9. (a) For any triangle ABC, find $\angle B$ given $\angle A = 41.02°$, $\angle C = 103.91°$
   (b) For any triangle ABC, find $\angle A$ given $\angle B = 1.023°$, $\angle C = 3.861°$
10. (a) $L = \frac{3}{20}G + 1$ for $G = 98$ (a leopard)
    (b) $L = \frac{3}{20}G + 1$ for $G = 37$ (a rabbit)

2 Order

Joe Trevino is bringing the company checkbook up to date. He started with a balance of $1829.42 and wrote checks for $213.69, $45.00, $184.20, and $54.55. He also made deposits of $309.00, $68.35, and $41.00. What did Joe calculate as the ending balance?

2-1 ORDER AMONG NUMBERS

One basket has 27 oranges in it and a second basket has 32 oranges. The second basket contains more oranges than the first; in mathematical notation,

$32 > 27$ $>$ means "is greater than"

The latest Mig jet cruises at Mach 2.3, whereas the latest Starfire jet cruises at Mach 2.1. Which airplane is faster? The Mig jet is faster, because

$2.3 > 2.1$ 2.3 is greater than 2.1

The side of a square is always shorter than its diagonal. Thus, for the square shown in Figure 2-1,

$1 < \sqrt{2}$ $<$ *means "is less than"*

FIGURE 2-1

These examples illustrate a basic fact about real numbers: there is an order among them. This order can be shown on a **number line,** as in Figure 2-2. The number line is like an infinite ruler, extending infinitely far in both directions, as indicated by the three dots (. . .) at each end. Each point on the line is assigned to a real number and every real number is assigned to a point on the line. The integers are assigned to equispaced points on the line. You may choose any scale you wish (an inch or a centimeter, for example) for the distance between consecutive integers. The rational numbers are assigned to appropriate points on the line; that is, $\frac{1}{2}$ is assigned to the point midway between 0 and 1; $\frac{8}{3}$ to the point two-thirds of the distance between 2 and 3; $-2\frac{3}{4}$ three-fourths of the distance between -2 and -3; and so on. The remaining points on the line are assigned irrational numbers, some of which are indicated on the number line in Figure 2-2.

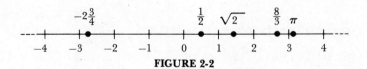

FIGURE 2-2

On any given number line, the number assigned to a point is called the **coordinate** of that point, and the point assigned to a number is called the **graph** of that number.

Problem 1 Points A, B, C, D, E, F, G, and H have the coordinates shown on the number line in Figure 2-3.

2-1: ORDER AMONG NUMBERS

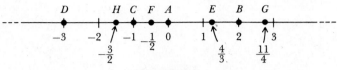

FIGURE 2-3

(a) Graph the set of numbers $\{-3, -1, -\frac{1}{2}, 0, 2, \frac{11}{4}\}$.

(b) Find the set of coordinates for the set of points $\{A, H, B, E, D\}$.

SOLUTION:

(a) The graph of the number -3 is the point D on the number line; the graph of -1 is point C; of $-\frac{1}{2}$ is point F; of 0 is point A; of 2 is point B; of $\frac{11}{4}$ is point G. Thus, the graph of the set $\{-3, -1, -\frac{1}{2}, 0, 2, \frac{11}{4}\}$ is the set of points $\{D, C, F, A, B, G\}$ on the number line in Figure 2-3.

(b) The coordinate of point A is the number 0; of H is $-\frac{3}{2}$; of B is 2; of E is $\frac{4}{3}$; of D is -3. Thus, the set of coordinates for the points $\{A, H, B, E, D\}$ is $\{0, -\frac{3}{2}, 2, \frac{4}{3}, -3\}$.

The point with coordinate 0 on a number line is called the **origin.** In Problem 1 above, point A is the origin.

Problem 2 Graph the set of numbers $S = \{0, \frac{1}{2}, 2, \frac{12}{5}, 3, \frac{10}{3}\}$ on a number line.

SOLUTION:

The graph consists of the heavy points (dots) on the number line in Figure 2-4.

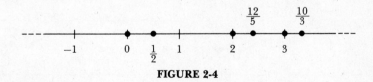

FIGURE 2-4

If on a horizontal number line the number 1 is to the right of 0, as in all of our examples so far, then we can tell the relative size of two numbers by looking at their graphs on the line. If the graphs of the numbers a, b, and c are the points A, B, and C, as shown in Figure 2-5, then

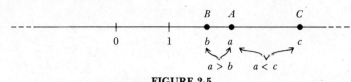

FIGURE 2-5

$a > b$ because A is to the right of B a is greater than b
$a < c$ because A is to the left of C a is less than c

Problem 3 Tell which number of each pair is greater by graphing them.

(a) $2, \dfrac{5}{2}$ (b) $-1, 0$ (c) $2.8, 3.2$

(d) $1, -4$ (e) $-2, -\dfrac{3}{2}$

SOLUTION:

From the number line in Figure 2-6,

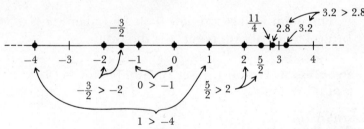

FIGURE 2-6

(a) $\dfrac{5}{2} > 2$ (b) $0 > -1$ (c) $3.2 > 2.8$

(d) $1 > -4$ (e) $-\dfrac{3}{2} > -2$

Quick Reinforcement

Tell which number in each pair is larger by graphing.

(a) $-2, -2.5$ (b) $\dfrac{5}{2}, 2$ (c) $\dfrac{21}{7}, \dfrac{19}{7}$

Answers (a) -2 (b) $\dfrac{5}{2}$ (c) $\dfrac{21}{7}$

EXERCISES 2-1

Graph each of the following sets of numbers on a number line.

1. (a) $\{8, 3, 5, -2\}$ (b) $\{6, -3, 1, 4\}$
2. (a) $\{8, -1, 0, -5\}$ (b) $\{-7, 2, 0, -3\}$
3. (a) $\{\frac{7}{2}, \frac{1}{3}, 5, 1\frac{1}{4}\}$ (b) $\{\frac{5}{2}, -\frac{3}{2}, 4, -2\}$
4. (a) $\{2.1, 3, -1.5, 1.3\}$ (b) $\{0.9, -0.3, 1.75, 0.2\}$
5. (a) $\{-\frac{2}{3}, \frac{2}{3}, 0, .75\}$ (b) $\{1.5, -\frac{7}{3}, 3, 5\}$
6. (a) $\{2\frac{3}{4}, -1.2, 3, -1\}$ (b) $\{-2, \frac{9}{4}, 1, \frac{5}{3}\}$
7. (a) $\{4.2, 1\frac{1}{2}, -\frac{1}{2}, 2.7\}$ (b) $\{-0.5, -1.75, -2\frac{1}{4}, 0\}$
8. (a) $\{10.5, 8.2, 11.9, 9.3\}$ (b) $\{-4, 1\frac{3}{4}, 2\frac{2}{3}, 0.6\}$

Find the set of coordinates for each of the following sets of points on the given number line.

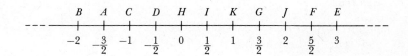

9. (a) $\{A, E, I\}$ (b) $\{B, F, J\}$
10. (a) $\{A, B, D, E, G\}$ (b) $\{B, C, E, F, H\}$
11. (a) $\{I, K, B, D\}$ (b) $\{J, A, C, E\}$
12. (a) $\{A, B, C, I, J, K\}$ (b) $\{C, D, E, F, G, H\}$
13. (a) $\{A, E, I, G, F\}$ (b) $\{B, C, H, K, J, E\}$

Tell which number of each pair is the greater by graphing each pair on a number line.

14. (a) $-3, 3$ (b) $-\frac{4}{3}, -1$
15. (a) $\frac{1}{2}, \frac{1}{3}$ (b) $0.75, \frac{2}{3}$
16. (a) $-1, -\frac{3}{2}$ (b) $-2, 0$
17. (a) $1.2, 2$ (b) $-\frac{2}{3}, -1$
18. (a) $5, 7$ (b) $-3, -4$
19. (a) $-10.5, -7$ (b) $9, 11$
20. (a) $\frac{5}{2}, 3$ (b) $-\frac{15}{4}, -5$
21. (a) $3.75, 4$ (b) $7, -2$
22. (a) $-9, 8$ (b) $0.75, -0.5$
23. (a) $2\frac{2}{3}, 2\frac{4}{5}$ (b) $-4, -\frac{8}{3}$
24. (a) $\frac{10}{3}, 3$ (b) $-\frac{10}{3}, -3$
25. (a) $-2.7, -3.2$ (b) $4.5, 5.4$

2-2 ORDER PROPERTIES

The six properties of real numbers given in Chapter 1 do not cover the ordering of numbers. Such properties, which have to do with the relative size of numbers, are called the **order properties.** They are listed below.

7. Transitive Property

For all real values of a, b, and c, if $a > b$ and $b > c$ are true, then $a > c$ is also true.

For example, $5 > 3$ and $3 > 1$ are true; therefore, $5 > 1$ is true.

8. Additive Property

For all real values of a, b, and c, if $a > b$ is true, then $a + c > b + c$ is also true.

For example, $9 > 7$ is true; therefore, $9 + 5 > 7 + 5$, or $14 > 12$, is also true. (The number 5 is chosen at random.)

9. Multiplicative Property

For all real values of a, b, and c, if $a > b$ and $c > 0$ are true, then $ac > bc$ is also true.

For example, $8 > 4$ and $3 > 0$ are true; therefore, $8 \cdot 3 > 4 \cdot 3$, or $24 > 12$, is also true.

10. Trichotomy Property

For all real values of a and b, one and only one of the following three statements is true:

(i) $a = b$ (ii) $a > b$ (iii) $b > a$

For example, given the pair 5, 8 of real numbers, either $5 = 8$, $5 > 8$, or $8 > 5$ is true. In this case, $8 > 5$ is true and the other two statements are false.

There is an order relation of *less than* ($<$) as well as greater than ($>$). By definition, $a < b$ is true whenever $b > a$ is true. For example, $21 < 25$ is true because $25 > 21$ is true. In terms of our horizontal number line, $a < b$ is true if a is to the left of b and false otherwise.

Properties (7) through (10) hold for "less than" as well as "greater than":

If $a < b$ and $b < c$ are true, then $a < c$ is also true. (*Transitive property* of $<$)

If $a < b$ is true, then $a + c < b + c$ is also true. (*Additive property* of $<$)

If $a < b$ and $c > 0$ are true, then $ac < bc$ is also true. (*Multiplicative property* of $<$)

One and only one of the following statements is true: (i) $a = b$ (ii) $a < b$ (iii) $b < a$. (*Trichotomy property* of $<$)

A number a is called **positive** if $a > 0$ and **negative** if $a < 0$. On a horizontal number line, the positive numbers are to the *right* of 0 and the negative numbers to the *left* of 0. According to the trichotomy property, given b the value 0, every real number a is either zero, positive, or negative.

The positive numbers are the **natural** numbers of arithmetic. You add and multiply them in the customary way, obtaining other positive numbers. That is, if $a > 0$ and $c > 0$, then $a + c > 0$ and $ac > 0$. For example, 7 and 5 are positive numbers; therefore, $7 + 5$, or 12, and $7 \cdot 5$, or 35, are positive numbers.

Quick Reinforcement

Name the property that makes each statement true.

(a) $-7 > 10$, therefore $(-7) + 3 > 10 + 3$

(b) $4 < 6$; therefore $4 \cdot 2 < 6 \cdot 2$

(c) $-2 < -1$ and $-1 < 0$; therefore $-2 < 0$

Answers (a) Additive property of $<$ (b) Multiplicative property of $<$ (c) Transitive property of $<$

EXERCISES 2-2

Which order property makes each of the statements below true?

1. (a) $5 > 3$ and $3 > 2$ are true; therefore, $5 > 2$ is true.
   (b) If $a > b$ is true, then $a + 7 > b + 7$ is true.
2. (a) $6 < 9$ and $9 < 13$ are true; therefore, $6 < 13$ is true.
   (b) $11 > 8$ and $3 > 0$ are true; therefore, $11 \cdot 3 > 8 \cdot 3$ is true.
3. (a) If $x > y$ is true, then $x + 2 > y + 2$ is true.
   (b) If $a > b$ is true, then $9a > 9b$ is true.

4. (a) $-2 > -5$ and $-5 > -10$ are true; therefore, $-2 > -10$ is true.
   (b) $-6 < 1$ and $1 < 6$ are true; therefore, $-6 < 6$ is true.
5. (a) If a is not less than 5, and a is not equal to 5, then a is greater than 5 is true.
   (b) If a is not less than 2, and a is not greater than 2, then a equals 2 is true.
6. (a) If $a > b$ is true, then $\frac{a}{2} > \frac{b}{2}$ is true.
   (b) $-2.3 < -1.2$ is true; therefore $-2.3 + 4 < -1.2 + 4$ is true.
7. (a) $-2.3 < -1.2$ and $4 > 0$ are true; therefore, $(-2.3)(4) < (-1.2)(4)$ is true.
   (b) If x is not equal to -3, and x is not greater than -3, then x is less than -3 is true.
8. (a) $12 > 8$ and $8 > 2$ are true; therefore, $12 > 2$ is true.
   (b) $5.7 < 9.2$ is true; therefore, $5.7 + 3 < 9.2 + 3$ is true.
9. (a) If $a < b$ is true, then $7a < 7b$ is true.
   (b) If $a > 6$ and $6 > c$ are true, then $a > c$ is true.
10. (a) If $a < 5$ and $5 < 9$ are true, then $a < 9$ is true.
    (b) $5 < 7$ and $10 > 7$ are true; therefore, $5 < 10$ is true.
11. (a) $11 > 4$ is true; therefore, $4 < 11$ is true.
    (b) $-8 < 8$ and $20 > 8$ are true; therefore, $-8 < 20$ is true.
12. (a) If $x < y$ is true, then $\frac{x}{4} < \frac{y}{4}$ is true.
    (b) $3 > -3$ and $7 > 0$ are true; therefore, $3 \cdot 7 > (-3) \cdot 7$ is true.

2-3 ADDITION AND SUBTRACTION OF REAL NUMBERS

You learned in arithmetic how to add positive numbers. But how do you add negative numbers, or positive and negative numbers? The answer to this question may be indicated by looking at temperatures at various times.

If the temperature now is 7°C and it rises 5°, the new temperature will be 12°C:

$7 + 5 = 12$ *Adding two positive numbers*

If the temperature now is -7°C and it drops 5°, the new temperature will be -12°C:

$(-7) + (-5) = -12$ *Adding two negative numbers*

If the temperature now is 7°C and it drops 5°, the new temperature will be 2°C:

$7 + (-5) = 2$ *Adding a positive and a negative number*

If the temperature starts at 7°C and drops 7°, the new temperature will be 0°C:

$7 + (-7) = 0$ *Adding a positive and a negative number*

Finally, if the temperature starts at 7°C and drops 12°, the new temperature will be −5°C:

$7 + (-12) = -5$ *Adding a positive and a negative number*

As these examples show, when you add numbers of the same sign, the sum also has that sign. However, when you add two numbers having different signs, the sum might be a positive number, zero, or a negative number. To help explain this amibguity of sign, let us consider the concept of absolute value.

Think of the set of real numbers as the set of points on a number line. Then the **absolute value** of a number a, designated by

$|a|$

is the distance between 0 and a on the line. For example, as shown by Figure 2-7,

$|5| = 5$
$|-5| = 5$

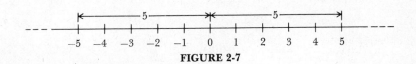

FIGURE 2-7

Other examples are

$|3.2| = 3.2, \qquad \left|-\frac{3}{2}\right| = \frac{3}{2}$

You can see from Figure 2-8 that if $a > 0$, then $|a| = a$, whereas if $a < 0$, then $|a| = -a$. In either case, $|a|$ is a positive number. It should also be noted that $|0| = 0$. For example,

$|7| = 7$ *Because $7 > 0$*
$|-7| = 7$ *Because $-7 < 0$ and $-(-7) = 7$*

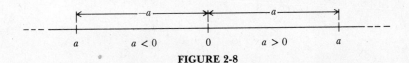

FIGURE 2-8

52 CHAPTER 2: ORDER

Returning to our temperature examples, we can now state the following rule.

To add numbers having the same sign, add absolute values and use their common sign for the answer. That is,

$a + b = |a| + |b|$ if $a > 0$ and $b > 0$
$a + b = -(|a| + |b|)$ if $a < 0$ and $b < 0$

For example,
$$4 + 22 = |4| + |22| = 26$$
$$(-4) + (-22) = -(|-4| + |-22|) \qquad |-4| = 4, |-22| = 22$$
$$= -(4 + 22)$$
$$= -26$$

Before looking at the sum of two numbers with opposite signs, we need to define the operation of **subtraction.** For all real numbers a, b, and c, $a - b = c$, provided $a = b + c$. (Read $a - b$ as "a minus b.") For example,

$14 - 8 = 6$ *Because* $14 = 8 + 6$
$23 - 30 = -7$ *Because* $23 = 30 + (-7)$
$-10 - 9 = -19$ *Because* $-10 = 9 + (-19)$

We can now state the following rule.

To add numbers having opposite signs, subtract absolute values, smaller from larger, and use the sign of the larger for the answer. That is,

$|a| > |b|$
$a + b = |a| - |b|$ if $a > 0$
$a + b = -(|a| - |b|)$ if $a < 0$

For example,

$15 + (-5) = |15| - |-5|$ *Because* $|15| > |-5|$ *and* $15 > 0$
$\qquad\qquad = 15 - 5$, or 10
$15 + (-25) = -(|-25| - |15|)$ *Because* $|-25| > |15|$ *and* $-25 < 0$
$\qquad\qquad = -(25 - 15)$, or -10

2-3: ADDITION AND SUBTRACTION OF REAL NUMBERS

A consequence of the definition of subtraction is that the equation

$$a + (-b) = a - b \quad \text{Because } a = a + b + (-b)$$

is true for all real numbers a and b. For example,

$28 + (-13) = 28 - 13 = 15$ Because $28 = 13 + 15$

$17 + (-20) = 17 - 20 = -3$ Because $17 = 20 + (-3)$

$(-16) + 7 = 7 + (-16) = 7 - 16 = -9$ Because $7 = 16 + (-9)$

You can also look at the equation

$$a - b = a + (-b)$$

as stating that subtracting b from a is the same as adding $-b$ to a. Thus, every subtraction problem can be converted to an addition problem. For example,

$5 - 10 = 5 + (-10) = -5$

$(-4) - 2 = (-4) + (-2) = -6$

$4 - (-3) = 4 + 3 = 7$ *The additive inverse of -3 is 3*

You add several numbers by adding them two at a time in any order. For example,

$$(-7) + 5 + 3 = [(-7) + 5] + 3$$
$$= (-2) + 3$$
$$= 1$$

and

$$-8 + 14 - 10 + 2 = (-8) + 14 + (-10) + 2$$
$$= (-8 - 10) + (14 + 2)$$
$$= -18 + 16$$
$$= -2$$

Quick Reinforcement

Find the value of each.

(a) $10 + (-24) =$

(b) $(-15) - 6 =$

(c) $(-21) + 38 + (-16) =$

(d) $10 - 3 + 7 - 14$

Answers (a) -14 (b) -21 (c) 1 (d) 0

EXERCISES 2-3

Evaluate each of the following.

1. (a) $|3|$ (b) $|-3|$
2. (a) $|9|$ (b) $|23|$
3. (a) $|-4|$ (b) $|2.4|$
4. (a) $|-2.2|$ (b) $-|6|$
5. (a) $-|-1|$ (b) $|-\frac{7}{2}|$
6. (a) $-|5.7|$ (b) $-|-11|$
7. (a) $(4) + (8)$ (b) $(3) + (7)$
8. (a) $(7) + (-2)$ (b) $(-9) + (6)$
9. (a) $(\frac{1}{4}) + (\frac{1}{2})$ (b) $(4.2) + (-2.3)$
10. (a) $(-6.5) + (3.2)$ (b) $(\frac{2}{3}) + (-\frac{5}{6})$
11. (a) $(-3) + (-10)$ (b) $(-11) + (-5)$
12. (a) $(-1.1) + (-5.6)$ (b) $(-\frac{8}{9}) + (-1\frac{1}{6})$
13. (a) $-11 + 8$ (b) $-16 + 7$
14. (a) $10.8 - 4.3$ (b) $-9.2 + 5.3$
15. (a) $-3\frac{1}{2} + 1\frac{3}{8}$ (b) $-\frac{11}{12} + \frac{5}{6}$
16. (a) $14 - 27$ (b) $33 - 29$
17. (a) $5.1 - 8.7$ (b) $4.3 - 7.5$
18. (a) $18 - 14$ (b) $26 - 38$
19. (a) $2 - 13$ (b) $13 - 62$
20. (a) $4\frac{3}{7} - 8\frac{1}{7}$ (b) $5\frac{2}{5} - 7\frac{4}{5}$
21. (a) $-9 - 22$ (b) $-3 - 15$
22. (a) $0 - 5$ (b) $0 - 12$
23. (a) $-3.7 - 8.1$ (b) $-5.2 - 10.4$
24. (a) $0 - \frac{3}{8}$ (b) $0 - 12.5$
25. (a) $(-9) + (-3) + (-5)$ (b) $11 + (-3) + (-5)$
26. (a) $(-17) + 12 + (-9)$ (b) $(-2.3) + (-3.2) + 0.5$
27. (a) $(-\frac{1}{2}) + (-\frac{1}{4}) + 1$ (b) $4 + (-\frac{5}{4}) + (-\frac{2}{3})$
28. (a) $6.9 + (-3.6) + (-4.3)$ (b) $(-102.5) + (-83.25) + (-14.25)$
29. (a) $21 + 15 - 11 - 5$ (b) $6 - 7 - 8 + 3$
30. (a) $-4 + 9 - 11 - 5 + 10$ (b) $7 - 2 - 5 + 4$
31. (a) $2.3 + 5.6 - 7.1 + 2.2$ (b) $8.2 - 5.5 - 3.2 - 2.1$
32. (a) $125 - 376 - 854 - 26$ (b) $300 - 500 + 100 - 750 - 350$
33. (a) $\frac{2}{3} + \frac{5}{6}$ (b) $2\frac{1}{4} - 3\frac{1}{2} - \frac{3}{4}$
34. (a) $8.2 - 9.4 + 1.1$ (b) $-5.3 - 4.9 - 0.8$
35. (a) $(5 - 7) + (2 - 5)$ (b) $(4.3 - 6.5) + (12.4 - 3.2)$

2-3: ADDITION AND SUBTRACTION OF REAL NUMBERS

36. (a) $5 - [2 + (-3 + 1)]$ (b) $1\frac{3}{5} - [4\frac{2}{3} + (-\frac{7}{15} + \frac{1}{3})]$
37. (a) $[3.2 + (-2.1 - 4.3)] + (-5.6)$ (b) $[-50 + (30 - 75)] + (-80)$
38. (a) $[-5 + (-7 + 3)] + (12 - 9)$ (b) $[(10 - 8) + (13 - 18) + (27 - 9)]$
39. (a) $(28 - 9) + [6 + (24 - 32)]$ (b) $(0.5 - 0.7) + [3.2 + (4.1 - 0.9)]$

Write and evaluate the mathematical expression for each of the following sentences.

40. (a) There are six employees in the sales office and four support employees for clerical work.
    (b) An English instructor used seven books in a literature class last semester but has chosen not to use two of them this semester.
41. (a) The car had 6 gallons of gasoline before filling up with 8 more gallons.
    (b) A full tank of 18 gallons of gasoline was purchased before a trip. The trip required 14 gallons of gasoline.
42. (a) My physics laboratory class requires 22 experiments finished during the semester. My laboratory partner and I have finished 16 of them.
    (b) A bank deposit of $250 was made, and then two withdrawals of $46 and $103 were made.
43. (a) A check for $56 against a checking account was written, followed by a check for $21 and a check for $75.
    (b) A fund-raising organization started out with $135 worth of expenses and then raised $375 on their first effort and $124 on their second effort.

Calculator Activities

Evaluate each of the following.

1. (a) $212 + (-571)$ (b) $(-1248) + (-10980)$
2. (a) $(-212) + (-571)$ (b) $986 + (-4089)$
3. (a) $(30.217) + (-23.996)$ (b) $-3.098 - 21.843 + 13.504$
4. (a) $(-6.702) + (-1.223) + (-5.031)$ (b) $1293 - 6.7 - 305.1$
5. (a) $(3.9 - 4.1) + [(-0.345) - 2]$ (b) $[489 - (3 - 4.81)] + 2089$

Write and evaluate a mathematical expression for each of the following sentences.

6. (a) The previous balance in Maria's checking account was $652.21. This month she had the following transactions: deposit of $43.25; check for $33.67; check for $102.99; check for $1.55; deposit of $23.40; check for $14.50; check for $12.20. What is her new balance after adding the deposits and subtracting the checks?
    (b) The trip across country required the following purchases of gasoline: 15.3 gal; 18.1 gal; 10.5 gal; 11.6 gal; 15.3 gal; 12.7 gal; 13.8 gal; 14.2 gal; 9.7 gal; 11.3 gal; 14.6 gal; and 13.7 gal. How much gas was used?

2-4 MULTIPLICATION AND DIVISION OF REAL NUMBERS

You know how to multiply positive numbers. For example,

$4 \times 7 = 28$

But do you multiply negative numbers, or positive and negative numbers? For example, what are $(-4) \times (-7)$ and $4 \times (-7)$? The key to answering these questions is in the observation that the *sum* of 4×7 and $4 \times (-7)$ is zero!

$4 \times 7 + 4 \times (-7) = 4 \times [7 + (-7)]$ *Distributive property*
$\qquad\qquad\qquad\qquad = 4 \times 0 \qquad\qquad 7 + (-7) = 0$
$\qquad\qquad\qquad\qquad = 0$

Since the sum of 4×7 and $4 \times (-7)$ is zero, each one of them is the negative of the other; thus,

$4 \times (-7) = -(4 \times 7)$

Hence,

$4 \times (-7) = -28$

Starting with $4 \times (-7)$ and $(-4) \times (-7)$ and using the same argument, you can show that $(-4) \times (-7)$ is the negative of $4 \times (-7)$. Thus,

$(-4) \times (-7) = -(-28) = 28 \qquad -(-a) = a$

Using absolute values, you can summarize the above results:

$4 \times 7 = |4| \times |7| = 28$
$4 \times (-7) = -(|4| \times |-7|) = -28$
$(-4) \times (-7) = |-4| \times |-7| = 28$

These results can be generalized with the following rules.

To multiply two numbers having the same sign, multiply their absolute values. That is,

$a \times b = |a| \times |b|$

if a and b have the same sign.

For example,

$(-5) \times (-12) = |-5| \times |-12|$
$\qquad\qquad\qquad = 5 \times 12, \text{ or } 60$

2-4: MULTIPLICATION AND DIVISION OF REAL NUMBERS

To multiply two numbers having opposite signs, multiply their absolute values and give the product the negative sign. That is,

$$a \times b = -(|a| \times |b|)$$

if a and b have opposite signs.

For example,

$$(-4) \times 16 = -(|-4| \times |16|)$$
$$= -(4 \times 16), \text{ or } -64$$

Quick Reinforcement

Find the value of each of the following.

(a) $(-5) \times 22$
(b) $5 \times (-80)$
(c) $(-9) \times (-11)$
(d) $(-12)(-12)$

Answers: (a) −110 (b) −400 (c) 99 (d) 144

You multiply several numbers by multiplying them two at a time in any order. Try to choose the order that will make the calculations easiest.

Problem 1 Find
(a) $(-5) \times 3 \times (-2)$
(b) $(-6) \cdot (-7) \cdot (-5)$
(c) $4 \cdot (-8) \cdot 10 \cdot 5$

SOLUTION:

(a) $(-5) \times 3 \times (-2) = [(-5) \times (-2)] \times 3$
$= (5 \times 2) \times 3$
$= 10 \times 3, \text{ or } 30$

(b) $(-6) \cdot (-7) \cdot (-5) = [(-5) \cdot (-6)] \cdot (-7)$
$= (5 \cdot 6) \cdot (-7)$
$= 30 \cdot (-7), \text{ or } -210$

(c) $4 \cdot (-8) \cdot 10 \cdot 5 = (4 \cdot 10) \cdot [(-8) \cdot 5]$
$= 40 \cdot (-40), \text{ or } -1600$

The operation of *division*, designated by ÷, is defined as follows:

$a \div b = c$ provided $a = b \times c$ *Notice the similarity to the definition of subtraction*

For example,

$12 \div 4 = 3$ *Because* $12 = 4 \times 3$
$60 \div 5 = 12$ *Because* $60 = 5 \times 12$

It is easy to divide negative as well as positive numbers. For example,

$42 \div (-6) = -7$ *Because* $42 = (-6) \times (-7)$
$(-36) \div 3 = -12$ *Because* $-36 = 3 \times (-12)$
$(-125) \div (-5) = 25$ *Because* $-125 = (-5) \times 25$

From these examples we can see the general rules for the division of real numbers.

To divide one number by another when both numbers have the same sign, divide their absolute values. That is,

$a \div b = |a| \div |b|$

if *a* and *b* have the same sign.

To divide one number by another when the numbers have opposite signs, divide their absolute values and give the quotient the negative sign. That is,

$a \div b = -(|a| \div |b|)$

if *a* and *b* have opposite signs.

Every rational number is a quotient of two integers. For example,

$\frac{3}{4} = 3 \div 4$

$-\frac{7}{12} = (-7) \div 12 \quad = 7 \div (-12)$

You multiply and divide two rational numbers as follows:

$\frac{a}{b} \times \frac{c}{d} = \frac{ac}{bd}$

$\frac{a}{b} \div \frac{c}{d} = \frac{a}{b} \times \frac{d}{c} = \frac{ad}{bc}$

Some examples of products and quotients of positive and negative rational numbers are given below.

2-4: MULTIPLICATION AND DIVISION OF REAL NUMBERS

Problem 2 Find

(a) $\frac{5}{2} \cdot \left(-\frac{3}{7}\right)$ (b) $\left(-\frac{5}{2}\right) \div \left(-\frac{1}{4}\right)$ (c) $\frac{2}{3} \times \left(-\frac{9}{4}\right)$

SOLUTION:

(a) $\frac{5}{2} \cdot \left(-\frac{3}{7}\right) = -\left(\frac{5}{2} \cdot \frac{3}{7}\right) = -\frac{15}{14}$

(b) $\left(-\frac{5}{2}\right) \div \left(-\frac{1}{4}\right) = \frac{5}{2} \cdot \frac{4}{1} = \frac{20}{2}$, or 10

(c) $\frac{2}{3} \times \left(-\frac{9}{4}\right) = -\left(\frac{2}{3} \times \frac{9}{4}\right) = -\frac{18}{12}$, or $-\frac{3}{2}$

Quick Reinforcement

Evaluate.

(a) $(-9) \div 3$ (b) $(-24) \div (-12)$ (c) $\left(-\frac{4}{3}\right) \cdot \frac{15}{8}$

(d) $\frac{2}{9} \div \left(-\frac{1}{3}\right)$ (e) $\left(-\frac{4}{7}\right) \div \left(-\frac{2}{21}\right)$

Answers (a) -3 (b) 2 (c) $-\frac{5}{2}$ (d) $-\frac{2}{3}$ (e) 6

Now you are ready to evaluate numerical expressions involving some or all of the four operations. Remember, you work within the innermost grouping symbols first and multiply and divide before you add and subtract, working from left to right.

Problem 3 Find

(a) $4 \cdot (-5) + 8 \div [(-2) + 6]$

(b) $3[4 - (-5)] \div [\frac{1}{2} \cdot (-4)] + \frac{1}{2}$

SOLUTION:

(a) $4 \cdot (-5) + 8 \div [(-2) + 6] = 4 \cdot (-5) + 8 \div 4$
$ = (-20) + 2$
$ = -18$

(b) $3[4-(-5)] \div \left[\frac{1}{2} \cdot (-4)\right] + \frac{1}{2} = 3(4+5) \div \left[-\left(\frac{1}{2} \cdot \frac{4}{1}\right)\right] + \frac{1}{2}$

$= 3(9) \div (-2) + \frac{1}{2}$

$= 27 \div (-2) + \frac{1}{2}$

$= -\frac{27}{2} + \frac{1}{2}$

$= -\frac{26}{2}$, or -13

Quick Reinforcement

Evaluate.

(a) $(5-9) \cdot 2 \div [(-3) + 1]$

(b) $\left(-\frac{4}{5}\right)\left(1 - \frac{3}{2}\right) + 1$

Answers (a) 4 (b) $\frac{7}{5}$

EXERCISES 2-4

Evaluate each of the following.

1. (a) 6×7 (b) 8×3
2. (a) $(-4) \times 3$ (b) $9 \times (-5)$
3. (a) $(-2) \times (-8)$ (b) $(-3) \times (-6)$
4. (a) $(-7) \times 5$ (b) $(-4) \times (-9)$
5. (a) $-(6 \times 8)$ (b) $-(4 \times 9)$
6. (a) $-[(-4) \times 7]$ (b) $-[5 \times (-6)]$
7. (a) $-[-(3 \times 5)]$ (b) $-[-(7 \times 1)]$
8. (a) $-[(-3) \times (-2)]$ (b) $-[(-7) \times (-8)]$
9. (a) $72 \div 9$ (b) $27 \div (-3)$
10. (a) $35 \div (-7)$ (b) $(-77) \div 11$
11. (a) $(-48) \div 8$ (b) $100 \div (-10)$
12. (a) $(-64) \div (-8)$ (b) $(-36) \div (-3)$
13. (a) $\frac{5}{8} \times \frac{4}{25}$ (b) $\frac{7}{15} \times \left(-\frac{5}{9}\right)$
14. (a) $\left(\frac{8}{21}\right) \times \left(-\frac{7}{16}\right)$ (b) $\left(-\frac{2}{3}\right) \times \left(\frac{9}{11}\right) \times \left(-\frac{22}{27}\right)$
15. (a) $\left(-\frac{1}{5}\right) \times \left(-\frac{2}{3}\right) \times \left(\frac{15}{22}\right)$ (b) $\frac{4}{7} \times \left(-\frac{3}{8}\right) \times \frac{1}{3}$

2-4: MULTIPLICATION AND DIVISION OF REAL NUMBERS

16. (a) $\frac{3}{7} \div \frac{5}{14}$ (b) $(\frac{9}{11}) \div (\frac{3}{4})$
17. (a) $(-\frac{3}{5}) \div (-\frac{9}{20})$ (b) $\frac{7}{9} \div (-\frac{14}{27})$
18. (a) $(-\frac{8}{15}) \div (\frac{4}{5})$ (b) $(-\frac{3}{8}) \div \frac{9}{4}$
19. (a) $(-2.3) \times (-1.5) \times 7$ (b) $(-0.9) \times (-1.2) \times [(-4.2) \times 8.0]$
20. (a) $6 \cdot (-3) + 12 \div [(-3) - 3]$ (b) $7[9 - (-5)]$
21. (a) $(-5) \times [8 - (-3)] \div (-6 - 5)$ (b) $-2[(-\frac{1}{2}) + \frac{3}{4}] \div (\frac{1}{4} - \frac{3}{8})$
22. (a) $(-1\frac{1}{2})(8 - 3\frac{1}{4}) \div (-2 + \frac{1}{2})$ (b) $-3[-(-7) + 2] - [(-5)(-5)]$
23. (a) $4(-5) + 8 \div [(-2) + 6]$ (b) $3[4 - (-5)] \div [\frac{1}{2} \cdot (-4)] + \frac{1}{2}$
24. (a) $\dfrac{4 \cdot (-2) \cdot 3}{(-8)(-5)}$ (b) $\dfrac{(-7) \cdot 9 \cdot (-10)}{(-15)(21)}$
25. (a) $\dfrac{(-\frac{1}{2})(-\frac{1}{4}) \cdot (-16)}{(-\frac{3}{5}) \cdot 25}$ (b) $\dfrac{(3.2) \cdot (-0.5)(5.0)}{(-2.25)(4.0)}$
26. (a) $4 - 2(3 - 5)$ (b) $8.3 - 2.5(12.3 - 4.3)$
27. (a) $[5 - (-2)][6 + (-11)]$ (b) $[13 - (-21)][-3 - (-9)]$
28. (a) $[3 \cdot (-9) - 2] - 4\{5 - [4 \div (-2)]\}$
    (b) $[25(-30 - \{-45\})][35 - (-5)(-20)]$

Write and evaluate a mathematical expression for each of the following sentences.

29. (a) The bookstore bought a dozen books at a cost of $4 each.
    (b) Jameson paid three bills, each for $15.
30. (a) The football team had five successive calls of 5 yards lost.
    (b) The football team had three successive calls of 7 yards gained.
31. (a) By the end of the second week of school, $\frac{1}{4}$ of the students in the chemistry class had dropped the class.
    (b) The partnership had losses of $3300 this last year, which had to be carried by the three partners. They each had an equal share.
32. (a) In another year, the partnership with three partners earned $9000, which they were able to share equally.
    (b) On four successive hands of blackjack, the player lost $10 each hand.
33. (a) The supermarket manager purchased five cases of toilet paper. Each case had 24 packages in it, and a package cost 30¢ wholesale.
    (b) A manufacturing plant had to deduct expenses from the profit of each item produced. Expenses on 350 items were $1.50 per item.

Calculator Activities

Evaluate each of the following to its best accuracy.

1. (a) $358 \cdot (-701)$ (b) $(-275.3) \cdot (-92.4)$
2. (a) $-(32.031 \times 3.926)$ (b) $-[(-1.032) \times (66.23)]$
3. (a) $33.011 \div (-12.395)$ (b) $(-1.23) \div (-5.32)$

62 CHAPTER 2: ORDER

4. (a) $(-21.77) \cdot (3.54) + (-6.58)$ (b) $789.1 - 12.0[3.5 \div (-2.0)]$
5. (a) $\frac{7}{20} + (-0.55 \times \frac{3}{100})$ (b) $481.0 - 15.0[4.8 \div (-6.0)]$
6. (a) $(67.91 - 53.83)(21.00 - 73.81)$ (b) $[\frac{3}{4} + (-0.50)] \div \frac{1}{40}$

2-5 POWERS AND SCIENTIFIC NOTATION

An important part of the development of mathematics has been the invention of a symbolic language of its own. In this language, words are replaced by symbols. For example:

WORDS	SYMBOLS
The square of 4	4^2, or $4 \cdot 4$
The cube of 4	4^3, or $4 \cdot 4 \cdot 4$
The positive number whose square is 5	$\sqrt{5}$
	Read "**square root** of 5"
The set of all numbers between -2 and 2	$\{x \mid -2 < x < 2\}$
	Read "the set of all real numbers x such that $-2 < x < 2$"

The following notation is very useful in mathematics. In this notation,

$6^1 = 6$
$6^2 = 6 \cdot 6$, or 36
$6^3 = 6 \cdot 6 \cdot 6$, or 216
$6^4 = 6 \cdot 6 \cdot 6 \cdot 6$, or 1296
$6^5 = 6 \cdot 6 \cdot 6 \cdot 6 \cdot 6$, or 7776

and so on. It is clear what 6^n means for every positive integer n. Using variables,

$x^n = \underbrace{x \cdot x \cdot x \cdot \ldots \cdot x}_{n\ x\text{'s}}$

The positive integer n is called the **exponent,** and x is called the **base.** The expression x^n is called the nth **power** of x. The second and third powers of x are given special names:

x^2 the **square** of x
x^3 the **cube** of x

2-5: POWERS AND SCIENTIFIC NOTATION

Exponents can be used to explain decimal notation. For example,

$87 = 8 \cdot 10^1 + 7$ $\qquad 10^1 = 10$

$436 = 4 \cdot 10^2 + 3 \cdot 10^1 + 6$ $\qquad 10^2 = 100$

$9652 = 9 \cdot 10^3 + 6 \cdot 10^2 + 5 \cdot 10^1 + 2$ $\qquad 10^3 = 1000$

$140{,}805 = 10^5 + 4 \cdot 10^4 + 8 \cdot 10^2 + 5$ $\qquad 100{,}000 + 40{,}000 + 800 + 5$

Special names are given to certain powers of 10:

$10^3 = 1000$ or one **thousand**

$10^6 = 1{,}000{,}000$ or one **million**

$10^9 = 1{,}000{,}000{,}000$ or one **billion**

$10^{12} = 1{,}000{,}000{,}000{,}000$ or one **trillion**

It is clear that we can express large numbers in a compact form by using exponents:

$1.28 \cdot 10^7$ meters *Diameter of the earth*

$3 \cdot 10^{11}$ dollars *Annual budget*

$1.5 \cdot 10^8$ kilometers *Distance to the sun*

The three numbers above are expressed in the form

$\boxed{A \cdot 10^n}$ *A greater than or equal to 1 and less than 10, n a positive integer*

A number in this form is said to be expressed in **scientific notation.**

Problem 1 Express each number in scientific notation.

(a) The speed of light is 300,000,000 meters per second.

(b) A light year is 9,460,000,000,000,000 meters (the distance traveled by a beam of light in one year).

(c) The distance between the sun and Mars is 229,000,000 kilometers.

SOLUTION:

(a) Move the decimal point from right to left until you have only one digit remaining to the left of the decimal point. The number of moves is the power of 10.

3.0̮.0̮.0̮.0̮.0̮.0̮.0̮.0̮. *Remaining number is*
8 moves   3.00000000, *or* 3
                 8 *moves, so* 10^8
300,000,000 = 3 · 10^8 *Scientific notation*

(b) 9.4̮.6̮.0̮.0̮.0̮.0̮.0̮.0̮.0̮.0̮.0̮.0̮.0̮.0̮.0̮.0̮. 9.46 *is remaining number*
              15 moves   15 *moves, so* 10^{15}
9,460,000,000,000,000 = 9.46 · 10^{15} *Scientific notation*

(c) 2.2̮.9̮.0̮.0̮.0̮.0̮.0̮.0̮. 2.29 *is remaining number*

      8 moves   8 *moves, so* 10^8
229,000,000 = 2.29 · 10^8 *Scientific notation*

Zero and negative exponents are also used in mathematics. We define the zero power of x as equal to 1:

$x^0 = 1$ *for all nonzero values of* x

For negative exponents,

$x^{-1} = \dfrac{1}{x}$

$x^{-2} = \dfrac{1}{x^2}$

$x^{-3} = \dfrac{1}{x^3}$

and so on. For example,

$2^{-3} = \dfrac{1}{2^3}$, or $\dfrac{1}{8}$

$5^{-4} = \dfrac{1}{5^4}$, or $\dfrac{1}{625}$

$10^{-1} = \dfrac{1}{10}$ or 0.1

$10^{-2} = \dfrac{1}{100}$, or 0.01

A number less than 1, such as 0.243, can be expressed in negative powers of 10:

0.243 = 0.2 + 0.04 + 0.003
     = 2 · 10^{-1} + 4 · 10^{-2} + 3 · 10^{-3}

2-5: POWERS AND SCIENTIFIC NOTATION

Any number expressed in decimal form can also be expressed in powers of 10:

$$82.79 = 8 \cdot 10^1 + 2 \cdot 10^0 + 7 \cdot 10^{-1} + 9 \cdot 10^{-2}$$
$$935.006 = 9 \cdot 10^2 + 3 \cdot 10^1 + 5 \cdot 10^0 + 6 \cdot 10^{-3}$$

Small numbers can be expressed in scientific notation by using negative exponents. Some examples are given below.

$0.0\,0\,0\,3 = 3 \cdot 10^{-4}$ *Move the decimal point from left to right until you have exactly one nonzero digit to the left of the decimal point*
4 moves

$0.0\,0\,8\,2 = 8.2 \cdot 10^{-3}$
3 moves

$0.0\,0\,0\,0\,0\,0\,7\,85 = 7.85 \cdot 10^{-7}$
7 moves

Some well-known physical constants are

$1.673 \cdot 10^{-24}$ gram: Mass of a hydrogen atom

$9.107 \cdot 10^{-28}$ gram: Mass of an electron

Problem 2 Express each decimal number in scientific notation.

(a) 1 pound = 0.45 kilogram
(b) 1 centimeter = 0.0328 foot
(c) Constant of gravitation = 0.0000000667 dyne.

SOLUTION:

(a) $0.45 = 4.5 \cdot 10^{-1}$
(b) $0.0328 = 3.28 \cdot 10^{-2}$
(c) $0.0000000667 = 6.67 \cdot 10^{-8}$

Quick Reinforcement

Express each number in scientific notation.

(a) 15680   (b) 0.00139

Answers: (a) 1.568×10^4 (b) 1.39×10^{-3}

EXERCISES 2-5

Express each integer in powers of 10.

1. (a) 93 (b) 79
2. (a) 104 (b) 112
3. (a) 267 (b) 304
4. (a) 9010 (b) 7112
5. (a) 3105 (b) 8893
6. (a) 114,000 (b) 201,000
7. (a) 63,980 (b) 75,505
8. (a) 938,325 (b) 407,032

Express each decimal in powers of 10.

9. (a) 0.201 (b) 0.972
10. (a) 6.03 (b) 9.22
11. (a) 4.791 (b) 8.838
12. (a) 11.52 (b) 13.377
13. (a) 81.963 (b) 60.002
14. (a) 27.001 (b) 38.305
15. (a) 27,000.02 (b) 530.901
16. (a) 380.07 (b) 5000.032

Express each number in scientific notation.

17. (a) 9,800,000 (b) 837,000,000
18. (a) 1200 (b) 5700
19. (a) 802,000 (b) 605,000
20. (a) 0.00053 (b) 0.000061
21. (a) 0.000000083 (b) 0.0000000095
22. (a) 0.00000077 (b) 0.0000000102
23. (a) 126,000,000,000 (b) 78,000,000,000,000
24. (a) 0.00000000001 (b) 0.00000000088

Express each number in scientific notation.

25. (a) The radius of the earth is approximately 6400 kilometers.
    (b) The moon is 386,000 kilometers from earth.
26. (a) Light travels at 30,000,000,000 centimeters per second in a vacuum.
    (b) The diameter of the orbit of the earth is 300,000,000 kilometers.
27. (a) The radius of the orbit of Venus is 109,000,000 kilometers.
    (b) A particle of matter is measured at 0.000000056 millimeter.

Calculator Activities

Use your calculator to express each number in scientific notation.

1. (a) 21^3 (b) 435^6
2. (a) 0.302^4 (b) 375^{-3}
3. (a) 6.8^9 (b) 3750^5
4. (a) $(51)^5 \div (51)^2$ (b) $(1.021)^4 \cdot (1.021)^2$
5. (a) 101^{-2} (b) $(2.2)^{-2} \cdot (3.1)^3$
6. (a) $[(-99) + 41]^{-2}$ (b) $(5.01 - 6.2)^3$
7. (a) $(5.31^2 + 12.3^3)^3$ (b) $(42.9^3 - 27.1^2)^4$
8. (a) $4.1 + 2(5 - 3.1)^2$ (b) $(-2.01)^2 + (4^8 \div 4^2)$
9. (a) $x^2 - 5$, when $x = 3.011$ (b) $12(x-1)^3$, when $x = 9$
10. (a) $2x^4$, when $x = -2.4$ (b) $7(3x - 2)^4$, when $x = 4.5$

2-6 SQUARE ROOTS AND APPROXIMATIONS

The opposite of squaring a number is taking its **square root.** For example, 9 is the square of 3; therefore, 3 is a square root of 9:

$$3^2 = 9$$

But 9 is also the square of -3; therefore, -3 is also a square root of 9:

$$(-3)^2 = 9$$

The symbol $\sqrt{}$, called a **radical sign,** is used to designate the positive square root of a positive number. Thus, $\sqrt{9}$ designates 3, the positive square root of 9:

$$\sqrt{9} = 3$$

The other square root of 9 is designated by $-\sqrt{9}$:

$\quad -\sqrt{9} = -3 \quad$ *There are two square roots of 9; 3 and -3*

We can now state another property of real numbers.

Every positive real number x has a positive square root $\sqrt{x}$.

For example,

$\quad \sqrt{144} = 12 \quad$ *Because $12^2 = 144$ and 12 is a positive number*

$\quad \sqrt{\dfrac{16}{25}} = \dfrac{4}{5} \quad \left(\dfrac{4}{5}\right)^2 = \dfrac{4}{5} \times \dfrac{4}{5}$, or $\dfrac{16}{25}$, and $\dfrac{4}{5} > 0$

Negative real numbers do not have real square roots; in fact, the square of every nonzero real number is a positive real number.

If you have a rectangular piece of land 300 meters by 400 meters (see Figure 2-9), how far is it from one corner of your land to the opposite corner? You can answer this question by using the **Pythagorean theorem,** which gives the relationship between the sides of a right triangle as

$$c^2 = a^2 + b^2$$

a and *b* are the lengths of the legs, *c* of the hypotenuse in Figure 2-10

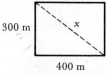

FIGURE 2-9

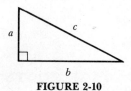

FIGURE 2-10

For the field shown in the figure,

$$x^2 = 300^2 + 400^2$$
$$= 90{,}000 + 160{,}000$$
$$= 250{,}000$$
$$x = \sqrt{250{,}000}, \text{ or } 500 \qquad 250{,}000 = 25 \cdot 10^4 = (5 \cdot 10^2)^2 = 500^2$$

Thus, the field is 500 meters from one corner to the opposite corner.

If the legs of a right triangle have lengths 5 and 12, as shown in Figure 2-11, then the length *c* of the hypotenuse is given by the equation

$$c^2 = 5^2 + 12^2$$

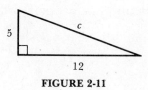

FIGURE 2-11

Therefore,

$$c = \sqrt{169}, \quad \text{or} \quad 13$$

2-6: SQUARE ROOTS AND APPROXIMATIONS

The lengths of the sides of a right triangle are not always integers as in the two preceding examples. Consider Figures 2-12 and 2-13:

$1^2 + 1^2 = c^2$ $1^2 + 2^2 = c^2$
$c^2 = 2$ $c^2 = 5$
$c = \sqrt{2}$ $c = \sqrt{5}$

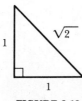

FIGURE 2-12

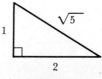

FIGURE 2-13

There is no integer whose square is 2, so $\sqrt{2}$ is not an integer. It is a real number, however. Similarly, $\sqrt{5}$ is not an integer.

The **perfect-square integers** are 0^2, 1^2, 2^2, 3^2, 4^2, and so on; that is,

0, 1, 4, 9, 16, 25, 36, 49, 64, 81, 100, . . .   *Infinitely many perfect square integers*

Stated another way, these are the integers whose square roots are also integers.

$\sqrt{4} = 2$, $\sqrt{64} = 8$, $\sqrt{121} = 11$

and so on. It may be shown that all the rest of the positive integers,

2, 3, 5, 6, 7, 8, 10, 11, . . .

have irrational square roots; that is,

$\sqrt{2}$, $\sqrt{3}$, $\sqrt{5}$, $\sqrt{6}$, $\sqrt{7}$, $\sqrt{8}$, $\sqrt{10}$, $\sqrt{11}$, . . .

are irrational numbers.

Some rational numbers are perfect-square rational numbers; for example,

$\frac{4}{9}, \frac{16}{25}, \frac{81}{49}, \frac{100}{121}$ $\frac{4}{9} = \left(\frac{2}{3}\right)^2, \frac{81}{49} = \left(\frac{9}{7}\right)^2$

As these examples illustrate, a rational number in simplest form (the numerator and denominator have no common factor other than 1 or -1) is a perfect square only if its numerator and denominator are perfect-square integers.

Irrational numbers such as $\sqrt{2}$, and $\sqrt{5}$ are not marked on a typical ruler. However, you can mark off $\sqrt{5}$ on a ruler by placing the ruler along the hypotenuse of a right triangle like the one shown in Figure 2-14, where one leg of the triangle measures 1 unit and the other leg measures 2 units. Other square roots can be marked in a similar fashion.

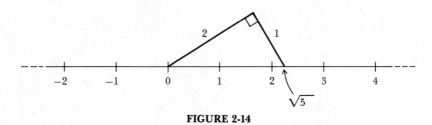

FIGURE 2-14

Once $\sqrt{5}$ is marked on a ruler, you can see its relationship to other marks already on the ruler. Suppose the ruler is marked off in meters, centimeters, and millimeters. By inspection,

$2 < \sqrt{5} < 3$ *Looking at the meter marks*
$2.2 < \sqrt{5} < 2.3$ *Looking at the centimeter marks*
$2.23 < \sqrt{5} < 2.24$ *Looking at the millimeter marks*

Thus, $\sqrt{5}$ is approximately 2 or 3; $\sqrt{5}$ is approximately 2.2 or 2.3; $\sqrt{5}$ is approximately 2.23 or 2.24. Clearly, a finer and finer scale on the ruler would allow a better and better approximation of $\sqrt{5}$. This example illustrates that any irrational number can be approximated by a rational number as accurately as you wish. In the example above, the rational number 2.23 (or 223/100) is a three-digit approximation of $\sqrt{5}$.

Problem 1 Find a three-digit approximation of $\sqrt{42}$.

SOLUTION:

$36 < 42 < 49$ *36 and 49 are consecutive perfect-square integers*
$6 < \sqrt{42} < 7$ *Take square roots*

Now find consecutive tenths between 6 and 7 whose squares surround 42. Use a calculator if you have one:

$6.2^2 = 38.44, \; 6.3^2 = 39.69, \; 6.4^2 = 40.96, \; 6.5^2 = 42.25$

2-6: SQUARE ROOTS AND APPROXIMATIONS

Thus,

$6.4^2 < 42 < 6.5^2$

or, taking square roots,

$6.4 < \sqrt{42} < 6.5$

Next, find consecutive hundredths between 6.4 and 6.5 whose squares surround 42.

$6.45^2 \approx 41.6$, $6.46^2 \approx 41.7$, $6.47^2 \approx 41.9$,
$6.48^2 \approx 41.99$, $6.49^2 \approx 42.1$

$\approx$ means "is approximately equal to"

Thus,

$6.48^2 < 42 < 6.49^2$

and

$6.48 < \sqrt{42} < 6.49$

We conclude that 6.48 or 6.49 is a three-digit approximation of $\sqrt{42}$. Since 6.48^2 is much closer to 42 than 6.49^2, you know 6.48 is a better approximation of $\sqrt{42}$ than 6.49.

Perhaps someone has in their home a square room whose area is 42 square meters. What is the length of one side of the room? From the formula $A = x^2$,

$42 = x^2$

and hence

$x = \sqrt{42}$

From the problem just solved,

$x \approx 6.48$

Thus, the length of each side is approximately 6.48 meters.

Quick Reinforcement

Find a two-digit approximation of each one.

(a) $\sqrt{200}$ (b) $\sqrt{53}$

Answers (a) 14 (b) 7.3

EXERCISES 2-6

Give the closest integers to each of the following square roots.

1. (a) $\sqrt{28}$ (b) $\sqrt{69}$
2. (a) $\sqrt{52}$ (b) $\sqrt{76}$
3. (a) $\sqrt{5}$ (b) $\sqrt{10}$
4. (a) $\sqrt{110}$ (b) $\sqrt{130}$
5. (a) $\sqrt{150}$ (b) $\sqrt{180}$
6. (a) $\sqrt{42}$ (b) $\sqrt{60}$
7. (a) $\sqrt{235}$ (b) $\sqrt{190}$
8. (a) $\sqrt{2}$ (b) $\sqrt{13}$

Find a two-digit approximation for each of the following square roots.

9. (a) $\sqrt{58}$ (b) $\sqrt{27}$
10. (a) $\sqrt{14}$ (b) $\sqrt{39}$
11. (a) $\sqrt{78}$ (b) $\sqrt{46}$
12. (a) $\sqrt{95}$ (b) $\sqrt{115}$
13. (a) $\sqrt{130}$ (b) $\sqrt{19}$
14. (a) $\sqrt{20}$ (b) $\sqrt{160}$
15. (a) $\sqrt{62}$ (b) $\sqrt{32}$
16. (a) $\sqrt{7}$ (b) $\sqrt{3}$
17. (a) $\sqrt{51}$ (b) $\sqrt{77}$
18. (a) $\sqrt{109}$ (b) $\sqrt{138}$
19. (a) $\sqrt{90}$ (b) $\sqrt{69}$
20. (a) $\sqrt{150}$ (b) $\sqrt{180}$

21. (a) Francis and Frances built a log cabin that is a perfect square. If the cabin has an area of 89 square meters, what is the length of one side of the cabin? Find the answer to the nearest centimeter.

    (b) Joshua decided to wallpaper one wall of his room. If the wall is a perfect square and has an area of 19 square meters, what dimensions does the wall have? Find the answer to the nearest centimeter.

22. (a) After placing all the air-conditioning equipment in the floor of the computer room, the workers needed to lay the false floor on top. If the computer room was a perfect square, and they ordered 205 square meters of tile, what were the dimensions of the computer room, to the nearest centimeter?

    (b) Carlos decided to build a birdhouse. If the front was a perfect square measuring 155 square centimeters, how wide did he make the birdhouse? Find the answer to the nearest millimeter.

KEY TERMS

Number line
Coordinate
Graph
Origin
Order properties
Positive
Negative
Absolute value
Square root
Exponent

Base
Power
Square
Cube
Scientific notation
Square root
Radical sign
Pythagorean theorem
Perfect-square integers

REVIEW EXERCISES

Graph each of the following sets of numbers on a number line.

1. (a) $\{5, 9, -2, -6\}$ (b) $\{11, 7, -5, -7\}$
2. (a) $\{\frac{4}{5}, \frac{7}{5}, -0.6, -2\}$ (b) $\{\frac{2}{3}, -0.75, 0, -1\}$

Find the set of coordinates for each of the following sets of points on the given number line.

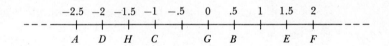

3. (a) $\{A, B, D, G\}$ (b) $\{C, D, H, E\}$
4. (a) $\{B, C, E, A\}$ (b) $\{F, A, E, G\}$

Tell which number of each pair is the greater by graphing each pair on a number line.

5. (a) $-7, -3$ (b) $5.4, -6.3$
6. (a) $-\frac{5}{4}, -\frac{4}{3}$ (b) $-2, -0.3$

Which order property makes each of the statements below true?

7. (a) $3 > 2$ and $2 > -3$ are true. Therefore, $3 > -3$ is true.
   (b) If $m < n$ is true, then $m + 8 < n + 8$ is true.
8. (a) If $a > b$ is true, and $5 > 0$ is true, then $5a > 5b$ is true.
   (b) If x is not less than 4, and x is not greater than 4, then $x = 4$ is true.

Evaluate each of the following.

9. (a) $|-3|$ (b) $-|7|$
10. (a) $(-6) + (-7.3)$ (b) $(-1.5) + 3 + (-0.5)$

11. (a) $7.3 - 6.2 + 5.7$ (b) $\frac{9}{2} + (-\frac{3}{4}) + \frac{1}{2}$
12. (a) $-5 + 7 - 11 - 2$ (b) $8.1 + 5.3 - 7.4 - 6.0$
13. (a) $(8 - 3) + (9 - 12)$ (b) $4 - [8 + (-3 + 5)]$
14. (a) $(-7) \cdot (12)$ (b) $-[8 \cdot (-3)]$
15. (a) $6[8 - (-4)]$
    (b) $(-3) \cdot [12 - (-2)] \div [10 + (-3)]$
16. (a) $\left(-\frac{10}{11}\right)\left(-\frac{22}{3}\right)\left(-\frac{9}{20}\right)$ (b) $\frac{(-12)(5)(-8)(-5)}{(-25)(6)(-2)}$

Write and evaluate the mathematical expression for each of the following sentences.

17. (a) A football team made 4 yards on the first play, lost 6 yards on the second play, but then gained 18 yards on the final play.
    (b) On 5 successive hands of blackjack, the player lost $15 each hand.
18. (a) With a starting balance of $320, a checking account showed checks written for $27, $63, $12, $5.50, and $72 and deposits made of $35 and $90.
    (b) A school business club started the year with $78.25 in its treasury and through fund raising earned $205.50 and $82.20. They put $30 aside for the national organization and paid $40 to a guest speaker.
19. (a) Forty people showed up to discuss taking the Sierra Club 15-mile hike. Only $\frac{5}{8}$ of those people actually took the hike.
    (b) A drugstore purchased 350 boxes of Band-aids at a wholesale cost of $0.78 per box and sold them all at a retail price of $0.90 per box.

Express each integer in terms of powers of 10.

20. (a) 205   (b) 1090
21. (a) 45,800   (b) 503,299

Express each decimal in terms of powers of 10.

22. (a) 2.34   (b) 0.863
23. (a) 7.9034   (b) 4003.0021

Express each number in scientific notation.

24. (a) 508,000,000   (b) 389,000,000,000
25. (a) 0.0000000092   (b) 0.00000000000075

Give the closest integers to each of the following square roots.

26. (a) $\sqrt{115}$ (b) $\sqrt{145}$
27. (a) $\sqrt{77}$ (b) $\sqrt{44}$

Find a two-digit approximation for each of the following square roots.

28. (a) $\sqrt{207}$ (b) $\sqrt{165}$

29. (a) $\sqrt{92}$ (b) $\sqrt{31}$
30. (a) Jennifer and Philip designed their own wedding invitations. The invitations were square and each had an area of 170 square centimeters. What were the dimensions of the invitations to the nearest millimeter?
   (b) Scott decided to build a square handball court. If the court had to be 60 square meters, what would its dimensions be to the nearest centimeter?

 Calculator Activities

Evaluate each of the following.

1. (a) $6523 + (-3427)$ (b) $(-2.043) + (-3.992)$
2. (a) $(-3.05) + (-5.12) + 7.40$ (b) $(3.9 - 4.2) + (5.7 - 2.6)$
3. (a) $(-832) \cdot (-521)$
   (b) $34.095 \cdot (-4.661) + (-1.483) \cdot (-0.101)$
4. (a) $\frac{9}{20} + [\frac{7}{50} \times (-0.62)]$
   (b) $(41.02 - 77.27)(-16.91 + 7.33)$

Use your calculator to express each number in scientific notation.

5. (a) 17^4 (b) 620^7
6. (a) 0.562^3 (b) 0.41^{-4}
7. (a) $32^3 \div 32^9$ (b) $(2.034)^5 \cdot (7.102)^3$
8. (a) $5.9 + 3(8.0 - 6.2)^3$ (b) $6(x^4 - 2)$, when $x = 203$

3 Equations in One Variable

The Hayashis are planning their route for their weekend ski trip to Lake Tahoe. They know that the distance of the first leg of the trip from San Francisco to Sacramento is 144 kilometers and that the second leg from Sacramento to Lake Tahoe is 210 kilometers. If they can average about 100 kilometers per hour on the first leg and 80 kilometers per hour on the second leg, what would be their expected time for the complete trip?

3-1 EQUATIONS AND THEIR SOLUTIONS

We study algebra to learn about various types of equations and develop methods of solving them. This is important because many real-world problems can be solved by solving related equations.

For example, suppose a manufacturer makes one item, hinges. His plant has a daily fixed cost of $1000 and a production cost of $3 per hinge. Thus, if he produces 200 hinges per day, his cost is

$3 \cdot 200 + 1000 = 600 + 1000 = \$1600$ per day

If he makes 1100 hinges, his cost is

$3 \cdot 1100 + 1000 = 3300 + 1000 = \$4300$ per day

How many hinges did he produce on Friday if his total cost that day was $5800? Since you do not know the number of hinges produced on Friday, you can refer to it as a variable x. Then

$$\begin{array}{rl} \$1000 & \text{Fixed costs} \\ \underline{3x} & \text{Production cost of } x \text{ hinges} \\ \$3x + 1000 & \text{Total costs on Friday} \end{array}$$

But you are told that the total cost is $5800. Therefore,

$3x + 1000 = 5800$

What value of x makes this equation true? You might answer this question as follows:

$3x = 5800 - 1000$ If $a + b = c$, then $a = c - b$

$3x = 4800$

$x = 4800 \div 3 = 1600$ If $ab = c$, then $a = c \div b$

Thus, 1600 hinges were made on Friday.

The equation $3x + 1000 = 5800$ is an example of an **equation in one variable** x. Other examples of equations in one variable are

$$3x + 1 = 5, \quad -2\left(y + \frac{1}{2}\right) = \frac{y}{3}, \quad 0.5x^2 - x + 1 = 0$$

If an equation involves only one variable, then any value of the variable that makes the equation true is called a **solution** of the equation. The set of all solutions is called the **solution set** of the equation. Given an equation in one variable and given a number, you can check to see if the number is a solution by replacing the variable by the number.

Problem 1 Is 5, -6, or 7 a solution of the following equation?

$2(y + 8) = 5(y - 1)$

SOLUTION:

Give the variable y the values 5, -6, and 7 in turn.

Let $y = 5$.   *This is a short way of saying "let y have the value 5"*

$2 \cdot (5 + 8) = 5 \cdot (5 - 1)$

$2 \cdot 13 = 5 \cdot 4$

$26 = 20$ *False: 5 is not a solution*

3-1: EQUATIONS AND THEIR SOLUTIONS

Let $y = -6$.

$$2 \cdot (-6 + 8) = 5 \cdot (-6 - 1)$$
$$2 \cdot 2 = 5 \cdot -7$$
$$4 = -35 \qquad \text{False: } -6 \text{ is not a solution}$$

Let $y = 7$.

$$2 \cdot (7 + 8) = 5 \cdot (7 - 1)$$
$$2 \cdot 15 = 5 \cdot 6$$
$$30 = 30 \qquad \text{True: } 7 \text{ is a solution of the given equation}$$

Problem 2 Which of the numbers -2, 2, and $\frac{1}{2}$ are solutions of the equation

$$2x(x + 3) = 2 + 3x$$

SOLUTION:

Let $x = -2$.

$$2 \cdot (-2) \cdot [(-2) + 3] = 2 + 3 \cdot (-2)$$
$$(-4) \cdot (1) = 2 + (-6)$$
$$-4 = -4 \qquad \text{True: } -2 \text{ is a solution}$$

Let $x = 2$.

$$2 \cdot 2 \cdot (2 + 3) = 2 + 3 \cdot 2$$
$$4 \cdot 5 = 2 + 6$$
$$20 = 8 \qquad \text{False: } 2 \text{ is not a solution}$$

Let $x = \frac{1}{2}$.

$$2 \cdot \frac{1}{2} \cdot \left(\frac{1}{2} + 3\right) = 2 + 3 \cdot \frac{1}{2}$$
$$1 \cdot \frac{7}{2} = 2 + \frac{3}{2}$$
$$\frac{7}{2} = \frac{7}{2} \qquad \text{True: } \frac{1}{2} \text{ is a solution}$$

80 CHAPTER 3: EQUATIONS IN ONE VARIABLE

Quick Reinforcement

Which of the numbers -1, 2, $\frac{3}{2}$, and 4 are solutions of the equation $x(x + 2) = 8(x - 1)$?

Answer 2, 4

EXERCISES 3-1

Find which values of the variable make each equation true.

1. (a) Which of the numbers 7, 8, 9, and 10 are solutions of
    $x + 3 = 12$
   (b) Which of the numbers 2, 3, 4, and 5 are solutions of
    $3x + 2 = 14$
2. (a) Which of the numbers 0, 1, -2, and 3 are solutions of
    $3y + 2 = y + 4$
   (b) Which of the numbers -1, 2, $\frac{1}{3}$, and 4 are solutions of
    $9x - 5 = 31$
3. (a) Which of the numbers 3, 5, -5, and 2 are solutions of
    $7z + 8 = 2z - 17$
   (b) Which of the numbers -4, 2, 0, and 4 are solutions of
    $x^2 + 3 = 19$
4. (a) Which of the numbers 12, 14, 16, and 18 are solutions of
    $2y + 4 = 4(y - 7)$
   (b) Which of the numbers -4, -3, -2, 1, and 3 are solutions of
    $6(m + 1) = 4m$
5. (a) Which of the numbers -4, $\frac{2}{3}$, 1, -3, and 3 are solutions of
    $x^2 + x = 12$
   (b) Which of the numbers -2, 0, 1, 2, and $-\frac{3}{4}$ are solutions of
    $x^2 = -4x - 4$
6. (a) Which of the numbers -1, 0, $\frac{1}{2}$, and $-\frac{1}{2}$ are solutions of
    $2y + 5 = 6$
   (b) Which of the numbers 2, -3, $\frac{1}{3}$, and $\frac{3}{4}$ are solutions of
    $4y + 2 = 3 + y$
7. (a) Which of the numbers 1, 2, 3, and 4 are solutions of
    $\frac{1}{2}y + 2 = 4$

(b) Which of the numbers $\frac{1}{2}$, -4, $\frac{3}{4}$, and 16 are solutions of
$$4 + \frac{1}{4}y = \frac{1}{2}y$$

8. (a) Which of the numbers 14, 7, -7, and -14 are solutions of
$$3(t+6) = 5(t-2)$$
(b) Which of the numbers -2, $-\frac{1}{2}$, -1, and 0 are solutions of
$$4(3x+2) = -2(x+3)$$

9. (a) Which of the numbers $\frac{7}{2}$, 0, 1, and -2 are solutions of
$$3(2x+7) = 42$$
(b) Which of the numbers -3, -2.4, -2.5, and 2 are solutions of
$$4y + 5 = 12y + 25$$

10. (a) Which of the numbers $\frac{1}{3}$, 4, $-\frac{1}{4}$, and $\frac{1}{4}$ are solutions of
$$5x - 3 = x - 2$$
(b) Which of the numbers $\frac{2}{3}$, -4, 0.75, and 1.5 are solutions of
$$8m - 5 = 1$$

Calculator Activities

Find which values of the variable make each equation true.

1. (a) Which of the numbers 1.2, -2, 2.61, and $\frac{1}{2}$ are solutions of
$$2x - 1.01 = 4.21$$
(b) Which of the numbers 4, 5.5, -10, and -16.1 are solutions of
$$0.2(x+1) = -3.02$$

2. (a) Which of the numbers -2.1, $\frac{1}{4}$, 0.01, and 3.102 are solutions of
$$1.62y + 1 = 1.125 + 1.12y$$
(b) Which of the numbers 0.012, 32.5, -2.01, and -15.04 are solutions of
$$0.5(y+1) - 1.01 = 0.495 + y$$

3. (a) Which of the numbers 205, 19, 51, and -5 are solutions of
$$255x - 1952 = 11053$$
(b) Which of the numbers 100, 85, -41, and -92 are solutions of
$$100{,}493 - 2145y = 453y + 339{,}509$$

4. (a) Which of the numbers 4.2, 3.2, -1, and -4 are solutions of
$$4.8(2.3x - 5.7) = 7.698$$
(b) Which of the numbers -3, -2, 1.1, and 3.4 are solutions of
$$1.93 + 0.82x = 0.07x - 0.32$$

82 CHAPTER 3: EQUATIONS IN ONE VARIABLE

5. (a) Which of the numbers 42, 50, -3, and -4 are solutions of
$$6302m + 9783 = -15{,}425$$
   (b) Which of the numbers 1.23, -1.23, 1.32, and -1.32 are solutions of
$$6.5y + 3.7 = y + 10.96$$

3-2 METHODS OF SOLVING EQUATIONS

You saw in the preceding section how to verify that a number is a solution of an equation involving one variable. In this section, you will be asked to *solve* such equations, that is, to *find* numbers that are solutions of given equations.

Two equations involving the same variable are said to be **equivalent** if they have the same solutions. If you solve one of two equivalent equations, you automatically solve the other. For example, if you are told that the two equations

$x + 3 = 7$ *Clearly, 4 is the solution*

and

$2(x - 1) = x + 2$

are equivalent, then you can easily solve the second equation; its solution is 4, because that is the solution of the first equation.

The additive property of equality suggests one way of forming a pair of equivalent equations. If an equation

$a = b$

is true, then so is the equation

$a + c = b + c$

for any value of c. For example, any value of x that makes the equation

$7 = x - 2$ *Its solution is 9*

true also makes the equation

$7 + 8 = (x - 2) + 8$ *Its solution is 9*

true. In other words, the two equations are equivalent.

We can state these observations as follows.

Addition Property for Equations

If you add the same number to each side of a given equation, the resulting equation is equivalent to the given one.

This property is used to solve the following equations. Adding the same number to each side of the equation enables us to isolate the variable on one side.

Problem 1 Solve the equation $x - 27 = 32$.

SOLUTION:

$$x - 27 = 32 \quad \text{Given equation}$$

equivalent $\quad x - 27 + 27 = 32 + 27 \quad$ Add 27 to each side of the equation to eliminate -27 from the left side

$$x + 0 = 59 \quad x \text{ is isolated on the left side}$$

$$x = 59$$

The equation $x = 59$ has 59 as its solution. This is also the solution of the equivalent given equation.

Problem 2 Solve the equation $14 = 19 + x$.

SOLUTION:

$$14 = 19 + x \quad \text{Given equation}$$

equivalent $\quad (-19) + 14 = (-19) + 19 + x \quad$ Add -19 to each side to eliminate 19 from the right side

$$-5 = 0 + x \quad x \text{ is isolated on the right side}$$

$$-5 = x$$

Thus, -5 is the solution of the given equation.

Another way to obtain equivalent equations is by the following property.

Multiplication Property for Equations

If you multiply each side of a given equation by the same nonzero number, the new equation is equivalent to the given one.

CHAPTER 3: EQUATIONS IN ONE VARIABLE

The following problems illustrate the use of the multiplication property.

Problem 3 Solve the equation $3y = 24$.

SOLUTION:

$\boxed{equivalent} \begin{cases} 3y = 24 & \text{Given equation} \\ \frac{1}{3} \cdot 3y = \frac{1}{3} \cdot 24 & \text{Multiply each side by } \frac{1}{3} \text{ to eliminate 3 from the left side} \\ 1 \cdot y = \frac{24}{3} & \text{y is isolated on the left side} \\ y = 8 & \end{cases}$

The given equation is equivalent to the final equation $y = 8$ by the multiplication property. Thus, 8 is the solution of the given equation.

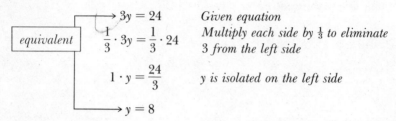

Problem 4 Solve the equation $\frac{4}{5}x = 20$.

SOLUTION:

$\boxed{equivalent} \begin{cases} \frac{4}{5}x = 20 & \text{Given equation} \\ \frac{5}{4} \cdot \frac{4}{5}x = \frac{5}{4} \cdot 20 & \text{Multiply each side by } \frac{5}{4}, \text{ the reciprocal of } \frac{4}{5}, \text{ to eliminate } \frac{4}{5} \text{ from the left side} \\ 1 \cdot x = \frac{100}{4} & \text{x is isolated on the left side} \\ x = 25 & \end{cases}$

Thus, the solution of the given equation is 25.

Both the addition and multiplication properties are used to solve the following problem.

Problem 5 Solve the equation $2a + 7 = 23$.

SOLUTION:

$\begin{aligned} 2a + 7 &= 23 & &\text{Given equation} \\ 2a + 7 + (-7) &= 23 + (-7) & &\text{Add } -7 \text{ to each side to eliminate 7 from the left side} \\ 2a + 0 &= 23 - 7 & &\text{2a is isolated on the left side} \\ 2a &= 16 & & \end{aligned}$

3-2: METHODS OF SOLVING EQUATIONS 85

$\frac{1}{2} \cdot 2a = \frac{1}{2} \cdot 16$ *Multiply each side by $\frac{1}{2}$ to eliminate 2 from the left side*

$1 \cdot a = \frac{16}{2}$ *a is isolated on the left side*

$a = 8$

The equations above are equivalent to each other. Thus, the given equation is equivalent to the last equation, $a = 8$, and 8 is the solution of the given equation.

√ CHECK:

$2a + 7 = 23$

Let $a = 8$.

$2 \cdot 8 + 7 = 23$

$16 + 7 = 23$ √   √ *means the equation is true*

You check the solution of an equation to make sure you did not make an arithmetical error while solving the equation. It is important to check your solution in the given equation, replacing the variable, each time it occurs, by your solution. If the resulting equation involving only numbers is true, your solution is correct; otherwise, you made a mistake somewhere and must go back and check your arithmetic.

In some equations, you can combine like terms on one or both sides of the equation before isolating the variable.

Problem 6 Solve the equation $10 + 6 = 3b + 2 - 1$.

SOLUTION:

$10 + 6 = 3b + 2 - 1$ *Given equation*

$16 = 3b + 1$ *Combine like terms*

$16 + (-1) = 3b + 1 + (-1)$ *Add -1 to each side to eliminate 1 from the right side*

$16 - 1 = 3b + 0$ *3b is isolated on the right side*

$15 = 3b$

$\frac{1}{3} \cdot 15 = \frac{1}{3} \cdot 3b$ *Multiply each side by $\frac{1}{3}$ to eliminate 3 from the right side*

$\frac{15}{3} = 1 \cdot b$ *b is isolated on the right side*

$5 = b$

Since the given equation is equivalent to $5 = b$, the given equation must have 5 as its solution.

✓ CHECK:

$$10 + 6 = 3b + 2 - 1$$

Let $b = 5$.

$$10 + 6 = 3 \cdot 5 + 2 - 1$$
$$10 + 6 = 15 + 2 - 1$$
$$16 = 16 \checkmark$$

Quick Reinforcement

Solve the following equations.

(a) $5 + x = -3$ (b) $4 = \dfrac{1}{3}x$ (c) $2x - 3 = 4$

Answers (a) -8 (b) 12 (c) $\dfrac{7}{2}$

EXERCISES 3-2

Solve each equation by using the addition property. Check all solutions.

1. (a) $x + 11 = 12$ (b) $y - 15 = 2$
2. (a) $10 = 3 + t$ (b) $5 = t - 3$
3. (a) $14 + 12 + x = 28$ (b) $6 - 13 = y - 7$
4. (a) $15 - x = 7$ (b) $32 - 11 + t = 9$
5. (a) $23 = t + 2 - 10$ (b) $x + 3 - 8 = 2$

Solve each equation by using the multiplication property. Check all solutions.

6. (a) $2x = 10$ (b) $3y = 9$
7. (a) $5y = 30$ (b) $7m = 28$
8. (a) $\tfrac{1}{2}x = 4$ (b) $\tfrac{1}{3}y = 2$
9. (a) $\tfrac{2}{3}t = \tfrac{5}{6}$ (b) $\tfrac{3}{5}x = 8$
10. (a) $\tfrac{1}{7}x = 7$ (b) $\tfrac{2}{9}y = \tfrac{4}{3}$

3-3: EQUATIONS IN THE REAL WORLD   87

Solve each equation by using the addition property, the multiplication property, or both. Check all solutions.

11. (a) $4 = 12 + 2x$ (b) $16 + 3x = 1$
12. (a) $3y + 5 = 11$ (b) $7y - 4 = 3$
13. (a) $4m + 20 - 6 = 34$ (b) $6 + 3t = 14 - 2$
14. (a) $3 = 5x + 7$ (b) $7y + 5 - 2 = 52$
15. (a) $27 + 6x = 15 - 6$ (b) $13 + 5 = 9x$
16. (a) $22 - 1 + 5 = 4y + 4$ (b) $12 - 4 + 2 = 5 + 6s$
17. (a) $7x - 5 = 13 - 31$ (b) $6 + 3 - 18 = 8y - 7$
18. (a) $42 + 6x = 12 - 6$ (b) $5t + 25 = 100 - 50$

 Calculator Activities

Solve each equation by using the addition property, the multiplication property, or both. Be sure to check your solutions. Round answers to the nearest thousandth.

1. (a) $21.1 = x - 0.009$ (b) $2.01x + 0.5 = 6.1$
2. (a) $-0.34x + 1.025 = -50.2$ (b) $1.4 - 0.029 = 1.3x + 5$
3. (a) $255x - 1952 = 11,053$ (b) $100,493 - 2598y = 339,509$
4. (a) $2x + 12.3 = 0.7$ (b) $13,933 - 1950 = 45p - 302$
5. (a) $\frac{t}{3} = -3.64$ (b) $\frac{y}{2} = 3.66$

3-3 EQUATIONS IN THE REAL WORLD

Real-world problems can sometimes be analyzed by translating them into algebraic language. The "hinge" problem at the beginning of this chapter is an example of such a problem. Another example is given below.

Problem 1 Cities A and B have a pollution count totaling 4215 units of particulates. City A has twice as many units of particulates as City B. Determine the pollution counts of the two cities.

SOLUTION:

In symbolic form, the first sentence says:

Pollution count of City B + Pollution count of City A = 4215

The second sentence says:

Pollution count of City A = 2 · pollution count of City B

The third sentence asks:

What is the pollution count of each city?

Because you do not know the pollution count of either City A or City B, you arbitrarily choose one of them to be a variable:

Let x = pollution count of City B

By the second sentence,

Pollution count of City $A = 2x$

By the first sentence,

$x + 2x = 4215$ *Equation relating the bits of information*

Solving this equation,

$3x = 4215$

$x = \dfrac{1}{3} \cdot 4215$

$x = 1405$

Therefore, City B has a pollution count of 1405 units, and City A has a pollution count of 2810 units.

✓ CHECK:

Their sum is $1405 + 2810 = 4215$ ✓

The diagram in Figure 3-1 is called a **flowchart.** Its purpose is to help you analyze a given applied problem. To use it, you follow the arrows beginning at "Start." There are work boxes, an input box, a solution box, and two decision boxes that contain questions having "yes" or "no" answers. How you answer each question determines your path out of the box. If you answer "no," you must go back to the start. This path is called a *loop.* If you answer "yes," you can continue down the flowchart. Eventually, you will solve the problem.

Problem 2 A soda pop machine accepts nickels, dimes, and quarters. One day, Mack the machine man collected 182 coins. Mack observed that there were 17 more nickels than dimes, and half as many quarters as dimes. How many of each coin were there?

SOLUTION:

182 coins

3-3: EQUATIONS IN THE REAL WORLD

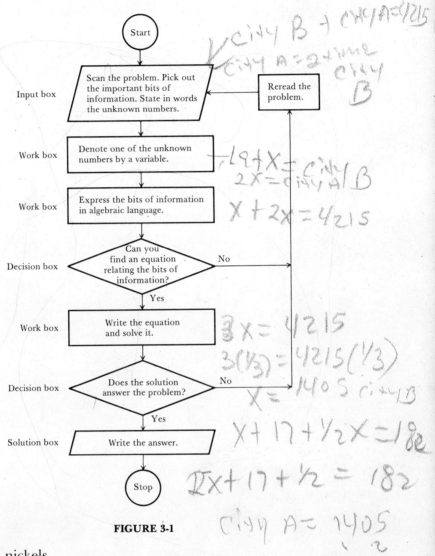

FIGURE 3-1

17 more nickels than dimes

Half as many quarters as dimes

Number of each kind of coin

Let x = number of dimes

Bits of information

Unknown number

It is easy to find the number of nickels and quarters if we know the number of dimes

90 CHAPTER 3: EQUATIONS IN ONE VARIABLE

182 coins

$17 + x$ = number of nickels   *Algebraic language*

$\frac{1}{2}x$ = number of quarters

$182 = x + (17 + x) + \frac{1}{2}x$ *The equation states that the total number of coins is the sum of the number of dimes, nickels, and quarters*

Solve this equation.

$182 = x + (17 + x) + \frac{1}{2}x$

■ $x + x + \frac{1}{2}x = \left(1 + 1 + \frac{1}{2}\right)x$

$= \frac{5}{2}x$

$182 = \frac{5}{2}x + 17$

$165 = \frac{5}{2}x$ *Add -17 to each side*

$\frac{2}{5} \cdot 165 = \frac{2}{5} \cdot \frac{5}{2}x$ *Multiply each side by $\frac{2}{5}$*

$66 = x$ $\frac{2}{5} \cdot 165 = 2 \cdot \frac{165}{5}$, or $2 \cdot 33$

There were 66 dimes, $17 + 66$, or 83, nickels, and $\frac{1}{2} \cdot 66$, or 33, quarters.

√ CHECK:

$66 + 83 + 33 = 182$ √

Problem 3 While visiting Julio and his nephew Dario, some friends asked how old Dario is. With a grin Julio said, "I am 3 times as old as he is. Three years ago the sum of our ages was 22 years." If you were their friends, could you calculate Dario's age from this information?

SOLUTION:

We want Dario's age

Julio's age is 3 times
Dario's age now   *Bits of information*

3-3: EQUATIONS IN THE REAL WORLD   91

3 years ago the sum
of their ages was 22

Let x = Dario's age now   *Variable*

A table can help show the relationships between the bits of information.

	Now	3 years ago
Dario	x	$x - 3$
Uncle Julio	$3x$	$3x - 3$
Sum of their ages		22

To get ages in the past, subtract the number of years from the current ages

We know from the given information that the ages in the column labeled "3 years ago" can be added to get the total at the bottom of the column.

$(x - 3) + (3x - 3) = 22$ *Equation relating bits of information*
$x - 3 + 3x - 3 = 22$
$4x - 6 = 22$
$4x = 28$
$x = 7$

Dario is 7 years old. (We also know that Uncle Julio is 21, since he is 3 times as old as Dario.)

✓ CHECK:

$(7 - 3) + (21 - 3) = 22$
$4 + 18 = 22$
$22 = 22$ ✓

Problem 4 Lorella bought 40 meters of fencing with which to enclose a rectangular vegetable garden. If her plans called for the length of the garden to be 2 meters more than the width, what would be the dimensions of her garden?

SOLUTION:

We want the length and width
Perimeter is 40 meters   *Bits of information*
The length is 2 more
than the width

92 CHAPTER 3: EQUATIONS IN ONE VARIABLE

Let x = width

Since the length is given in terms of the width, it is easier to let the width be our variable

In any problem involving geometric shapes, a diagram like the one in Figure 3-2 is extremely helpful.

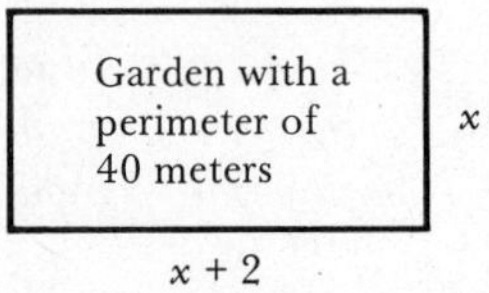

```
┌─────────────────┐
│ Garden with a   │
│ perimeter of    │  x
│ 40 meters       │
└─────────────────┘
        x + 2
```

The length is 2 more than the width, or $x + 2$

FIGURE 3-2

We recall that the formula for the perimeter of a rectangle is

$P = 2l + 2w$

Substituting $x + 2$ for the length l, and x for the width w we get

$40 = 2(x + 2) + 2(x)$
$40 = 2x + 4 + 2x$
$40 = 4x + 4$ *Subtract 4 from both sides*
$36 = 4x$ *Divide both sides by 4*
$9 = x$

The width is 9 meters and the length is $9 + 2$, or 11, meters.

✓ CHECK: $9+2 = 11$

$40 = 2(11) + 2(9)$
$40 = 22 + 18$
$40 = 40$ ✓

Problem 5 Donna Delaney wants to invest enough money now so that she will have $15,500 in her account one year from now. The rate of interest on her investment is 9.5% per year. How much should Donna invest?

SOLUTION:

Donna wants to invest some money
Rate of interest is 9.5% per year   *Bits of information*
Donna wants $15,500 in her account one year from now

Let x = dollars invested now   *Variable*

$0.095x$ = dollars of interest one year from now

$15{,}500 = x + 0.095x$ *Equation relating bits of information*

$15{,}500 = 1 \cdot x + 0.095x$ $x = 1 \cdot x$

$15{,}500 = (1 + 0.095)x$

$15{,}500 = 1.095 \cdot x$

$\dfrac{15{,}500}{1.095} = x$ *Divide each side by* 1.095

$x \approx 14{,}155$ *Using calculator*

Donna needs to invest \$14,155 now to reach her goal of having \$15,500 one year from now.

Quick Reinforcement

Can John's dad be three times as old as John if the sum of their ages equals 56? Determine the age of each for this to be possible.

Answers 14, 42

EXERCISES 3-3

1. (a) A father is three times as old as his son, and the difference in their ages is 24 years. How old is each?
   (b) There are two brothers, Sam and Mike. If the sum of their ages is 16, and three times Sam's age is Mike's age, how old are the brothers?
2. (a) Two students shared in the purchase of an \$86 electronic calculator. One paid \$26 more than five times what the other student paid. What did each pay?
   (b) Helen, trying to get out of debt, paid \$750 to two creditors. If one creditor demanded four times as much as the other, what did she pay to each one?
3. (a) A rope is 25 meters long. For Joseph's purpose, he must cut it into two pieces, with one piece 6 meters longer than the other. What will be the lengths of the two pieces?

94 CHAPTER 3: EQUATIONS IN ONE VARIABLE

(b) As office manager, Judy is responsible for setting up the partitions in the new travel agency. The office has 250 square meters of floor space and the partitions must divide the room into two areas—the reception room and the private offices. Seven times the area of the offices should equal twice the area of the reception room. What is the area of each?

4. (a) During the first act at a Las Vegas nightclub, a person in the audience tells the performing mind reader that he has 18 coins worth $3.60 in his pockets, made up of quarters and dimes. Within seconds, the mind reader tells him how many of each coin he has. Can you?

(b) Late for work, Jim finds he is out of change. Since the bus only takes exact change, he borrows all the money from his son's piggy bank, leaving an I.O.U. He finds that the bank contains only dimes and nickels. If the total value of the 15 coins in the bank is $1.10, how many of each type of coin did Jim borrow?

5. (a) The Environmental Protection Agency tells Company B to reduce its pollution particulate count to 1050 units. This will meet present air standards. This is 9500 units less than twice its present count. How much pollution is Company B currently pouring into the atmosphere?

(b) While preparing foods in the hospital kitchen, the dietician found that each meal needed 144 units made up of three nutritional components. If each meal needed four times as many units of component B as component A, and seven times as many units of component C as component A, how many units of each component would there be in each meal?

6. (a) In business, we compute revenue earned by multiplying the number of items sold by the difference between the selling price of each item and the cost of each item, and then subtracting fixed costs. If the fixed costs on a certain item are $2400, the cost per item is $1, and the selling price is $3 per item, how many of these items must be sold if a revenue of $15,400 is desired?

(b) To meet cost-of-living increases, a person must earn $20,900 next year if that person is earning $19,000 now. What is the cost-of-living increase as a percent?

7. (a) As a reward for running a marathon, T-shirts will be given to the participants. A total of 1050 runners are expected. If you order the same amount of small and extra-large sizes, twice as many medium as small, and twice as many large as extra-large, how many of each size will you order?

(b) In preparing for a backpack trip, a group decided to take freeze-dried packages of food, since such food was both light and nutritional. They decided that they needed twice as many packages of desserts as main courses (the dessert packages were small!), three times as many vegetable packages as main courses, and four more starch packages than

3-3: EQUATIONS IN THE REAL WORLD 95

dessert packages. If they were going to carry 44 packages of freeze-dried food, how many of each type were they going to carry?

8. (a) In fencing a rectangular vegetable garden, Manuel noticed that the length was exactly twice the width and that the perimeter of the garden was 9.6 meters. What were the dimensions of the vegetable garden?

(b) A stained-glass window is in the shape of a triangle. Two of the sides of the window are equal in length, and the third side, the base of the window, is 50 centimeters shorter than the others. If the perimeter of the window is 340 centimeters, how long are the sides?

Nostalgia Problems

1. (a) My horse and saddle are worth $100, and my horse is worth 7 times as much as my saddle. What is the value of each?

(b) A farmer has 4 times as many cows as horses and 5 times as many sheep as cows, and there are 100 animals altogether. How many horses does he have?

2. (a) A girl had 120 pins and needles, and she had seven times as many pins as needles. How many did she have of each?

(b) A teacher said that her school consisted of 64 scholars and that there were 3 times as many in arithmetic as in algebra and 4 times as many in grammar as in arithmetic. How many were there in each study?

Calculator Activities

1. (a) To meet the continual cost-of-living increases, Shirley computed that she must earn $21,406 next year just to stay even with inflation. If she is currently earning $19,250, what is the cost-of-living percentage?

(b) The Barchas family was designing a new home. They decided that they needed four and a half times as much square footage for the bedrooms and bathrooms as for the kitchen. They also wanted two and a quarter times as much room for the family room as for the kitchen and three and a half times as much room for the living room and dining room as for the kitchen. If they were planning on a 2250-square-foot house, what would be the size of each room?

2. (a) You are told that your offer on a new housing project is $1.1 million above your competitor's, yet both bids together total $3.25 million. How much did your competitor bid?

(b) You must keep a certain amount of money in a savings account that receives $5\frac{1}{4}\%$ interest. To beat inflation you decide to invest 9.2 times as much money in a mutual fund that earns 15.5%. If you have $18,000 to invest, how much should be invested in the mutual fund?

3. (a) The Lucas family decided to divide their money into three parts and invest the parts separately. They wanted to have an annual income of

$2500 from the investments. One part would be invested at 9%; the second part, which would be $500 less than the first part, would be invested at 9.5%; and the third part, which would be double the second, would be invested at 10.2%. How much money would they invest at each rate (to the nearest penny), and what would be the total amount invested?

(b) Lucia Monroe, Lee Sang, and Mel Zucker were planning to go into partnership to establish a natural foods store. Lee Sang's share was to be 0.8 of Lucia Monroe's share, but Mel Zucker's share was to be 1.4 of Lucia Monroe's share. If the total amount of starting capital was to be $21,840, how much did each of the partners contribute?

3-4 MORE SOLVING

Here are more problems involving equations in one variable. Remember, the goal is to isolate the variable on one side of the equation.

Problem 1 Solve the equation $5x - 1 = 3x + 17$.

SOLUTION:

$$5x - 1 = 3x + 17$$ — *Given equation*

$$(-3x) + 5x - 1 = (-3x) + 3x + 17$$ — *There are x-terms on both sides; add $-3x$ to each side to eliminate $3x$ from the right side*

$$(-3 + 5)x - 1 = 0 + 17$$ $-3x + 5x = (-3)x + 5x,$ or $(-3 + 5)x$

$$2x - 1 = 17$$

$$2x = 18$$

$$\frac{1}{2}(2x) = \frac{1}{2}(18)$$

$$1 \cdot x = \frac{18}{2}$$

$$x = 9$$

The solution is 9.

√ CHECK:

$$5x - 1 = 3x + 17$$

Let $x = 9$.

$$5 \cdot 9 - 1 = 3 \cdot 9 + 17$$
$$45 - 1 = 27 + 17$$
$$44 = 44 \checkmark$$

Problem 2 Solve the equation $3t + 23 = 3 - 2t$.

SOLUTION:

$$3t + 23 = 3 - 2t$$
$$3t + 23 + 2t = 3 - 2t + 2t \qquad \text{Add } 2t \text{ to each side to eliminate } -2t \text{ from the right side}$$
$$5t + 23 = 3$$
$$5t + 23 - 23 = 3 - 23 \qquad \text{Add } -23 \text{ to each side to eliminate } 23 \text{ from the left side}$$
$$5t = -20$$
$$\frac{1}{5} \cdot 5t = \frac{1}{5}(-20) \qquad \text{Multiply each side by } \frac{1}{5} \text{ to eliminate the numerical factor in } 5t$$
$$t = -4$$

The equation $3t + 23 = 3 - 2t$ has the solution -4.

CHECK:

$$3t + 23 = 3 - 2t$$

Let $t = -4$.

$$3(-4) + 23 = 3 - 2(-4)$$
$$-12 + 23 = 3 + 8$$
$$11 = 11 \checkmark$$

Problem 3 Solve the equation $3(a + 1) + 4 = 5a - 1$.

SOLUTION:

First eliminate the parentheses, then solve as before.

$$3(a + 1) + 4 = 5a - 1 \qquad \text{Given equation}$$
$$3a + 3 + 4 = 5a - 1$$
$$-3a + 3a + 7 = -3a + 5a - 1 \qquad \text{Add } -3a \text{ to each side}$$
$$7 = 2a - 1$$
$$8 = 2a \qquad \text{Add 1 to each side}$$
$$4 = a \qquad \text{Multiply each side by } \frac{1}{2}$$

98 CHAPTER 3: EQUATIONS IN ONE VARIABLE

✓ CHECK:
$$3(a+1)+4 = 5a-1$$
Let $a = 4$.
$$3(4+1)+4 = 5 \cdot 4 - 1$$
$$3 \cdot 5 + 4 = 20 - 1$$
$$15 + 4 = 19$$
$$19 = 19 \checkmark$$

Thus, 4 is the solution of the given equation.

Problem 4 Solve the equation $\frac{2}{3}y + 2 = \frac{5}{6} + y$.

SOLUTION:

When an equation involves fractions, you can get a simpler equivalent equation by multiplying each side by the least common multiple (LCM) of the denominators. The new equation will involve only integers and the variable.

$\frac{2}{3}y + 2 = \frac{5}{6} + y$	Given equation
$6\left(\frac{2}{3}y + 2\right) = 6\left(\frac{5}{6} + y\right)$	Multiply each side by 6, the LCM of the denominators
$6 \cdot \frac{2}{3}y + 6 \cdot 2 = 6 \cdot \frac{5}{6} + 6y$	
$4y + 12 = 5 + 6y$	
$12 = 5 + 2y$	Add $-4y$ to each side
$7 = 2y$	Add -5 to each side
$\frac{7}{2} = y$	Multiply each side by $\frac{1}{2}$

✓ CHECK:
$$\frac{2}{3}y + 2 = \frac{5}{6} + y$$

Let $y = \frac{7}{2}$.
$$\frac{2}{3} \cdot \frac{7}{2} + 2 = \frac{5}{6} + \frac{7}{2}$$
$$\frac{7}{3} + 2 = \frac{5}{6} + \frac{7}{2}$$

$$\frac{14}{6} + \frac{12}{6} = \frac{5}{6} + \frac{21}{6}$$

$$\frac{26}{6} = \frac{26}{6} \checkmark$$

Thus, $\frac{7}{2}$ is a solution of the given equation.

The following problem involves fractions and grouping symbols.

Problem 5 Solve the equation $\frac{1}{2}(x-2) + 1 = \frac{1}{4} - x$.

SOLUTION:

$$\frac{1}{2}(x-2) + 1 = \frac{1}{4} - x \qquad \text{Given equation}$$

$$4\left[\frac{1}{2}(x-2) + 1\right] = 4\left[\frac{1}{4} - x\right] \qquad \text{Multiply each side by 4, the LCM of the denominators}$$

$$4 \cdot \frac{1}{2}(x-2) + 4 \cdot 1 = 4 \cdot \frac{1}{4} - 4x \qquad \text{Distributive property}$$

$$2(x-2) + 4 = 1 - 4x$$

$$2x - 4 + 4 = 1 - 4x$$

$$2x = 1 - 4x$$

$$6x = 1 \qquad \text{Add } 4x \text{ to each side}$$

$$x = \frac{1}{6} \qquad \text{Multiply each side by } \frac{1}{6}$$

✓ CHECK:

$$\frac{1}{2}(x-2) + 1 = \frac{1}{4} - x$$

Let $x = \frac{1}{6}$.

$$\frac{1}{2}\left(\frac{1}{6} - 2\right) + 1 = \frac{1}{4} - \frac{1}{6}$$

$$\frac{1}{2}\left(-\frac{11}{6}\right) + 1 = \frac{3}{12} - \frac{2}{12}$$

$$-\frac{11}{12} + 1 = \frac{1}{12}$$

$$\frac{1}{12} = \frac{1}{12} \checkmark$$

Thus, $\frac{1}{6}$ is a solution of the given equation.

Quick Reinforcement

Solve each of the following equations.

(a) $3x - 1 = 5 - 2x$
(b) $5 + x = 2(1 - x) + 1$
(c) $\frac{x+1}{2} = \frac{x}{4} - 1$

Answers (a) $\frac{6}{5}$ (b) $-\frac{2}{3}$ (c) -6

EXERCISES 3-4

Solve each of the following equations. Check each solution in the given equation.

1. (a) $3x - 2x + 4 = 3$ (b) $6x + 3 = 5x - 7$
2. (a) $9y + 3 = 8y + 6$ (b) $6z - 7 = 7z - 8$
3. (a) $10x + 5 = 9x - 11$ (b) $2m + 9 = m - 1$
4. (a) $5x + 22 - 2x = 31$ (b) $3t + 12 = 7t + 102$
5. (a) $10x - 6x + 14 = 62$ (b) $5x - 10 = 3x + 12$
6. (a) $3y - 20 = -y - 4$ (b) $4m + 45 = 7m - 30$
7. (a) $\frac{3x}{4} = 25$ (b) $\frac{x}{5} + 25 = 30$
8. (a) $\frac{x}{4} + 3 = 7$ (b) $4x - 3 = x - 3$
9. (a) $4x - 2 = x + 7$ (b) $x - 1 = 4x - 91$
10. (a) $3t + 2 = t + 8$ (b) $4m - 12 = m + 6$
11. (a) $2(x - 11) = 20$ (b) $3(x + 4) = 5$
12. (a) $5x + 22 - 2x = 31$ (b) $y = 4(y - 21)$
13. (a) $3x - 20 = -(x + 4)$ (b) $\frac{3}{5}(x - 3) = 3$
14. (a) $2y + 4 = 4(y - 7)$ (b) $5(2 + 2t) = 0$
15. (a) $-3(m - 2) = 4(m + 4)$ (b) $4(m - 1) = 3(m - 2)$
16. (a) $3(x - 1) = 3(x + 4)$ (b) $5(-x + 3) = (x + 10) - 5$
17. (a) $5(m + 1) + 6(m + 2) = 6(m + 7)$
    (b) $3(t + 1) + 4(t + 2) = 6(t + 3)$
18. (a) $7(x + 3) = 45 + 4(3x - 16)$
    (b) $\frac{6}{7}(x + 5) = \frac{6}{7}$

19. (a) $\frac{x}{2} - 3 + \frac{x}{3} = 2$

    (b) $\frac{5x}{8} + \frac{1}{4} = \frac{11}{6} + \frac{7x}{12}$

20. (a) $\frac{x}{3} - \frac{x}{4} + 2 = 3$

    (b) $\frac{x}{4} + \frac{x}{8} - \frac{x}{6} = \frac{5}{12}$

21. (a) $\frac{1}{2}x + \frac{1}{3}x + \frac{1}{4}x = 39$

    (b) $\frac{11x - 80}{6} - \frac{8x - 5}{15} = 0$

22. (a) $5(x + 2) + 3 = 4(2x - 3) + 1$

    (b) $18 - 3(1 - x) + 5 = 9$

23. (a) $2(y - 3) = 5(y + 1) - 11$

    (b) $t - 2 = -3t + 2(t + 5)$

24. (a) $z - 2 = 4(z + 1) - 5z$

    (b) $3(x + 2) = 7(x - 1) + 13$

25. (a) $3(2x - 1) + 8(2x - 1) = 12(2x - 1)$

    (b) $6(x + 3) - x = x + 5(x + 3) + 2$

26. (a) $5(z - 1) = 5(z + 2) - 15$

    (b) $4y - 3(2 - y) = 6 + 12y$

Calculator Activities

Solve each of the following and check. Round answers to thousandths.

1. (a) $0.2x - 1 = 3.15 - 0.05x$ (b) $0.5 - 0.01x = 2(1 + x)$
2. (a) $1.3(0.1x + 0.51) = -1.3$ (b) $0.025 - 3(1 + 0.1x) = -5.1$
3. (a) $1 - 6.7(0.4x - 1) = 3.67x - 0.3$ (b) $\frac{5}{9}x - 0.23 = \frac{1.02x - 1}{3}$
4. (a) $\frac{1}{8} - \frac{2(x - 0.1)}{3} = 0.52 - \frac{x}{6}$ (b) $\frac{x - 1}{0.5} + 2 = 1 - \frac{x}{4}$
5. (a) $100{,}493 - 2145y = 453y + 339{,}509$

    (b) $492(105x - 41) = 23{,}192x + 10{,}000$

6. (a) $\frac{524 - 34x}{121} - 1015 = \frac{24}{11} - 42x$

    (b) $\frac{x}{9} + 216(52 - 13x) = 412x - \frac{1}{6}$

3-5 MORE APPLIED PROBLEMS

Two important types of applied problems are distance-rate-time problems and mixture problems.

Problem 1 At noon, the Fast Express is speeding along the track at 120 km/hr, 60 kilometers behind a freight train traveling at 80 km/hr. At what time will the Fast Express catch up with the freight train?

SOLUTION:

120 km/hr = speed of Express

80 km/hr = speed of freight   *Bits of information*

At noon, Express is 60 kilometers behind freight

Let t = time, in hours, it takes Express to catch freight   *Variable*

Note that an important formula states that rate multiplied by time equals distance.

Rate × Time = Distance

Express	120	t	$120t$
Freight	80	t	$80t$

$120t = 60 + 80t$ *The equation states that in t hours, Express goes 60 kilometers farther than the freight*

$120t - 80t = 60$

$40t = 60$

$t = \dfrac{3}{2}$

Thus, the Fast Express catches up with the freight train in $1\frac{1}{2}$ hours, that is, at 1:30 p.m.

√ CHECK:

In $1\frac{1}{2}$ hours, the Express goes 180 kilometers, the freight 120 kilometers. The Express goes 60 kilometers farther than the freight, and hence will catch up with the freight. √

Problem 2 The Elixir of life consists of a total of 3 liters containing two solutions, Magic and Triple E. Magic contains 2% of Elixir's

Secret ingredient, and Triple E 5%. The final Elixir contains 4% of Secret. How much of Magic and Triple E are contained in the Elixir?

SOLUTION:

Let $x =$ the number of liters of Magic in Elixir

$3 - x =$ the number of liters of Triple E in Elixir

$0.02x =$ the number of liters of Secret in Magic $\quad$ 2% of x is Secret

and so on. The following table gives the relationships between the various ingredients.

Magic + Triple E = Elixir

Liters of solution	x	$3 - x$	3
Liters of Secret	$0.02x$	$0.05(3 - x)$	$0.04(3)$

This table yields the equation

$0.02x + 0.05(3 - x) = 0.04(3)$

$0.02x + 0.15 - 0.05x = 0.12$ $\qquad$ *Distributive property*

$(0.02 - 0.05)x = 0.12 - 0.15$

$-0.03x = -0.03$

$3x = 3$ $\qquad$ *Multiply each side by -100, then divide by 3*

$x = 1$

Thus, Elixir is made up of 1 liter of Magic and 2 liters of Triple E.

√ CHECK:

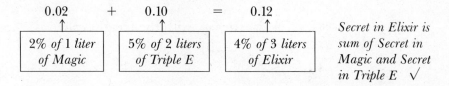

Secret in Elixir is sum of Secret in Magic and Secret in Triple E √

Problem 3 Alex drives 19 blocks to work along the path shown in Figure 3-3. The distance BC is twice as many blocks as AB, CD is one more block than AB, DE is the same number of blocks as AB, and EF is 4 times as many blocks as AB. Describe how many blocks there are in each part of Alex's journey.

104 CHAPTER 3: EQUATIONS IN ONE VARIABLE

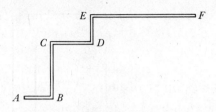

FIGURE 3-3

SOLUTION:

Alex drives 19 blocks	
BC is twice AB	
CD is 1 block more than AB	*Bits of information*
DE equals AB	
EF is 4 times AB	
The number of blocks in AB, BC, CD, DE, and EF	*Unknown numbers*
$x =$ number of blocks in AB	*Variable*
$2x =$ number of blocks in BC	
$1 + x =$ number of blocks in CD	*Algebraic language: each part of the journey is described in terms of AB*
$x =$ number of blocks in DE	
$4x =$ number of blocks in EF	
$x + 2x + 1 + x + x + 4x = 19$	*Equation: total trip of 19 blocks is sum of various parts of trip*

$$1 + 9x = 19$$
$$(⅑)9x = 18 (⅑)$$
$$x = 2$$

Hence, AB is 2 blocks and the journey is as shown in Figure 3-4.

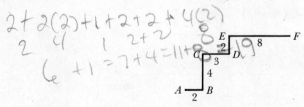

FIGURE 3-4

Problem 4 A square house has a yard on three sides as shown in Figure 3-5. The yard has an area of 315 square meters. Find the dimensions of the house.

3-5: MORE APPLIED PROBLEMS

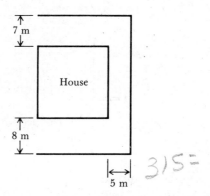

FIGURE 3-5

SOLUTION:

House is square	
Yard has an area of 315 square meters	*Bits of information*
Shape of yard as shown	
Dimensions of house	*Unknown numbers*
Let x = length of house, in meters	*Variable*

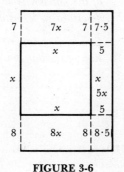

Algebraic language: yard is divided into five rectangles (Figure 3-6)

FIGURE 3-6

$$8x + (8 \cdot 5) + 5x + (7 \cdot 5) + 7x = 315$$

Equation: area of the yard is sum of areas of five rectangular pieces

$$8x + 40 + 5x + 35 + 7x = 315$$
$$20x + 75 = 315$$
$$20x = 315 - 75$$
$$20x = 240$$
$$x = 12$$

The house is 12 meters by 12 meters. You check it!

CHAPTER 3: EQUATIONS IN ONE VARIABLE

Problem 5 Five years ago the Chiang family moved to California and bought a house in the Santa Cruz Mountains. Along with the house was a guesthouse, which, they were told, was 3 times as old as the house and of some historic importance. A friend recently told them that now the guesthouse is only twice as old as the main house. To decide if the guesthouse is truly of historic importance, they need to know just how old it currently is. Would you be able to help them?

SOLUTION:

Want to find the age of the guesthouse

5 years ago the guesthouse was 3 times the age of the house   *Bits of information*

Now the guesthouse is twice the age of the house

Let x = age of the guesthouse now   *Variable*

Recall that a well-labeled table can be extremely helpful.

	Now	5 years ago
Guesthouse	x	$x - 5$
House	$\frac{1}{2}x$	$\frac{1}{2}x - 5$

If the guesthouse is twice the age of the house, then the house is one-half the age of the guesthouse!

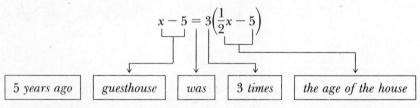

$x - 5 = \frac{3}{2}x - 15$ Subtract x from both sides and add 15 to both sides

$10 = \frac{1}{2}x$ Multiply both sides by 2

$20 = x$

The guesthouse is 20 years old now. Therefore, the house is now 10 years old.

✓ CHECK:

$20 - 5 = 3(10 - 5)$

$15 = 15$ ✓

3-5: MORE APPLIED PROBLEMS

Problem 6 There are two sections of an algebra course containing 56 students altogether. The teacher of section A complained about having most of the students in his section. He observed that he had only 7 fewer than twice as many students as were in section B. How many students were in each section?

SOLUTION:

56 students in two sections	
Section A contains 7 fewer than twice as many students as section B	*Bits of information*
Number of students in each section	*Unknown numbers*
Let $x =$ number of students in section A	*Variable*
$56 - x =$ number of students in section B	
$2(56 - x) =$ twice as many students as in section B	
$2(56 - x) - 7 =$ seven fewer than twice as many students as in section B	*Algebraic language*
$x = 2(56 - x) - 7$	*Equation relating bits of information*
$x = 112 - 2x - 7$	
$x = 105 - 2x$	
$3x = 105$	
$\frac{1}{3} \cdot 3x = \frac{1}{3} \cdot 105$	
$x = 35$	

Thus, section A has 35 students and section B has $56 - 35$, or 21, students.

√ CHECK:

$35 = 2 \cdot 21 - 7$ *Section A contains 7 fewer than*
$35 = 42 - 7$ *twice as many students as section B*
$35 = 35$ √

Quick Reinforcement

Two trains leave the station at the same time traveling in opposite directions. The faster train rolls along at 90 km/hr and the slower train at 65 km/hr. How long will it be before they are 2170 kilometers apart? (Analyze as in Problem 1)

Answer 14 hours

108 CHAPTER 3: EQUATIONS IN ONE VARIABLE

EXERCISES 3-5

1. (a) May Loomis had a 350-kilometer drive to return to college after the mid-semester break. In the first 4 hours of the trip, she was able to maintain a constant speed, but for the last hour of the trip, over a country road, her speed was reduced to $\frac{2}{3}$ of her previous rate. What was her original speed?

   (b) In training for a cross-country bicycle trip, Steve put in a practice ride of 28 kilometers. The first part of the trip was somewhat uphill and took him two hours at a fairly constant speed. He clocked the last leg of the trip at one hour, going at $1\frac{1}{2}$ times his uphill rate. What was his speed on the uphill part of the ride?

2. (a) The perimeter of a square, in centimeters, is 7 times the length of a side, decreased by 12. Find the length of a side.

   (b) The width of a rectangular flower garden is two-thirds its length. If the perimeter of the garden is five times the length, decreased by five, what are the dimensions of the garden?

3. (a) From a point on a straight road, Shirley and Pete begin a bicycle race in opposite directions. Shirley rides at 12km/hr and Pete rides at 10 km/hr. How long will it take for them to be 55 kilometers apart?

   (b) Two cars are 400 kilometers apart. Starting at 1 p.m. and traveling toward one another, they plan on meeting at 5 p.m. What must be the average speed of each car if one plans to go 20 km/hr faster than the other?

4. (a) Joe and Martha were arguing over the difference between their ages. Joe said that he was three times as old as Martha. And Martha claimed that three years ago the sum of their ages was 42! If they are both right, how old are they now?

   (b) Helen is five years older than her sister Alicia. In five years Helen will be one and one-half times as old as Alicia. What are their current ages?

5. (a) The club treasurer reports his dues to the coin-collecting executive committee in riddle form. He says that they have $11.50 in quarters, dimes, and nickels. They have the same number of dimes as nickels, but four times as many quarters as nickels. How many of each type do they have?

   (b) The freshman class goal is to sell 405 boxes of candy before Valentine's day so they can gross the $603.75 necessary to fund their class dance. They have two types of candy, one selling at $1.25 per box, the other at $1.75. How many boxes of each should they sell?

6. (a) Due to his financial difficulty, a man had to pay off a debt of $290 in three different payments. The second payment was to be $30 more than the first, and the third payment had to be twice as much as the second. What must be the sizes of his three payments?

3-5: MORE APPLIED PROBLEMS

(b) On a given day, a store owner paid bills totaling $2800 to three different suppliers. If she paid the second supplier $500 less than the first, and the third supplier $900 more than the first, what were the amounts paid to each supplier?

7. (a) In planning your rectagular house lot, it becomes apparent that the length is to be twice the width, and the difference between the length and the width is to be 28 meters. What must be the dimensions?

(b) Horace Peabody is 67 years old. He has three children named Algernon, Lancelot, and Felicity. Algernon is three years older than Felicity, and Lancelot is five years younger than Felicity. In eight years, the sum of the children's ages will equal their father's age *now*. How old are the children now?

8. (a) Willy Deal sells rental-income units for Really Deal Real Estate Company. His salary consists of an $8000 base salary plus a commission of 1.5% on the selling price of each rental unit he sells. How much must Willy Deal sell in order to gross $38,000 in wages for the year?

(b) The interior decorator you hired had a flair for triangular windows. The living room had such a window, and you could tell that two of its angles were the same size, while the third measured twice the sum of the other two. What are the measurements of each angle of this window? (Reminder: The angles of a triangle add up to 180°.)

9. (a) In a tiny ecosystem, restrictions are imposed so that the total predator-prey population is 341 and the number of prey is fifty more than one-half the number of predators. What must be the predator population and the prey population?

(b) Two different types of bacteria are going to be studied. The professor asks that there be a total of 1090 bacteria in this study with the number of type A bacteria equaling twice the difference of the two amounts. How much of each should be brought for the study?

10. (a) A family had $8000 available for investment. They decided to put part of the money in a 7% term account and leave the rest fairly fluid in a 5% passbook account. They wanted the income from interest to be $550 for the year. How much money are they leaving fluid in the passbook account?

(b) A drug company had 800 liters of cough medicine containing 5% alcohol. How many liters of cough medicine containing only 2% alcohol must be added to end up with the cough medicine having an alcohol content of only 4%?

Nostalgia Problems

1. (a) In a certain orchard, one-half of the trees are apple, one-fourth are peach, and one-sixth are plum. Besides these trees, there are 100 cherry trees and 100 pear trees. How many trees are in the orchard?

(b) A clerk spends $\frac{2}{3}$ of his salary for room and board, $\frac{2}{3}$ of the remainder for clothes, and saves $150 a year. What is his salary?

2. (a) A young man who came into possession of a fortune spent $\frac{3}{8}$ of it the first year and $\frac{4}{5}$ of the remainder the next year, at which time he had $1420 left. What was his original fortune?

(b) A carpenter agrees to the conditions that he be paid $5.50 for every day he works, while he must pay $6.60 for every day he does not work. At the end of 30 days, he finds he has paid out as much as he has received. How many days did he work?

3. (a) A shepherd was met by a band of robbers who plundered him of half of his flock and half a sheep over. Afterward, a second party met him and took half of what he had left and half a sheep over; and soon after this, a third party met him and treated him in like manner; and then he had 5 sheep left. How many sheep did he start out with?

(b) A man bought a horse and chaise for $341. If $\frac{3}{8}$ of the price of the horse is subtracted from twice the price of the chaise, the remainder will be the same as if $\frac{5}{7}$ of the price of the chaise is subtracted from 3 times the price of the horse. What is the price of each?

Calculator Activities

1. (a) Since last year, meat prices have climbed 24%. You now pay $4.55 a pound for rib eye steak. What was the "never to be seen again" price per pound last year?

(b) Two swimmers finish a race only 0.42 second apart. Their combined time is 2.1 minutes. Determine the finishing time of each swimmer to the nearest hundredth of a second.

2. (a) The city track club is organizing a marathon. Entrants will pay $4 if they want a club T-shirt, $2 otherwise. If the club plans for 420 runners desiring T-shirts and 80 not, and if the club's cost is $2.90 per T-shirt, how much money is available for prizes and refreshments from the proceeds of the T-shirt and entry money?

(b) A headline in the newspaper reads "Investment Fever Hits!!" The article goes on to say that real estate is returning 55% per year on investments while stocks are returning only 9% and mutual funds are yeilding 12%. You are told that to realize a $5280 annual investment income, the same amount should be invested in mutual funds as in the stock market, and that 3 times the total of these two should go to real estate. How much do you need for each investment?

3-6 MORE FORMULAS

Equations relating real-world variables are called **formulas.** For example,

$d = rt$ *Distance = rate × time*

$A = (1 + r)P$ *Principal P at rate r per year yields amount A after one year*

If you travel for 3 hours at an average speed of 90 km/hr, then, according to the formula above, you will travel a distance of

$d = 90 \cdot 3$ $r = 90, t = 3$

$= 270$ *kilometers*

If you invest $1200 for one year at 10% per year, at the end of 1 year you will have

$A = (1 + 0.10) \cdot 1200$ $P = 1200, r = 0.10$ (*the decimal equivalent of* 10%)

$= 1.1 \cdot 1200$

$= \$1320$

Often, a formula can be solved for any one of its variables in terms of the remaining ones. Here are some examples of how this is done.

Problem 1 Solve the formula $d = rt$ for r in terms of d and t.

SOLUTION:

$d = r \cdot t$ *Given formula*

$d \cdot \dfrac{1}{t} = r \cdot t \cdot \dfrac{1}{t}$ *Try to isolate r on one side of the equation*

$\dfrac{d}{t} = r \cdot 1$

$r = d \div t$ *Desired formula*

If, for example, you drive 350 kilometers in $3\frac{1}{2}$ hours, your average speed r was

$r = 350 \div \dfrac{7}{2}$

$= 350 \cdot \dfrac{2}{7}$

$= 100$

Thus, your average speed was 100 km/hr.

Problem 2 Solve the formula $A = (1 + r)P$ for r in terms of A and P.

SOLUTION:

$A = (1 + r)P$ *Given formula*

$A \cdot \dfrac{1}{P} = (1 + r)P \cdot \dfrac{1}{P}$ *Try to isolate r on one side of the equation*

$$\frac{A}{P} = (1+r) \cdot 1$$

$$\frac{A}{P} = 1 + r$$

$$r = \frac{A}{P} - 1 \qquad \text{Desired formula}$$

For example, what rate of interest must you receive to build up your nest egg from $640 to $720 in one year? By the formula above,

$$r = \frac{720}{640} - 1 \qquad \text{Using } A = 720,\ P = 640$$

$$r = 1.125 - 1 \qquad \text{Using calculator}$$

$$r = 0.125$$

Thus, you must receive 12.5% interest for the year.

Quick Reinforcement

(a) Solve the formula $P = 2(x+y)$ for y in terms of P and x. (P is perimeter, x and y are lengths of adjacent sides of a rectangle.)

(b) Solve the formula $A = xy$ for x in terms of A and y. (A is area of a rectangle, x and y are lengths of adjacent sides.)

Answers (a) $y = \frac{P - 2x}{2}$ (b) $x = \frac{A}{y}$

EXERCISES 3-6

For each of the following formulas, solve for the indicated variable.

1. (a) $I = p \cdot r \cdot t$, for t \qquad (b) $E = I \cdot R$, for R
2. (a) $P = 2(l + w)$, for w \qquad (b) $P = 2(l + w)$, for l
3. (a) $A = \frac{1}{2}bh$, for b \qquad (b) $A = \frac{1}{2}bh$, for h
4. (a) $P = S - C$, for C \qquad (b) $P = S - C$, for S
5. (a) $A = \frac{a+b}{2}$, for a \qquad (b) $A = \frac{1}{2}(a+b)$, for b
6. (a) $A = \frac{1}{2}h(b_1 + b_2)$, for b_2 \qquad (b) $A = \frac{1}{2}h(b_1 + b_2)$, for h
7. (a) $C = 2\pi r$, for r \qquad (b) $L = 2\pi r + 3s$, for s
8. (a) $v = -32t + k$, for t \qquad (b) $v = -32t + k$, for k

9. (a) $L = \frac{3}{20}G + 1$, for G (b) $A = P(1+r)$, for r

10. (a) $s = \dfrac{a}{1-r}$, for r (b) $s = \dfrac{1}{2}at^2$, for a

11. (a) The density d of a substance is given by the formula $d = \dfrac{m}{v}$, where m is the mass and v is the volume. Solve for v.
    (b) The number of prey (y) and the number of predators (d) in a small ecosystem are related by the formula $y = -2d + 100$. Solve for d.

12. (a) The number of hours (H) of sleep required by a growing child at age A is given by the formula $H = 17 - (A/2)$.
    (i) How much sleep is required by a 10-year-old child?
    (ii) Solve for A.
    (iii) If the child required 14 hours of sleep, how old is it?
    (b) The total cost in business is the number of items n times the cost c per item, plus the fixed costs f, given by the formula $T = nc + f$. Solve for n.

Calculator Activities

1. (a) The population P of township A is related to the number of births b and deaths d by the formula
   $$P = 0.83(b + 1235) - d$$
   Solve for b.
   (b) A company's profits P are determined by its gross income I and costs C, using the formula $P = I - C$. If $I = 1.5x$ and $C = 0.35(y + 1575)$, with x denoting the number of items sold and y denoting the number of items produced, solve the profit formula for y.

2. (a) If P money is invested at $r\%$ interest rate and compounded once at the end of the year, then the amount of money accumulated (A) is given by $A = P + rP$.
   (i) Solve for r.
   (ii) What interest rate must you have if you want \$2250 to grow to \$3000 by the end of one year?
   (b) A small animal weighs w grams after it is t weeks old according to the formula $w = 100(1 + 0.2t)$.
   (i) Solve for t.
   (ii) If the current weight of the animal is 300 grams, how old is it in weeks?

3. (a) If $w = 3000(1 + 3t)$ shows the gallons of waste w dumped into a river in t days, solve for t.
   (b) A bacterial colony grows to a size n according to $n = 1000(3t + 1)$, where t is time in hours.

114 CHAPTER 3: EQUATIONS IN ONE VARIABLE

(i) Solve for t.

(ii) If the current number of bacteria is 5500, how long has the culture been growing?

KEY TERMS

Equation in one variable
Solution
Solution set
Equivalent

Addition property for equations
Multiplication property for equations
Flowchart
Formulas

REVIEW EXERCISES

For each equation and each set of numbers, find which values of the variable selected from the set of numbers make the equation true.

1. (a) Which of the numbers $\{3, -5, 6, 7\}$ are solutions of

    $3x - 5 = 2x + 1$

    (b) Which of the numbers $\{\frac{7}{2}, -\frac{7}{2}, -7, 0\}$ are solutions of

    $8(y + 4) = 3y - 3$

2. (a) Which of the numbers $\{-3, 0, 3, 9\}$ are solutions of

    $x^2 + 3 = 12$

    (b) Which of the numbers $\{-1, 0, 0.5, 1.5\}$ are solutions of

    $5(x + 1) = 3(x + 2)$

Solve each of the equations by using the addition property. Check all solutions.

3. (a) $x - 14 = 6$ (b) $t + 11 = 25$
4. (a) $3 + x + 12 = 21$ (b) $8 + 7 - 5 = t + 1$

Solve each equation by using the multiplication property. Check all solutions.

5. (a) $14x = 28$ (b) $9y = 243$
6. (a) $\frac{2}{5}x = \frac{3}{5}$ (b) $\frac{1}{4}x = \frac{3}{8}$

Solve each equation. Check all solutions.

7. (a) $19 = 5 + 2y$ (b) $5z + 11 = -14$
8. (a) $9 = -8x - 7$ (b) $4x - 4 = -4$
9. (a) $7x + 3 = 5x - 9$ (b) $14 - 3t + 6t = 8t - 12$
10. (a) $-7m - 9m = 32$ (b) $2y + 11 - 8 = 4y + 2$

11. (a) $5(x-3) = 12 - 4x$ (b) $7(x-4) = x + 26$
12. (a) $-(3x-7) + 2 = x + 9 - 4x$ (b) $-(3x-4) - 3(4-x) = 4(x+1)$
13. (a) $\frac{y}{3} - 3 = \frac{y}{4} - 2$ (b) $\frac{y+1}{2} = \frac{2y}{3}$
14. (a) $6(n+2) - 3n = 2n - 3$ (b) $2(y-3) + 3y = y - 16$
15. (a) A friend came to visit the Bernstein family and asked how old the children were. Irwin, being something of a mathematical puzzle-player, said that he was 11 years older than his little sister Isabel, but that in 7 years he will be twice as old as she. Can you solve Irwin's puzzle?
    (b) Two families living next door to each other leave on vacation going in the same direction. However, the Samuels family in their Mercedes travel at twice the speed of the Bower family in their old station wagon. If at the end of three hours they are 105 kilometers apart, what are their speeds?
16. (a) Jeremy decided to count the coins in his bank. He told his brother Joshua that he had as many dimes as quarters and three times as many half-dollars as dimes. He calculated that he had managed to save $35.15. Joshua was able to tell him how many coins of each type he had; can you?
    (b) In a growing suburban area, an old-time landowner sold one-tenth of his land last month and 38 acres this month. He still owns four acres less than one-third of his original homestead. What was the size of his homestead before the sales?
17. (a) The cost of renting a car for one day is $32 plus an additional $0.21 per mile. On a business trip to Los Angeles, Sam Barronson rented a car and drove 95 miles. What was the total cost of the day's rental?
    (b) A ball is tossed upward into the air with an initial velocity of 110 feet per second. If at time t (in seconds) its velocity is given by the equation $v = 110 - 32t$, when will the ball be at its maximum height? (Note: When the ball is at its maximum height, the velocity has slowed down to zero.)
18. (a) The chemistry laboratory needed a 40% salt water solution for a certain experiment. How many liters of a 20% solution should be added to 30 liters of a 50% solution to have the correct percentage for the experiment?
    (b) In doing a needlepoint design in the shape of a square, Norma decided to increase the length of each side by 3 centimeters to have enough room for the complete pattern. If this gave the design a new perimeter of 108 centimeters, what were the original dimensions of the design?

In each of the following formulas, solve for the indicated value.

19. (a) $a = \dfrac{w_1 - w_2}{t}$, for t (b) $a = \dfrac{w_1 - w_2}{t}$, for w_2
    (Formula for angular acceleration)

20. (a) $C = \dfrac{KA}{4\pi d}$, for A (b) $C = \dfrac{KA}{4\pi d}$, for d

(Formula for capacitance of a parallel plate condenser of area A, distance apart d.)

21. (a) The current I in a simple circuit with an external resistance R, internal resistance r, and a power source in volts E is shown by the formula $I = \dfrac{E}{R + r}$, where the resistances are in ohms and the current is measured in amperes.

   (i) Solve for E.

   (ii) How many volts would be necessary if the current I is 4 amperes, the resistance R is 1 ohm, and the resistance r is $\frac{1}{2}$ ohm?

(b) In a mercury thermometer, the mercury expands as the temperature increases. The space, or volume (V), of the mercury in a thermometer can be given as

$$V = 0.2(T + 10) + V_b$$

where T is the current temperature and V_b is the starting volume of the mercury in the bulb in cubic centimeters.

   (i) Solve for T.

   (ii) If the starting volume V_b is 0.1 cubic centimeter, the final volume V is 3 cubic centimeters, what is the temperature?

Calculator Activities

Solve the following equations, checking solutions. Give answers to thousandths.

1. (a) $0.21x - 5.1 = 6.011$ (b) $0.3(1.1 - x) = 4.9x + 1.88$

2. (a) $\dfrac{y}{0.31} - 2.1 = 0.051 + y$ (b) $292(26 - 42x) = 15{,}401 + 414x$

3. (a) $19.2x + 0.007 = 2(2.1x + 0.161)$ (b) $\dfrac{8.2y + 2.02}{0.5} = 100.8$

4. (a) Assume that revenue equals income minus costs as given by the equation $R = n(p - c) - f$, where n is the number of items produced, p is the price per item, c is the cost per item, and f is the fixed costs. If each item costs $0.85 and sells for $1.95, and there are fixed costs of $1490, how many items must be produced to have a revenue of $14,600?

(b) In the problem above, if fixed costs rise by $300 and cost per item rises by $0.05, what would be the new selling price in order to keep the same revenue and number of items produced? Round your answer to the nearest penny.

4 Inequalities in One Variable

Ske Taft is trying to determine how many cartons of books can be stored in the available space in the company stockroom. If the available space is 900 square meters of floor space with shelving, and seven layers of shelving, determine the number of cartons that can be stored if each carton requires 2.5 square meters of space.

4-1 INEQUALITIES AND THEIR SOLUTIONS

The mathematical statement

$$3 + 2x > 3x - 5 \quad \begin{array}{l} 3 + 2x \text{ is the left side,} \\ 3x - 5 \text{ is the right side} \end{array}$$

is an example of an inequality. An **inequality** is a statement consisting of two mathematical expressions, called the *sides*, joined by an inequality sign such as $<$ or $>$. Other inequality signs are $\geq$, which means "greater than or equal to," and $\leq$, which means "less than or equal to." By definition,

$a \geq b$ if either $a > b$ or $a = b$

and

$a \leq b$ if either $a < b$ or $a = b$

117

Inequalities occur all the time in our everyday living. Here are some common expressions involving inequalities:

1. Keep costs C under $10,000: $C < 10,000$
2. To make a profit, income I must be greater than costs C: $I > C$
3. Take at least 500 units (U) of vitamin C per day: $U \geq 500$
4. Buy no more than 5 suits:

$s \leq 5$ *s is number of suits you buy*

Some inequalities are true, such as

$7 > 2$

$5 \leq 5$

Some are false, such as

$3 \geq 7$

$9 < 9$

And others are neither true nor false, such as

$3 + 2x > 3x - 5$

An inequality involving variables, such as the one above, becomes either true or false when values are given to the variables. For example, give x the values 0, 5, 8, and 11 in the inequality

$3 + 2x > 3x - 5$

Let $x = 0$.

$3 + 2 \cdot 0 > 3 \cdot 0 - 5$

$\quad 3 > -5$ *True*

Let $x = 5$.

$3 + 2 \cdot 5 > 3 \cdot 5 - 5$

$3 + 10 > 15 - 5$

$\quad 13 > 10$ *True*

Let $x = 8$.

$3 + 2 \cdot 8 > 3 \cdot 8 - 5$

$3 + 16 > 24 - 5$

$\quad 19 > 19$ *False*

Let $x = 11$.

$3 + 2 \cdot 11 > 3 \cdot 11 - 5$

$3 + 22 > 33 - 5$

$25 > 28$ *False*

If an inequality involves one variable, then a **solution** of the inequality is any value of the variable that makes the inequality true. The set of all solutions is called the **solution set.**

Problem   Find two solutions for each of the following inequalities.

(a) $7 - 2x < 3$ (b) $5t + 7 \leq 8$ (c) $x - 5 \geq 15 - x$

SOLUTION:

Since you have not yet learned how to solve inequalities, you must resort to testing various numbers. It is easy to test 0. Then test some positive values of the variable. Then, if you still haven't found any solutions, test some negative values.

(a)

x	$7 - 2x < 3$	
0	$7 - 2 \cdot 0 < 3$	$7 < 3$ *is false*
2	$7 - 2 \cdot 2 < 3$	$3 < 3$ *is false*
3	$7 - 2 \cdot 3 < 3$	$1 < 3$ *is true*
5	$7 - 2 \cdot 5 < 3$	$-3 < 3$ *is true*

Thus, 3 and 5 are solutions.

(b)

t	$5t + 7 \leq 8$	
0	$5 \cdot 0 + 7 \leq 8$	$7 \leq 8$ *is true*
1	$5 \cdot 1 + 7 \leq 8$	$12 \leq 8$ *is false*
3	$5 \cdot 3 + 7 \leq 8$	$22 \leq 8$ *is false*
-2	$5 \cdot (-2) + 7 \leq 8$	$-3 \leq 8$ *is true*

Thus, 0 and -2 are solutions.

(c)

x	$x - 5 \geq 15 - x$	
0	$0 - 5 \geq 15 - 0$	$-5 \geq 15$ *is false*
5	$5 - 5 \geq 15 - 5$	$0 \geq 10$ *is false*
10	$10 - 5 \geq 15 - 10$	$5 \geq 5$ *is true*
20	$20 - 5 \geq 15 - 20$	$15 \geq -5$ *is true*

Thus, 10 and 20 are solutions.

Each inequality above has many more solutions than the two given. For example, any value of x greater than 10 is a solution of part (c)!

Quick Reinforcement

(a) Is the inequality $2 - 5 < \frac{1}{2}(6 + 2)$ true or false?

(b) Find two solutions for the inequality $3 + 2y > 7$.

(c) Put into symbols: The number of bacteria B must be fewer than 250.

Answers (a) True (b) 5 and 10 (c) $B > 250$

EXERCISES 4-1

State whether each inequality is true or false.

1. (a) $6 \leq 6$ (b) $7 + 2 > 9 - 1$ TRUE
2. (a) $14 - 3 \leq 9 + 1$ 10 (b) $6 - 4 \geq -3 + 1$ TRUE
3. (a) $3(4 + 1) \geq 4(3 + 2)$ 5·4 =20 (b) $(-3)(-2) \leq (3)(2)$ FALSE
4. (a) $6 - 9 \leq (-1)(4)$ = -4 (b) $\frac{1}{3}(17 + 4) > \frac{2}{3}(6)$
5. (a) $(7)(-3)(-2)(0) > \frac{1}{2}$ (b) $4.2(11 + 6 - 2) \leq 92$
   $-21 = -42(0) > 1/2$

Find two solutions for each of the following inequalities.

6. (a) $x > -3$ (b) $y + 3 \leq 4$
7. (a) $z - 2 < 5$ (b) $x - 7 \geq 3$
8. (a) $2x + 3 < 6$ (b) $4y - 2 > 6$
9. (a) $2(t + 3) > t - 8$ (b) $3(x + 4) \geq 11 + 2x$
10. (a) $5(2 + x) \leq 25$ (b) $3y \leq y + 2y$
11. (a) $2(3y - 2) > 7y - 2$ (b) $4x > 5(x - 1)$
12. (a) $3(x + 1) < 5x$ (b) $2 - 3z < 3(1 - z)$

Rewrite each of the following expressions using symbols.

13. (a) The number of predators D must be less than twice the number of prey P. $2(P) - D$
    (b) The pollution count C of a city must not exceed 1525.
14. (a) The laboratory must order at least 300 cubic centimeters of sodium chloride (s). $S \leq 300$
    (b) The student must earn at least an 87 on the final exam (f) in order to receive an A. $300 = x + y$

15. (a) The sum of Patty's salary (s_1) and her husband Mark's salary (s_2) must be at least $2300 per month for them to cover all their expenses.
    (b) The number of hours allocated to labor for product A must be less than twice those allocated to product B.
16. (a) The selling price S of a new line of calculators is computed as no less than twice the cost C plus $10.
    (b) It was calculated that the number b of bacteria will be at least 2300 more than twice the number of hours t since the experiment began.
17. (a) A saleswoman estimated that her salary s this year (including her commissions) will be at least 4% of $250,000 plus $9000.
    (b) The water skier estimated that if he practiced t times he would manage to do his best trick correctly no more than 60% of the time.
18. (a) The sum of a number and 20 must be less than one-half the number.
    (b) The quotient of a number and 30 is greater than twice the number plus three.

Calculator Activities

State whether each inequality is true or false.

1. (a) $152 - 4914 < 12(54 - 181)$ (b) $2.5 - 3.01 \geq -1.021 + 4$
2. (a) $0.5(4 - 2.9) < 3.21 - 0.0115$ (b) $519(288 - 149) > (121)^2$

Find two solutions for each of the following inequalities.

3. (a) $x - 3.5 < 4.2$ (b) $519 - x \geq 1089$
4. (a) $3.26 + 0.5 < 1 - 0.2x$ (b) $149(x - 12) < 587 + x$
5. (a) $3.79 - x \geq 2.1(x + 1)$ (b) $1905(15 + x) > 2222x + 354$

4-2 METHODS OF SOLVING INEQUALITIES

Two inequalities having the same solutions are said to be *equivalent*. We can often solve a complicated inequality by finding a less complicated equivalent inequality to solve.

If an inequality $a > b$ is true, then the inequality $a + c > b + c$ is also true for any number c by the order property for addition (see Section 2-2). This property also applies to inequalities involving the other order signs, $<$, $\geq$, and $\leq$. We can summarize as follows.

Addition Property for Inequality

If you add the same number to each side of a given inequality, the resulting inequality is equivalent to the given one.

122 CHAPTER 4: INEQUALITIES IN ONE VARIABLE

The following problems can be solved by using this property. You solve each inequality by isolating the variable on one side of the inequality sign and everything else on the other side.

Problem 1 Solve the inequality $x + 8 > 11$.

SOLUTION:

	$x + 8 > 11$	*Given inequality*
equivalent	$x + 8 + (-8) > 11 + (-8)$	*Add -8 to each side, to eliminate the 8 on the left side*
	$x + 0 > 11 - 8$	
	$x > 3$	*x is isolated on the left side*

The solution set of the inequality $x > 3$ is the set of all real numbers greater than 3. Hence, this is the solution set of the given inequality. The notation

$\{x \mid x > 3\}$ *The set of all real values of x greater than 3*

is used for this solution set.

Problem 2 Solve the inequality $4 + 3x \leq 2x + 5$.

SOLUTION:

	$4 + 3x \leq 2x + 5$	*Given inequality*
	$4 + 3x - 2x \leq 2x + 5 - 2x$	*Add $-2x$ to each side to eliminate $2x$ on the right side*
equivalent	$4 + (3 - 2)x \leq 0 + 5$	
	$4 + x \leq 5$	
	$-4 + 4 + x \leq -4 + 5$	*Add -4 to each side to eliminate 4 from the left side*
	$0 + x \leq 5 - 4$	
	$x \leq 1$	*x is isolated on the left side*

The solution set of the given inequality is the set of all real numbers less than or equal to 1, that is,

$\{x \mid x \leq 1\}$

Included in this set are 1, all positive numbers between 0 and 1, 0, and all negative numbers.

The order property for multiplication (Section 2-2) states that if $a > b$ is a true inequality, then $ac > bc$ is also true for every positive number c. If you multiply each side of a true inequality by a negative number, however, you must *reverse* the inequality sign in order to obtain a true inequality. For example,

$4 > 3$

is a true inequality. If you multiply each side by -2 and reverse the inequality sign, you obtain the true inequality

$-8 < -6$ $<$ *is the reverse of* $>$

shown in Figure 4-1.

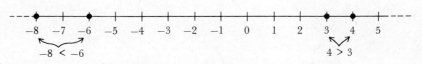

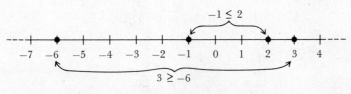

FIGURE 4-1

Another example is the inequality

$-1 \leq 2$ *True inequality*

If you multiply each side by -3, you must also reverse the inequality sign to obtain the true inequality

$3 \geq -6$ $\geq$ *is the reverse of* $\leq$

shown in Figure 4-2.

FIGURE 4-2

These examples illustrate the following property.

Multiplication Property for Inequalities

(a) If you multiply each side of a given inequality by the same positive number, the new inequality is equivalent to the given one.

(b) If you multiply each side of a given inequality by the same negative number and reverse the inequality sign, the new inequality is equivalent to the given one.

Problem 3 Solve the inequality $2x > 5$.

SOLUTION:

$\boxed{equivalent}$
$2x > 5$ — Given inequality

$\frac{1}{2} \cdot 2x > \frac{1}{2} \cdot 5$ — Multiply each side by the positive number $\frac{1}{2}$ to eliminate the 2 on the left side

$1 \cdot x > \frac{5}{2}$

$x > \frac{5}{2}$

Thus, the solution set of the given inequality is the set of all real numbers greater than $\frac{5}{2}$, or

$\left\{ x \mid x > \frac{5}{2} \right\}$

Problem 4 Solve the inequality $5 - 3x \leq 14$.

SOLUTION:

$\boxed{equivalent}$
$5 - 3x \leq 14$

$-5 + 5 - 3x \leq -5 + 14$ — Add -5 to each side

$-3x \leq 9$

$\left(-\frac{1}{3}\right) \cdot (-3x) \geq \left(-\frac{1}{3}\right) \cdot 9$ — Multiply each side by the negative number $-\frac{1}{3}$ and reverse the inequality sign

$1 \cdot x \geq -\frac{9}{3}$

$x \geq -3$

Thus, the solution set of the given inequality is

$\{x \mid x \geq -3\}$ The set of all real values of x that are greater than or equal to -3

4-2: METHODS OF SOLVING INEQUALITIES

Problem 5 Solve the inequality $3x - 7 \leq 5(x + 1)$.

SOLUTION:

$3x - 7 \leq 5(x + 1)$	Given inequality
$3x - 7 \leq 5x + 5$	Distributive property
$-7 \leq 2x + 5$	Add $-3x$ to each side
$-12 \leq 2x$	Add -5 to each side
$-6 \leq x$	Multiply each side by $\frac{1}{2}$
$x \geq -6$	By definition, $a > b$ if $b < a$

Thus, the solution set of the given inequality is

$\{x \mid x \geq -6\}$

Problem 6 Four less than three times a number is less than the number increased by six. Describe the number as best you can.

SOLUTION:

Let x designate the number. Then

$3x - 4$	Four less than three times a number
$x + 6$	Number increased by six
$3x - 4 < x + 6$	$3x - 4$ is less than $x + 6$
$2x - 4 < 6$	Add $-x$ to each side
$2x < 10$	Add 4 to each side
$x < 5$	Multiply each side by $\frac{1}{2}$

The number in question is less than 5. That's all you can say.

Quick Reinforcement

(a) Solve the following inequalities:

   (i) $4 + x > 12$ (ii) $-4x \geq 13$ (iii) $3x - 1 < 4$

(b) The sum of 1 and a number is more than twice the number. What can you say about the number?

Answers: (a) (i) $x > 8$ (ii) $x \leq -13/4$ (iii) $x > 5/3$ (b) $x < 1$

126 CHAPTER 4: INEQUALITIES IN ONE VARIABLE

EXERCISES 4-2

Solve each of the following inequalities.

1. (a) $x + 3 < 11$ (b) $y - 7 > 9$
2. (a) $y - 9 \geq 32$ (b) $x + 5 < 2$
3. (a) $z + 7 > 12$ (b) $y - 4 < 2$
4. (a) $w - 3 < 1$ (b) $w - 5 \geq -1$
5. (a) $5x > 35$ (b) $9y \leq -54$
6. (a) $7x \leq 14$ (b) $21x > 49$
7. (a) $9y > -63$ (b) $-2x < 14$
8. (a) $-x < -2$ (b) $-3x > -9$
9. (a) $-4t \geq 6$ (b) $-t < 7$
10. (a) $5x - 2 > -17$ (b) $2x + 3 > 9$
11. (a) $9y + 7 < 34$ (b) $-4y + 2 \leq -6$
12. (a) $3y - 5 + 17 \leq 0$ (b) $2z + 8 > 3z - 1$
13. (a) $2x - 7 > 4x$ (b) $10m + 3 > 20m + 1$
14. (a) $3(x + 4) > 2(x - 1)$ (b) $-2x + 3(x + 1) < 0$
15. (a) $5p - 30 \geq 2p$ (b) $36 - 7y \leq 2y - 18$
16. (a) $6(x - 7) + 30 < 0$ (b) $8x + 6 < 5x + 27$
17. (a) $7w - 12 < 28 + 5w$ (b) $3(2x + 1) > 25$
18. (a) $5(3x - 4) > -5$ (b) $6(y - 3) \leq 21$
19. (a) $-9(4 + x) \leq 36$ (b) $4 + 3(5 - x) > -5$
20. (a) $0.2(x - 12) < 5.6$ (b) $0.3(y + 5) > 28.5$

Write the inequality that describes each of the following sentences. Solve the inequality.

21. (a) The number of female students in the class was 14. The sum of the number of male and female students was less than 30.
    (b) Five times the number of bunsen burners in the chemistry laboratory is still less than the number of petri dishes, which is 150.
22. (a) The base $12 cost for a toaster oven plus the profit margin desired must be less than $19 for the toaster oven to sell.
    (b) The difference between the number of games won and lost must be greater than 6, when 5 games are lost, for the coach to be secure in his job.
23. (a) If a car is to maintain a speed of 50 kilometers per hour for no more than four hours, what is the maximum distance the car can travel?
    (b) Seven times the sum of a number plus 9 must be no more than 55; what is the range of values for the number?

24. (a) The gambler found that if he lost no more than $6 a day for a week he could manage not to suffer financially. What is the maximum he can afford to lose?
    (b) A family found that they could buy a house if the sum of the mortgage payment and the $75 monthly tax bill was less than the difference between their combined incomes of $2100 per month (after taxes) and their expenses of $1400. What is the maximum mortgage payment they can make?

25. (a) In designing a house, the architect decided that the length of one room would be 14 meters and the minimum area of the room would be 154 square meters. What is the minimum width of the room?
    (b) The board of trustees for Ozooza State College decided that the minimum perimeter of a new track for the athletic field was to be 300 meters. If the width of the track was to be 15 meters, what was the minimum length of the track? (Assume a rectangular track.)

Calculator Activities

Solve each of the following inequalities.

1. (a) $x - 1089 < 4896$ (b) $0.021 + x > 43.2$
2. (a) $y + 14.52 > -90.805$ (b) $49x < 52.1$
3. (a) $0.15x < 5.2 - 1.8$ (b) $-11x + 5 \geq 512$
4. (a) $1.41 - 2.9 + 0.2x \leq 5$ (b) $528x - 49 + 381 < 0$

4-3 GRAPHING INEQUALITIES

An inequality of the form

$$Ax + B < Cx + D$$

$A, B, C,$ and D are real numbers; the inequality sign $<$ can be replaced by $>$, $\leq$, or $\geq$

can be solved using the methods shown in the preceding section.

Problem 1 Solve the inequality $5x + 7 < 3x + 4$.

SOLUTION:

You solve such an inequality by isolating the variable x on one side of the inequality and everything else on the other side.

$$5x + 7 < 3x + 4 \qquad \text{Given inequality}$$
$$5x + 7 - 3x < 3x + 4 - 3x \qquad \text{Add } -3x \text{ to each side}$$
$$(5-3)x + 7 < (3-3)x + 4$$
$$2x + 7 < 4$$
$$2x + 7 - 7 < 4 - 7 \qquad \text{Add } -7 \text{ to each side}$$
$$2x < -3$$
$$\frac{1}{2} \cdot 2x < \frac{1}{2} \cdot (-3) \qquad \text{Multiply each side by } \frac{1}{2}$$
$$x < -\frac{3}{2}$$

Thus, the solution set of the given inequality is

$$\left\{ x \mid x < -\frac{3}{2} \right\}$$

You can see what an inequality in one variable looks like by graphing its solution set on a number line. In the problem above, the graph consists of all points to the left of $-\frac{3}{2}$, as indicated in Figure 4-3.

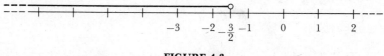

FIGURE 4-3

A part of a line having one endpoint A and extending infinitely far away from A is called a **ray** (Figure 4-4). If the endpoint is excluded, as indicated below by the hollow dot at A, then the ray is called an **open ray.** For example, the graph of $5x + 7 < 3x + 4$ is an open ray.

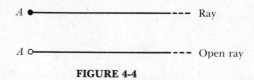

FIGURE 4-4

Problem 2 Graph the inequality $2x - 3 \leq 4x - 5$.

SOLUTION:

$$2x - 3 \leq 4x - 5 \qquad \text{Given inequality}$$
$$(-2x) + 2x - 3 \leq (-2x) + 4x - 5 \qquad \text{Add } -2x \text{ to each side}$$
$$-3 \leq 2x - 5$$
$$-3 + 5 \leq 2x - 5 + 5 \qquad \text{Add 5 to each side}$$
$$2 \leq 2x$$
$$\frac{1}{2} \cdot 2 \leq \frac{1}{2} \cdot 2x \qquad \text{Multiply each side by } \frac{1}{2}$$
$$1 \leq x \qquad \text{Equivalent to } x \geq 1$$

Thus, the graph of the given inequality is the ray with endpoint at 1, as shown in Figure 4-5.

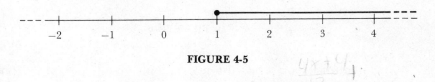

FIGURE 4-5

Problem 3 Graph the inequality $\dfrac{x + 4}{4} + 2 < \dfrac{2x + 5}{3} + 3$.

SOLUTION:

The least common multiple of the denominators 4 and 3 is 12. If you multiply each side by 12, you obtain an inequality having no fractions.

$$\frac{x + 4}{4} + 2 < \frac{2x + 5}{3} + 3 \qquad \text{Given inequality}$$
$$12\left(\frac{x + 4}{4} + 2\right) < 12\left(\frac{2x + 5}{3} + 3\right)$$
$$12\left(\frac{x + 4}{4}\right) + 12 \cdot 2 < 12\left(\frac{2x + 5}{3}\right) + 12 \cdot 3 \qquad \text{Distributive property}$$
$$3(x + 4) + 24 < 4(2x + 5) + 36$$
$$3x + 12 + 24 < 8x + 20 + 36$$
$$3x + 36 < 8x + 56$$

CHAPTER 4: INEQUALITIES IN ONE VARIABLE

$$3x + 36 - 36 < 8x + 56 - 36 \quad \text{Add } -36 \text{ to each side}$$
$$3x < 8x + 20$$
$$-8x + 3x < -8x + 8x + 20 \quad \text{Add } -8x \text{ to each side}$$
$$-5x < 20$$
$$\left(-\frac{1}{5}\right) \cdot (-5x) > \left(-\frac{1}{5}\right)(20) \quad \text{Multiply each side by } -\frac{1}{5} \text{ and reverse the inequality sign}$$
$$x > -4$$

Thus, the solution set of the given inequality is

$$\{x \mid x > -4\}$$

The graph is an open ray with endpoint at -4, as shown in Figure 4-6.

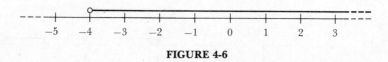

FIGURE 4-6

Sometimes you encounter problems involving more than one inequality. Consider the following examples.

Problem 4 A manufacturer has costs of $1.5x + 1000$ dollars and income of $5x$ dollars when it produces x items. The company wants to keep costs below $4000 and income greater than $9000. How many items must the company produce?

SOLUTION:

By what is given,

$$1.5x + 1000 < 4000 \quad \textit{Costs less than } \$4000$$
$$5x > 9000 \quad \textit{Income greater than } \$9000$$

You are asked to find values of x that are solutions of *both* inequalities. The first inequality is solved as follows.

$$1.5x + 1000 < 4000 \quad \textit{Given inequality}$$
$$1.5x < 3000 \quad \textit{Add } -1000 \textit{ to each side}$$
$$x < 2000 \quad \textit{Multiply each side by } \tfrac{2}{3}$$

The second inequality is solved as follows.

$5x > 9000$ *Given inequality*

$x > 1800$ *Multiply each side by $\frac{1}{5}$*

Thus, any number greater than 1800 but less than 2000 is a solution of the given problem. For example, 1927 is a solution.

In the language of sets, the solution of the problem above is the *intersection* of the solution sets of the two inequalities.

$\{x \mid 1800 < x < 2000\} = \{x \mid x > 1800\} \cap \{x \mid x < 2000\}$

The graph of the solution set is given in Figure 4-7.

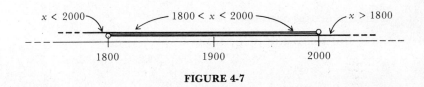

FIGURE 4-7

Problem 5 Find all values of x that are solutions either of the inequality

$2x + 3 < x + 5$

or of the inequality

$3x + 1 > x + 7$

or of both.

SOLUTION:

Solve each inequality in turn.

$2x + 3 < x + 5$	First inequality
$x + 3 < 5$	Add $-x$ to each side
$x < 2$	Add -3 to each side to obtain solution of first inequality
$3x + 1 > x + 7$	Second inequality
$2x + 1 > 7$	Add $-x$ to each side
$2x > 6$	Add -1 to each side
$x > 3$	Multiply each side by $\frac{1}{2}$ to obtain solution of second inequality

Every number less than 2 *or* more than 3 is a solution of the problem. In set notation, the solution of the problem is

$$\{x \mid x < 2\} \cup \{x \mid x > 3\}$$

This is the *union* of the solution sets of the two given inequalities. Its graph is shown in Figure 4-8.

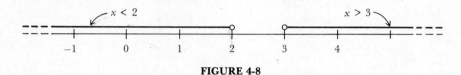

FIGURE 4-8

Quick Reinforcement

Solve the following inequalities.

(a) $5x - 5 < 100$

(b) $x + 10 > 50$

(c) $5x - 5 < 100 \quad \text{or} \quad x + 10 > 50$

Answers (a) $\{x \mid x < 21\}$ (b) $\{x \mid x > 40\}$ (c) $\{x \mid x < 21\} \cup \{x \mid x > 40\}$

EXERCISES 4-3

Solve each of the following inequalities and draw its graph.

1. (a) $\dfrac{x}{-3} < -7$ (b) $\dfrac{x}{-5} > -2$

2. (a) $\dfrac{y}{-4} \geq 5$ (b) $-\dfrac{3x}{2} \leq -15$

3. (a) $3 - \tfrac{3}{4}x > \tfrac{1}{2}(x + 5)$ (b) $y + 2 > \tfrac{1}{3}y - 2$

4. (a) $5(k - 2) + 3 \leq 2(k - 3) + 2k$
   (b) $-4x + 2(x - 3) \geq 5x - (3 + 6x) - 7$

5. (a) $4t > 5(t - 1)$ (b) $2x - 4 \geq x - 5$

6. (a) $\dfrac{x}{5} - 3 < \dfrac{3}{5} - x$ (b) $\dfrac{x - 3}{2} - 1 > \dfrac{x}{4}$

7. (a) $\dfrac{4y + 3}{9} > \dfrac{3y + 5}{7}$ (b) $\dfrac{z + 3}{2} - \dfrac{z}{3} < 4$

8. (a) $3 - \dfrac{2x - 3}{3} \geq \dfrac{5 - x}{2}$ (b) $1 - \dfrac{3x - 1}{6} > \dfrac{2 - x}{3}$

9. (a) $\frac{1}{2}z + 1 \leq z + 2$ (b) $(1 - x) - 2x \geq x + 1$
10. (a) $2(x - 3) + 5 < 5 - x$ (b) $3 - (2 + x) > -9$
11. (a) $\frac{x + 2}{3} - \frac{x - 3}{4} + 2 \leq x - \frac{x - 1}{2}$ (b) $\frac{4x - 2}{11} - \frac{3x - 5}{13} \geq 1$
12. (a) $\frac{x}{5} - \frac{x - 2}{3} \geq -\frac{x}{2} + \frac{13}{3}$ (b) $\frac{3x}{4} - \frac{x - 1}{2} < 6x - \frac{20x + 13}{4}$
13. (a) $-1 \leq \frac{2}{3}y + 5$ (b) $\frac{1}{4}(x - 1) > \frac{1}{5}(x + 1)$
14. (a) $2 - \frac{1}{3}y < \frac{1}{6}(y - 2)$ (b) $\frac{3}{5}(y + 3) < \frac{1}{3}(y + 1)$
15. (a) $(1 - x) - 2x < x + 1$ (b) $\frac{x - 1}{3} - 1 \leq \frac{x}{2}$
16. (a) $5 - 6x < 23$ and $3x + 5 \leq 7 - 2x$
    (b) $5x - 7 < 3x + 4$ or $x > 10$
17. (a) $4(x + 2) > 16$ or $3(x - 5) < -21$
    (b) $3x + 9 \geq -3(2x + 3)$ and $4x + 2 \leq -2(x + 2)$
18. (a) $5(2x - 3) > 7x$ or $2(3x + 4) < 8(2 + x)$
    (b) $-2(3x + 9) - 6 > -3(3x + 11) - 6$ and $-3(5x + 24) + 2 \leq 2(3 - 2x) - 10$
19. (a) A manufacturing firm can rent a warehouse that contains 5000 square meters of floor space. The product they manufacture requires 4 square meters of floor space per unit for storage. Determine the inequality that must be satisfied by the number of units x that can be stored on the warehouse floor. Graph the solution.
    (b) If the warehouse (problem above) can provide three layers of shelving for these products (not counting the floor itself), determine the inequality that shows the maximum number of units x that can be stored in the warehouse. Graph the solution.
20. (a) A grocer can purchase canned soup from a wholesaler at a discount if she buys at least 120 cans. However, the cans are boxed in cartons of 24 cans each. Determine the inequality that must be satisfied by the number of cartons x the grocer must purchase. What is the least number of cartons the grocer must purchase to receive the discount? Graph the solution.
    (b) The grocer in the previous problem will also receive a discount on purchases of packaged frankfurters if she buys at least 60 dozen packages. If the frankfurters come in cases of thirty-six packages to a case, what is the least number of cases of frankfurters that the grocer must buy to receive her discount? Graph the solution.
21. (a) The Forestry Department wants to replant some trees in an area that suffered fire two years ago. They have found that to prevent crowding of the trees, for every 100 square meters the number of trees x must be such that three times the number of trees less twenty-two is less than sixty-eight. The department also knows that they must plant at

least 4 trees for every 100 square meters for the trees to properly reseed themselves. Determine the inequalities that represent the Forestry Department's situation and graph the solution.

(b) The local nursery is trying to decide how many cuttings to take of a variety of plants. They know they must take at least four times as many rose cuttings R as peach cuttings P, but no more than 6 times as many rose cuttings as peach cuttings. If they take 20 peach cuttings, find the inequalities that describe the number of rose cuttings they should take, and graph the solution.

Calculator Activities

Solve each of the following inequalities and draw its graph.

1. (a) $1.2x + 0.5 < 3.01x$ (b) $589x - 486 \geq 1000 - 261x$
2. (a) $41(x - 2) \leq 5(19x - 1)$ (b) $0.2(x - 0.051) < 3.1x - 4.81$
3. (a) $210x - 5(21x + 10) > 4 - 50x$ (b) $0.95(x - 0.2) < 4 - 0.1(5 - x)$
4. (a) $1.2x + 0.5 < 3.01x$ and $0.2(x - 0.051) > 3.1x - 4.81$
   (b) $589x - 486 \geq 1000 - 261x$ or $41(x - 2) \leq 5(19x - 1)$

KEY TERMS

Inequality
Additive property for inequality
Multiplicative property
   for inequalities
Ray
Open ray

REVIEW EXERCISES

State whether each inequality is true or false.

1. (a) $15 - 4 \geq -2(5)$ (b) $-3(8 - 2) < -12$
2. (a) $4(14)(-2) > 16(-2)$ (b) $(-2)(-\frac{1}{3})(-6) \geq 3$

Find two solutions for each of the following inequalities.

3. (a) $x - 5 > -2$ (b) $4y - 8 < 2$
4. (a) $6x + 2 \geq 2x - 4$ (b) $3z + 5 > 3(2z - 1)$

Rewrite each of the following expressions using symbols.

5. (a) The number of sopranos in the choir must be at least half again as many as the number of tenors.
   (b) The mobile library unit takes at least twenty more than twice as many novels as nonfiction books to the local hospitals.
6. (a) The amount of profit planned for is at least 15% of the selling price.
   (b) It has been found that the number of people willing to go back-packing is less than two-thirds of those who enjoy hiking.

Solve each of the following inequalities and draw its graph.

7. (a) $7(3x) > 42$ (b) $-2(4x) < -32$
8. (a) $x - 7 \leq 2x + 4$ (b) $3y \geq 2(y + 8)$
9. (a) $7(y + 3) - 4(3y - 16) \geq 45$ (b) $3x + 3 \geq 6(x + 3) - 4(x + 2)$
10. (a) $\dfrac{x + 5}{2} + \dfrac{5(x + 5)}{6} > 3(x + 5) - 20$
    (b) $\dfrac{11x - 80}{6} - \dfrac{8x - 5}{15} \leq 0$
11. (a) $5(x + 1) + 6(x + 2) > 6(x + 7)$ (b) $\dfrac{3x}{4} + 16 \leq \dfrac{x}{2} + \dfrac{x}{8} + 17$
12. (a) $2 - 3x < 3(1 - x)$ (b) $2x + 3(x + 2) > 4(x - 1) + 10$
13. (a) $2x + 3 > 9$ or $-2x + 3 > 9$ (b) $5x - 3 \geq 7x - 8$ and $x \geq 0$
14. (a) $3x + 5 > -10$ and $x < 3$
    (b) $4x + 3 \leq 6x - 6$ or $\dfrac{3x}{2} + \dfrac{1}{3} < \dfrac{5x}{6} - \dfrac{2}{3}$
15. (a) The sum of the ages of three siblings, Frank, Mark (7 years old), and Melissa (9 years old), is less than three times Frank's age. How old can Frank be?
    (b) The amount of profit in dollars is the same for two items. In figuring out what that profit should be, the manufacturer found that the difference between twice the profit and the number six was less than or equal to the difference between the number three and the profit. What is the range of the profit? Graph it.
16. (a) In a university's School of Engineering, there are at least $12\frac{1}{2}$ times as many male graduates as female graduates, but not more than 15 times as many. If this year there are six women graduating, what is the probable range of graduating men?
    (b) A first-grade teacher found that his pupils felt successful about their reading ability when there were no more than three sentences per page in a beginning reader. If the reader he chose had 32 pages, what is the maximum number of sentences that it could contain to satisfy him? If the minimum number of sentences was set at 70 sentences, graph the possible solutions.

Calculator Activities

State whether each inequality is true or false.

1. (a) $382 - 8863 < 14(550 + 40)$ (b) $2.7 + 9.9(-1.1) \geq -2.1 - 6.09$
2. (a) $0.7(3.2 + 8.7) > -0.3(-42.5)$ (b) $903(112 - 5638) < -60{,}000(85)$

Find two solutions for each of the following inequalities.

3. (a) $x + 4.2 < 9.75$ (b) $845 - x > 721$
4. (a) $8.21 + 1.9(x - 1) \leq -1.1x$ (b) $7.8x - 5.1 > 3(x - 0.3)$
5. (a) $5905 + 926 - 2x > 6x$ (b) $452x - 90{,}428x < -347{,}221$
6. (a) $320(x + 4) < 92(2x - 3)$ (b) $4.22 - 3.1(x + 5) \geq 6$
7. (a) $2.1(x - 1) > 6.3 \quad \text{and} \quad x < 5.2$
   (b) $52x - 3427 < 23x + 12{,}233 \quad \text{or} \quad x > 670$

5 Equations in Two Variables and Their Graphs

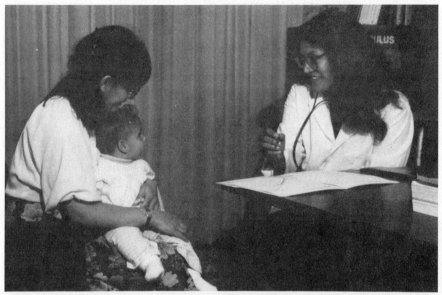

On a visit to the doctor, Kay Barchas and her daughter, Shana, listen while their pediatrician explains how Shana's current weight and height are plotted as a part of her growth curve and how she compares to the normal curve for babies her age.

5-1 EQUATIONS IN TWO VARIABLES

Equations involving two variables occur frequently in the mathematical formulation of real-world problems. For example, the equation

$$C = 1.5x + 1000$$

was used earlier to describe the cost C to produce x items. This is an equation in two variables x and C. When values are given to these variables, the resulting equation is either true or false. For example,

x	C	$C = 1.5x + 1000$	
100	1500	$1500 = 1.5 \cdot 100 + 1000$	
		$1500 = 1150$	False
2000	4000	$4000 = 1.5 \cdot 2000 + 1000$	
		$4000 = 4000$	True

137

Values of x and C that make the equation true are called *solutions* of the equation. Thus,

$x = 2000$, $\quad C = 4000$

$x = 600$, $\quad\; C = 1900$

are solutions of the equation $C = 1.5x + 1000$. You can add many more solutions to this list by giving x a value, multiplying it by 1.5, and adding 1000 to get the value of C.

It is convenient to denote the solution $x = 2000$, $C = 4000$ by $(2000, 4000)$ and $x = 600, C = 1900$ by $(600, 1900)$. We call $(2000, 4000)$ an **ordered pair** of numbers. Each ordered pair consists of two numbers. The first number is a value of x; the second number is a value of C. Thus, 2000 is the first number and 4000 is the second number in the ordered pair $(2000, 4000)$; and 600 is the first number and 1900 is the second number in the ordered pair $(600, 1900)$.

As another example, consider the following equation in the two variables x and y.

$3x + y - 4 = 0$

$x = 1, y = 1$, or $(1, 1)$ is a solution $\qquad$ *True:* $3 \cdot 1 + 1 - 4 = 0$

$x = 0, y = 4$, or $(0, 4)$ is a solution $\qquad$ *True:* $3 \cdot 0 + 4 - 4 = 0$

$x = 4, y = 0$, or $(4, 0)$ is not a solution $\quad$ *False:* $3 \cdot 4 + 0 - 4 = 0$

Observe that the first number in each ordered pair is an x-value, the second a y-value.

The equation can be solved for y in terms of x.

equivalent $\quad\begin{array}{ll} 3x + y - 4 = 0 & \textit{Given equation} \\ y - 4 = -3x & \textit{Add } -3x \textit{ to each side} \\ y = -3x + 4 & \textit{Add 4 to each side} \end{array}$

Every solution of the final equation, and hence of the given equation, has the form

$(x, -3x + 4)$

For example,

$(-3, 13) \qquad x = -3, y = (-3) \cdot (-3) + 4 = 13$

$(-1, 7) \qquad\; x = -1, y = (-3) \cdot (-1) + 4 = 7$

$(2, -2) \qquad\; x = 2, y = (-3) \cdot 2 + 4 = -2$

are solutions of the given equation. Actually,

$\{(x, -3x + 4) \mid x \text{ any real number}\}$

is the set of *all* solutions of the equation $3x + y - 4 = 0$. This means that there are infinitely many different solutions of this equation.

Problem 1 Describe the solution set of the equation $2x + y = 10$.

SOLUTION:

You can express variable y in terms of variable x as follows:

$2x + y = 10$ — *Given equation*

$-2x + 2x + y = -2x + 10$ — *Add $-2x$ to each side to isolate y on the left side*

$y = -2x + 10$ — *Equivalent to given equation*

Every solution is an ordered pair of numbers; the first is a value of x, and the second is the corresponding value of y. For example,

SOLUTION:

when $x = 4$, $y = -2 \cdot 4 + 10 = -8 + 10$, or 2 $(4, 2)$

when $x = 0$, $y = -2 \cdot 0 + 10 = 0 + 10$, or 10 $(0, 10)$

when $x = -3$, $y = (-2) \cdot (-3) + 10 = 6 + 10$, or 16 $(-3, 16)$

For any value of x, the corresponding value of y is $-2x + 10$. Thus, the solution set of the given equation is

$\{(x, -2x + 10) \mid x \text{ any real number}\}$

Problem 2 Describe the solution set of the temperature equation

$5F - 9C - 160 = 0$ *F is Fahrenheit, C is Celsius*

SOLUTION:

Each solution is an ordered pair of numbers; the first is a Celsius temperature and the second is the corresponding Fahrenheit temperature. When $C = 0$, $5F - 160 = 0$ and $F = 32$. When $C = 100$, $5F - 900 - 160 = 0$, $5F = 1060$, and $F = 212$. Thus, two solutions of the temperature equation are

$C = 0$, $F = 32$, or $(0, 32)$ *Freezing point*

$C = 100$, $F = 212$, or $(100, 212)$ *Boiling point*

You can also solve the given equation for F in terms of C:

$5F - 9C - 160 = 0$ — *Given equation*

$5F = 9C + 160$ — *Add $9C + 160$ to each side*

$F = \frac{9}{5}C + 32$ — *Multiply each side by $\frac{1}{5}$ to obtain an equivalent equation*

Now you can list more solutions of the temperature equation:

	SOLUTION:
$C = -10$, $F = -18 + 32$, or 14.	$(-10, 14)$
$C = -5$, $F = -9 + 32$, or 23.	$(-5, 23)$
$C = 5$, $F = 9 + 32$, or 41.	$(5, 41)$
$C = 10$, $F = 18 + 32$, or 50.	$(10, 50)$

Observe that a 5° change in C causes a 9° change in F. The solution set of the temperature equation is

$$\left\{ \left(C, \frac{9}{5}C + 32\right) \,\middle|\, C \text{ any real number} \right\}$$

Problem 3 Describe the solution set of the equation $3x - y = 2x + 3y + 2$.

SOLUTION:

It is easy to solve this equation for x in terms of y:

$3x - y = 2x + 3y + 2$	Given equation
$x - y = 3y + 2$	Add $-2x$ to each side
$x = 4y + 2$	Add y to each side to obtain an equivalent equation

The solution set consists of all ordered pairs (x, y) such that $x = 4y + 2$, that is,

$$\{(4y + 2, y) \mid y \text{ any real number}\}$$

Here are some specific solutions:

when $y = 0$, solution is $(2, 0)$
when $y = 5$, solution is $(22, 5)$
when $y = -2$, solution is $(-6, -2)$

Problem 4 Find the solution set of the equation $y - x^2 + 1 = 0$.

SOLUTION:

You solve this equation for y as follows:

| $y - x^2 + 1 = 0$ | Given equation |
| $y = x^2 - 1$ | Add $x^2 - 1$ to each side |

Thus, the solution set is

$\{(x, x^2 - 1) \mid x \text{ any real number}\}$ *Set of all ordered pairs (x, y) where $y = x^2 - 1$*

Some solutions are

$(0, -1)$	$0^2 - 1 = -1$
$(1, 0)$	$1^2 - 1 = 0$
$(-1, 0)$	$(-1)^2 - 1 = 0$
$(4, 15)$	$4^2 - 1 = 15$
$(-4, 15)$	$(-4)^2 - 1 = 15$

Quick Reinforcement

Find (a) the solution set and (b) three solutions of $3x + y = 2$.

Answers (a) $\{(x, 2 - 3x) \mid x \text{ any real number}\}$ (b) $(1, -1), (0, 2), (-1, 5)$

EXERCISES 5-1

Find the solution set for each of the following equations by solving for x. Find three ordered pairs for each.

1. (a) $x + 3y - 3 = 0$ (b) $x - 2y + 11 = 0$
2. (a) $x - \frac{2}{3}y + 1 = 0$ (b) $x + \frac{1}{2}y - \frac{5}{9} = 0$
3. (a) $2x - 8y + 4 = 0$ (b) $3x + 12y - 9 = 0$
4. (a) $3x + 15y - 2 = 0$ (b) $2x - 10y - 7 = 0$
5. (a) $x - 3y^2 + 1 = 0$ (b) $x - 2y^2 - 2 = 0$
6. (a) $2x - 6y^2 + 3 = 0$ (b) $3x - 6y^2 + 1 = 0$
7. (a) $1.2x - 3.6y = 4.8$ (b) $3.5x + 17.5y = 7$
8. (a) $5x + 20y^2 = -15$ (b) $3x + y^2 - 9 = 0$
9. (a) $3y + 2x - 3 = x + 5y$ (b) $4x - 2y = 6x + 3y - 1$
10. (a) $2x + 3y^2 = y^2 - 2x + 8$ (b) $7y^2 + 5x + 3 = 3x + 5y^2 - 1$

Find the solution set for each of the following equations by solving for y. Find three ordered pairs for each.

11. (a) $3x + y - 5 = 0$ (b) $2x - y + 7 = 0$
12. (a) $-4x + y = -7$ (b) $y - 8x = -3$

13. (a) $0.5y + 0.2x = 1$ (b) $0.3x + 1.2y - 0.5 = 0$
14. (a) $9 - 6x - 3y = 0$ (b) $5 - 2y + 8x = 0$
15. (a) $xy = 1$ (b) $xy = 5$
16. (a) $3xy + 2 = 0$ (b) $4xy - 6 = 0$
17. (a) $2x^2 + 3y = 6$ (b) $5y - 15x^2 = 10$
18. (a) $2.5y + 6.25x^2 - 7.5 = 0$ (b) $1.3y - 2.6x^2 = 6.5$
19. (a) $2xy + 3 = 1$ (b) $5xy - 2 = 8$
20. (a) $2x + 7y = 16$ (b) $-x + 5y = 25$
21. The relationship of the number of hours of sleep H needed by a growing child of age A is given as $2H + A - 34 = 0$
    (a) Solve for H
    (b) Find three ordered pairs in the solution set and interpret their meaning.
22. The relationship of the weight of a small animal in grams w to its age in weeks t is shown by $w - 20t = 100$.
    (a) Solve for t.
    (b) Find three ordered pairs in the solution set and interpret their meaning.
23. The percentage of the population in college in relationship to a base year of 1940 is shown by the equation $P - 1.25t - 15 = 0$, where P is the percentage in college the tth year after 1940.
    (a) Solve for P.
    (b) Find three ordered pairs in the solution set and interpret their meaning.
24. The braking distance of a car going at a fixed speed for a given set of regular tires is dependent on the air temperature measured in degrees Fahrenheit (F). If the braking distance is in feet, then the relationship is shown by $D - 2F = 115$.
    (a) Solve for F.
    (b) Find three ordered pairs in the solution set and interpret their meaning.

5-2 COORDINATE PLANES

A **grid** on a flat surface, or plane, consists of equidistant horizontal and vertical lines as shown in Figure 5-1. The points at which the horizontal lines intersect with the vertical lines are called the **grid points.**

    A moon satellite takes a picture of a crater on the moon and sends it back to earth in the form of a set of ordered pairs of numbers. These numbers are **plotted** on a grid, each ordered pair of numbers plotting a point. The set of dots on the grid reproduce the picture of the crater on the moon!

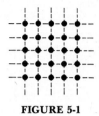

FIGURE 5-1

Basic to the process described above is the concept of a **coordinate system on a plane.** To describe such a system, select a grid point in the center of a grid and call it the **origin.** The horizontal grid line through the origin is arbitrarily called the **x-axis,** the vertical grid line the **y-axis.** We assume the x- and y-axes to be number lines, with numbers assigned to points in the usual way and 0 assigned to the origin. The grid in Figure 5-2 has integers assigned to the grid points on the axes.

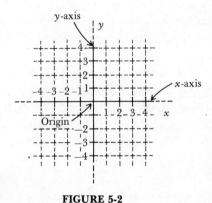

FIGURE 5-2

Every number a can be assigned to a point on the x-axis and every number b to a point on the y-axis. Therefore, every ordered pair (a, b) can be assigned to the point on the plane that is directly above or below a on the x-axis and directly to the right or to the left of b on the y-axis. You can see how this system works from Figure 5-3, which shows four plotted points. The ordered pair $(3, 5)$ is assigned to the point directly above 3 on the x-axis and to the right of 5 on the y-axis; the ordered pair $(2, -4)$ is assigned to the point directly below 2 on the x-axis and to the right of -4 on the y-axis; the ordered pair $(-2, -6)$ is assigned to the point directly below -2 on the x-axis and

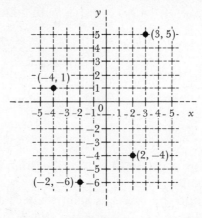

FIGURE 5-3

to the left of -6 on the y-axis; and the ordered pair $(-4, 1)$ is assigned to the point directly above -4 on the x-axis and to the left of 1 on the y-axis.

The association of ordered pairs of numbers with points on a plane is called a **coordinate system** on the plane, and any plane having a coordinate system is called a **coordinate plane.** The numbers associated with a point on a coordinate plane are called the **coordinates** of the point; the first number is the **x-coordinate,** the second the **y-coordinate.** For example, the point with coordinates $(2, -4)$ has x-coordinate 2 and y-coordinate -4. The origin has coordinates $(0, 0)$. Each point on the x-axis has coordinates of the form $(a, 0)$, and each point on the y-axis has coordinates of the form $(0, b)$.

The axes separate a coordinate plane into four parts, called **quadrants.** The quadrants are numbered as shown in Figure 5-4. The quadrant in which a point lies is determined by the sign of each of its coordinates.

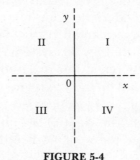

FIGURE 5-4

QUADRANT OF (a, b)	SIGN OF COORDINATES
I	$a > 0, b > 0$
II	$a < 0, b > 0$
III	$a < 0, b < 0$
IV	$a > 0, b < 0$

Quick Reinforcement

(a) Plot the points $(5, 1)$ and $(0, -2)$ on a grid.

(b) Find the coordinates of points A and B in the figure shown here.

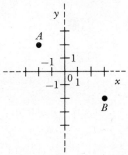

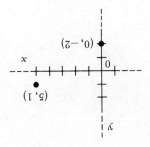

Answers: (a)

(b) Point A has coordinates $(-2, 2)$, point B has coordinates $(3, -2)$.

Every point in a coordinate plane has a unique ordered pair of coordinates; and each ordered pair of numbers locates a unique point in the plane. This unique point is called the **graph** of the ordered pair of numbers. In turn, the graph of a set of ordered pairs of numbers consists of the graphs of the individual ordered pairs in the set.

Problem 1 A satellite observed a strange formation on the moon. It sent back to earth the following set of ordered pairs:

$\{(1,1), (2,1), (3,1), (1,2), (3,2), (1,3), (2,3), (3,3)\}$

Graph the set on a grid and describe it.

SOLUTION:

The graph consists of eight points as shown in Figure 5-5. This graph forms a square.

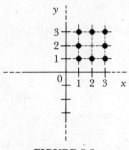

FIGURE 5-5

Problem 2 At a different location, the satellite sent to earth the following picture:

$\{(5,0), (6,0), (7,0), (6,1), (5,2), (4,3), (3,4), (2,5), (1,6),$

$(0,7), (0,6), (0,5), (1,4), (2,3), (3,2), (4,1)\}$

Show it on a grid.

SOLUTION:

These sixteen points are plotted in Figure 5-6. They form two "ridges" of peaks—or whatever your imagination sees!

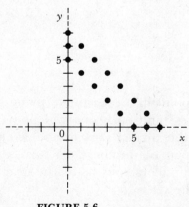

FIGURE 5-6

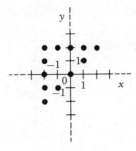

FIGURE 5-7

Problem 3 The space center wants to send back to the satellite a picture of the right triangle shown in Figure 5-7. What set of ordered pairs should they send?

SOLUTION:

They should send the following set:

$\{(0,0), (1,1), (2,2), (1,2), (0,2), (-1,2), (-2,2), (-2,1), (-2,0),$
$(-2,-1), (-2,-2), (-1,-1)\}$

Problem 4 From Figure 5-8, determine the coordinates of point A and explain what they mean.

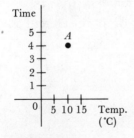

FIGURE 5-8

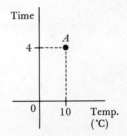

FIGURE 5-9

SOLUTION:

As shown in Figure 5-9, the coordinates of A are $(10, 4)$. That is, it was 10°C at 4 a.m.

Quick Reinforcement

(a) Graph the set of ordered pairs and tell its "shape":

$\{(4,0), (2,2), (2,-2), (0,-4), (0,4), (-2,2), (-2,-2), (-4,0)\}$

(b) Find the set of ordered pairs having this figure as its graph.

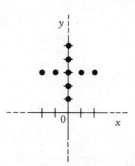

(c) (i) Plot the following information about the stock on the coordinate system below:

Day	Price ($)
1	12.5
2	14
3	10.2
4	12.5

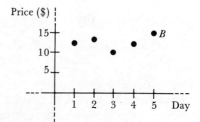

(ii) Find the coordinates of point B and give the associated information.

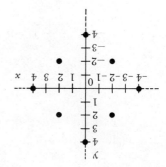

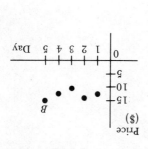

Answers (a) Square (see the figure). (b) {(0,1), (0,2), (0,3), (0,4), (0,5), (−2,3), (−1,3), (1,3), (2,3)} (c) (i) See the figure (ii) B has coordinates (5, 15). That is, the price of the stock on the fifth day was $15.

EXERCISES 5-2

Plot the following points on a grid.

1. (a) $(3, 4)$ and $(-1, 4)$ (b) $(2, 1)$ and $(4, -3)$
2. (a) $(-1, -3)$ and $(-4, 1)$ (b) $(6, 2)$ and $(-6, -2)$
3. (a) $(3, -3)$ and $(-3, 3)$ (b) $(-4, -2)$ and $(2, 4)$
4. (a) $(8, -2)$ and $(-3, -4)$ (b) $(-4, -5)$ and $(-2, -4)$
5. (a) $(4, 0)$ and $(0, 5)$ (b) $(0, 6)$ and $(-2, 0)$
6. (a) $(0, 0)$ and $(-3, 0)$ (b) $(0, -2)$ and $(-2, 0)$

Find the coordinates of each of the following points.

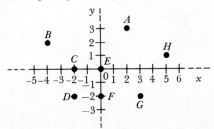

7. (a) B (b) D
8. (a) A (b) E
9. (a) C (b) F
10. (a) H (b) G

Graph each set of ordered pairs. Is there a shape to the set of points?

11. (a) $\{(-1, 1), (0, 0), (1, 1)\}$ (b) $\{(3, 0), (3, 3), (3, 6)\}$
12. (a) $\{(0, 0), (2, 0), (2, 2), (4, 2)\}$ (b) $\{(3, 5), (4, 4), (4, 6), (5, 5)\}$
13. (a) $\{(1, 1), (2, 2), (3, 3)\}$ (b) $\{(4, 2), (3, 4), (4, 3), (5, 4)\}$
14. (a) $\{(-4, -2), (3, -3), (0, 5)\}$
    (b) $\{(-2, -2), (-1, -1), (0, -3), (-1, -4)\}$
15. (a) $\{(1, 0), (2, 1), (3, 0), (0, 3), (4, 3)\}$
    (b) $\{(3, 1), (2, 3), (3, 2), (4, 3), (3, 3)\}$

Find the set of ordered pairs having the following figures as graphs.

16. (a) 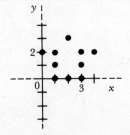 (b)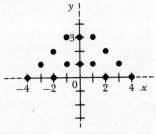

17. (a) 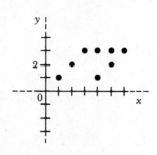 (b)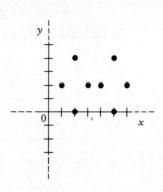

18. The following set of points represents temperatures taken in New York City during the month of February.

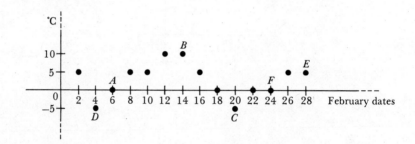

(a) Find the coordinates of points $A, D,$ and E and interpret their meaning.
(b) Find the coordinates of points $B, C,$ and F and interpret their meaning.

19. The following set of points represents the grades earned by an algebra student on the ten quizzes given during the term.

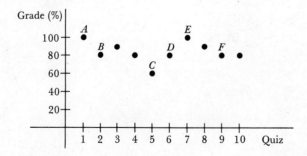

(a) Find the coordinates of points $A, B,$ and C and interpret their meaning.
(b) Find the coordinates of points $D, E,$ and F and interpret their meaning.

20. The following set of points represents the population changes in the town of Whoville in the ten-year span 1970–1979.

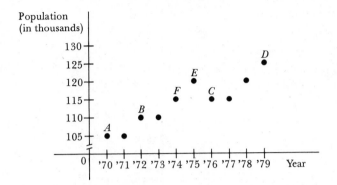

(a) Find the coordinates of points D, E, and F and interpret their meaning.
(b) Find the coordinates of points A, B, and C and interpret their meaning.

5-3 LINEAR EQUATIONS IN TWO VARIABLES

Consider the table of values below, where x designates a company's costs and y represents the company's income in thousands of dollars.

x	y
0	0
1	2
2	4
3	6
4	8
⋮	⋮

Clearly, the pattern shown in the table is described by the equation

$y = 2x$ *Income is twice the cost*

This equation is an example of a **linear equation in two variables**. Note that each of the ordered pairs $(0, 0)$, $(1, 2)$, $(2, 4)$, $(3, 6)$, and $(4, 8)$ is a *solution* of the equation $y = 2x$. Why it is called *linear* will soon become clear.

Every equation in two variables has a set of ordered pairs as its solution set. As a set of ordered pairs, the solution set has a graph in a coordinate plane. The graph of an equation in two variables is the graph of the solution set of the equation. In the example above,

152 CHAPTER 5: EQUATIONS IN TWO VARIABLES AND THEIR GRAPHS

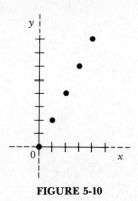

FIGURE 5-10

the table of values has a graph consisting of five points, as shown in Figure 5-10. Observe that the five points lie on a line. The equation $y = 2x$ has many more solutions than the five listed in the table. For example,

$(0.5, 1)$, $(1.25, 2.5)$, $(-2, -4)$ *y is twice x*

are also solutions of $y = 2x$, as shown in Figure 5-11. Actually, the solution set of $y = 2x$ is the set of all points on the line passing through the five given points. Because the graph forms a straight line, the equation is called a linear equation.

As another example, suppose the number y of prey in an ecosystem is two more than the number x of predators; that is,

$y = x + 2$

Thus, when $x = 0$, $y = 2$; when $x = 2$, $y = 4$; when $x = 5$, $y = 7$; and when $x = 8$, $y = 10$. These four solutions of the equation $y = x + 2$

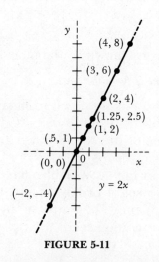

FIGURE 5-11

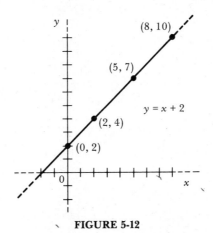

FIGURE 5-12

are plotted in Figure 5-12. Clearly, the four points lie on a straight line. In fact, every solution of the equation $y = x + 2$ lies on the line, and every point on the line is a solution of the equation. Because the graph of $y = x + 2$ is the line shown in the figure, the equation is a linear equation.

It can be shown that every equation in two variables x and y of the form

$ax + by + c = 0$ a, b, and c are numbers (a or $b \neq 0$)

has a line as its graph. The case where $a = 1$, $b = 1$, and $c = 1$ is shown in Figure 5-13. Of course, every equation that is equivalent to an equation of the given form also has a line as its graph. You call every such equation a linear equation in two variables. The two earlier equations,

$y = 2x$ Equivalent to $-2x + y = 0$

and

$y = x + 2$ Equivalent to $-x + y + (-2) = 0$

are examples of linear equations. Further examples follow.

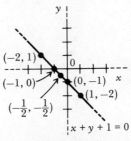

FIGURE 5-13

Problem 1 Sketch the graph of the equation $2x + y - 1 = 0$.

SOLUTION:

This equation has the form

$$ax + by + c = 0 \qquad a = 2, b = 1, c = -1$$

Thus, it is linear. To find solutions of the given equation, we first solve the equation for y:

$$2x + y - 1 = 0 \qquad \textit{Given equation}$$
$$-2x + 1 + 2x + y - 1 = -2x + 1 \qquad \textit{Add } -2x + 1 \textit{ to each side}$$
$$y = -2x + 1 \qquad \textit{Equivalent equation}$$

To find solutions of this equation, we give x values and solve for y. If $x = 0$, then $y = 1$; if $x = 3$, then $y = -5$; if $x = -2$, then $y = 5$. These solutions are plotted in Figure 5-14. Observe that they lie on a line L. (Hereafter, "line" always means "straight line.") Thus, L is the graph of the given linear equation.

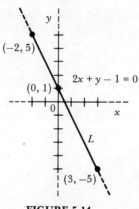

FIGURE 5-14

Problem 2 Sketch the graph of the equation $4x + 3y = 1$.

SOLUTION:

This equation is linear, since it is equivalent to the linear equation $4x + 3y - 1 = 0$. Thus, the graph of the given equation is a line L. You can determine L by finding two points on it. Aren't the equations

$$4 \cdot 1 + 3 \cdot (-1) = 1 \qquad x = 1, y = -1$$
$$4 \cdot (-2) + 3 \cdot 3 = 1 \qquad x = -2, y = 3$$

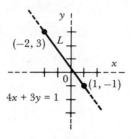

FIGURE 5-15

true? Therefore, $(1, -1)$ and $(-2, 3)$ are two solutions of the given equation. Thus, the graph of the given equation is the line L containing the points $(1, -1)$ and $(-2, 3)$, as shown in Figure 5-15.

Problem 3 Sketch the graph of the linear equation $2x - 3y + 5 = 0$.

SOLUTION:

You only need to find two solutions. The graph is a line, and you can draw a line if you are given two points on it. One such point can be found by giving x the value 0 in the equation and solving the resulting equation for y.

Let $x = 0$.

$$2 \cdot 0 - 3y + 5 = 0$$
$$-3y + 5 = 0$$
$$3y = 5$$
$$y = \frac{5}{3} \quad (0, \tfrac{5}{3}) \text{ is a solution}$$

Similarly, we can give y the value 0 and solve for x.

Let $y = 0$.

$$2x - 3 \cdot 0 + 5 = 0$$
$$2x + 5 = 0$$
$$2x = -5$$
$$x = -\frac{5}{2} \quad (-\tfrac{5}{2}, 0) \text{ is a solution}$$

Thus, the points $(0, \tfrac{5}{3})$ and $(-\tfrac{5}{2}, 0)$ are on the graph. Hence, the graph is the line L of Figure 5-16. You can check to see if L is the graph of the given equation by finding one more solution of the equation. For example, if x has the value 2, then

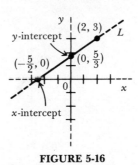

FIGURE 5-16

$$2 \cdot 2 - 3y + 5 = 0$$
$$3y = 9$$
$$y = 3$$

Thus, $(2, 3)$ is a solution of the given equation. The point $(2, 3)$ appears to be on L.

The line L of the problem above intersects the x-axis at $(-\frac{5}{2}, 0)$ and the y-axis at $(0, \frac{5}{3})$. We call $-\frac{5}{2}$ the **x-intercept** and $\frac{5}{3}$ the **y-intercept** of line L, as indicated in the figure above.

Problem 4 Find the intercepts of the graph of the equation

$$y - 2x = 1$$

SOLUTION:

You can find the intercepts by finding the value of each variable when the other variable has the value 0.

(a) To find the x-intercept, let $y = 0$ in the equation $y - 2x = 1$:

$$0 - 2x = 1$$
$$-2x = 1$$
$$x = -\frac{1}{2}$$

Thus, $-\frac{1}{2}$ is the x-intercept of the graph.

(b) To find the y-intercept, let $x = 0$ in the equation $y - 2x = 1$:

$$y - 2 \cdot 0 = 1$$
$$y - 0 = 1$$
$$y = 1$$

Thus, 1 is the y-intercept. Using the intercepts, you can sketch the graph of the given equation as shown in Figure 5-17.

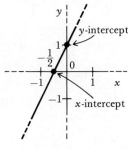

FIGURE 5-17

Not all lines have both an x-intercept and a y-intercept, as the following examples show.

Problem 5 Describe the graph in a coordinate plane of the equation $x = 2$.

SOLUTION:

The equation $x = 2$ can be considered a linear equation in variables x and y:

$x + 0y = 2 \qquad 0y = 0$ *for every value of y*

This is the equation whose graph you are asked to describe. Some solutions of this equation are

$(2, 5) \qquad 2 + 0 \cdot 5 = 2$ *is true*
$(2, -3) \qquad 2 + 0 \cdot (-3) = 2$ *is true*
$(2, 0) \qquad 2 + 0 \cdot 0 = 2$ *is true*

In fact, (a, b) is a solution for any number b as long as a is 2. Thus, the graph of $x = 2$ is the vertical line L of Figure 5-18. The line L has an x-intercept, 2, but no y-intercept.

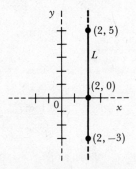

FIGURE 5-18

158   CHAPTER 5: EQUATIONS IN TWO VARIABLES AND THEIR GRAPHS

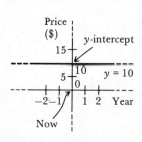

FIGURE 5-19   FIGURE 5-20

A horizontal line has only a y-intercept. For example, if the price of a stock is constant, say $10, then you can graph price versus year as shown in the figure. The price is the graph of the equation $y = 10$ (Figure 5-19).

Problem 6 What information is associated with the point A in Figure 5-20?

SOLUTION:

From the figure, point A has coordinates $(1973, 3.5)$. This means that the company's profits in 1973 were $3.5 million.

Quick Reinforcement

(a) Sketch the graph of the equation $x - 2y = 2$ and check your answer.

(b) Find the intercepts of the graph of the equation $9x + 2y - 18 = 0$, and sketch the graph.

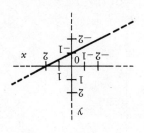 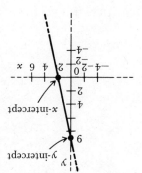

Answers (a)   (b) The x-intercept is 2, the y-intercept is 9.

EXERCISES 5-3

Sketch the graph of each equation, using the given values for x or y.

1. (a) $x = y$, with $x = 1, 2,$ and 3 (b) $y = x + 1$, with $x = 0, 1,$ and 2
2. (a) $x = y + 1$, with $y = 0, 2,$ and 4 (b) $x + y = 3$, with $y = 0, 2,$ and 4
3. (a) $2x - y = 3$, with $x = -\frac{1}{2}, 0,$ and $\frac{1}{2}$
   (b) $y = 3x + 2$, with $x = \frac{1}{3}, \frac{4}{3},$ and $\frac{7}{3}$
4. (a) $2x = 3y + 1$, with $y = -1, 1,$ and 3
   (b) $0.5y = 5 - x$, with $y = -2, 0,$ and 2
5. (a) $y = -x$, with $y = -1, 3,$ and 4
   (b) $y = x - 1$, with $y = -5, -1,$ and 0
6. (a) $3y + 1 = x + 2$, with $x = 1, 2,$ and 3
   (b) $4x = y + x - 2$, with $x = -\frac{2}{3}, \frac{2}{3},$ and 0
7. (a) $0.2x = 0.1y + 0.1$, with $y = -2, 1,$ and 3
   (b) $0.5y - 0.5x = 1.5$, with $y = -3, 0,$ and 3
8. (a) $4x + 7y = 0$, with $x = 0, 0.75,$ and 2.5
   (b) $x - 5y = 1$, with $x = -4, 0,$ and 6
9. (a) $x = 5$, with $y = -2.5, 1,$ and 3.2
   (b) $x = -2$, with $y = -\frac{1}{2}, 0,$ and $3\frac{1}{2}$
10. (a) $y = 7$, with $x = -2, -1,$ and 4 (b) $y = -3$, with $x = 5, 7,$ and 9

Sketch the graph of each equation by choosing values for x or y. Plot three points on each graph.

11. (a) $x + y = 4$ (b) $x + y = 6$
12. (a) $y - x = -1$ (b) $x - y = 2$
13. (a) $\frac{1}{2}x - \frac{1}{2}y = -\frac{3}{2}$ (b) $\frac{1}{4}x + \frac{1}{2}y = 0$
14. (a) $0.8x + 0.8y = 0$ (b) $0.4x + 0.1y = 0.2$
15. (a) $5x - y + 1 = 0$ (b) $x + 2y - 2 = 0$
16. (a) $3x - y + 1 = 0$ (b) $x + 3y - 6 = 0$
17. (a) $x = -7$ (b) $y = 4$
18. (a) $y + 1 = 0$ (b) $x - 3 = 0$
19. (a) $4y + 2x = 12$ (b) $x - 5y = 15$
20. (a) $7x + 3y = 21$ (b) $2y - 9x = 18$

Sketch the graph of each equation by determining the x- and y-intercepts.

21. (a) $x - y = 5$ (b) $y + x = 2$
22. (a) $2x + y = 4$ (b) $3x - y = 3$
23. (a) $x + 5y = 10$ (b) $x - 3y = -6$
24. (a) $2x - 3y = 6$ (b) $4x + 5y = 20$

25. (a) $4y - x = 2$ (b) $\frac{1}{2}x + \frac{1}{4}y = 1$
26. (a) $0.2x + 0.3y = 0.6$ (b) $0.4y - 0.2x = 1.2$
27. Raggs, Ltd., a clothing manufacturer, has determined that its revenue R from the sale of x suits is $80 per suit. This is shown by the equation

    $R = 80x$

    If you label the y-axis R, you have an equation you can graph.
    (a) Choose three values for x (why must they all be positive?) and find the corresponding values for R. Sketch the graph.
    (b) By reading your graph, estimate what the revenue would be if 45 suits were sold.

28. As a person descends into the depths of the ocean, the water pressure p increases with depth d according to the equation

    $p = \frac{15}{33}d + 15$

    (a) If d takes on the values 0, 33, and 66 feet, find the corresponding pressures p.
    (b) Letting d be on the x-axis and p be on the y-axis, sketch the graph. Why would d not take on any negative values? What does it mean when $d = 0$?

29. A typewriter manufacturer is considering introducing a new, moderately-priced electric typewriter with some innovative features. Their market research department gave the management a demand-price equation based on their research. They found that price p and demand d had the following relationship:

    $d + 60p = 12{,}000$

    Consider d as representing values on the y-axis and p as representing values on the x-axis.
    (a) Find the x- and y-intercepts and sketch the graph.
    (b) Find the estimated demand d for prices of $70, $120, and $160. Which one would bring in a greater total revenue? (The revenue can be figured by multiplying the price per item by the estimated demand.)

30. The equation that shows the comparison of a person's body weight w with a person's brain weight b is

    $b = 0.025w$

    (a) Choose three values for w and find the corresponding values for b. Using w on the x-axis and b on the y-axis, sketch the graph.
    (b) Explain why so many practical applications of graphing use only the first quadrant of the graph.

Calculator Activities

1. The normal weight w of an adult in pounds is related to his or her height h in inches, according to the equation

    $w = 5.5h - 220$

    (a) Given heights of 62 inches, 65 inches, 74 inches, and 81 inches, find the corresponding weights. Use these points to sketch the graph, with h plotted on the x-axis and w plotted on the y-axis.

    (b) Use your graph to estimate the appropriate weight for someone measuring 70 inches; 60 inches. Since the equation is for a normal adult, does the value of the y-intercept have any real meaning? Why or why not?

2. A frisbee manufacturer has computed the price p of frisbees as related to the demand d according to the equation

    $20{,}000p + 0.5d - 90{,}000 = 0$

    (a) If p is related to the x-axis and d to the y-axis, find the x- and y-intercepts and sketch the graph.

    (b) With a price of $2.75, approximately what would be the demand?

3. A local theater owner knows that the price of his tickets and the number of people who attend his productions are very definitely related. If the price is p and the number attending is n, then their relationship is given by

    $p + 0.01n = 11$

    (a) Find the x- and y-intercepts and sketch the graph, plotting n on the x-axis and p on the y-axis. Interpret the meaning of the intercepts.

    (b) If the owner decided he needed 900 people to attend a certain production, what price ticket should he sell?

5-4 THE SLOPE OF A LINE

The slope of a road is defined as shown in Figure 5-21. For example, if a road has a rise of 20 meters for a run of 120 meters, then it has the slope shown in Figure 5-22.

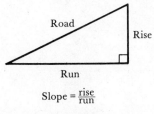

FIGURE 5-21

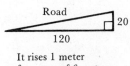

FIGURE 5-22

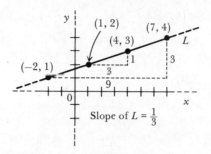

Slope of $L = \frac{1}{3}$

FIGURE 5-23

The concept of slope can be carried over to lines in a coordinate plane. The line L of Figure 5-23 has a rise of 1 for a run of 3 going from point $(1, 2)$ to point $(4, 3)$ on L. We say L has a **slope** of $\frac{1}{3}$. If you choose two other points on L and calculate the rise over run in going from one point to the other, you will still get a slope of $\frac{1}{3}$. For example, going from point $(-2, 1)$ to point $(7, 4)$, the rise is 3 and the run is 9. Hence, the slope is $\frac{3}{9}$, or $\frac{1}{3}$.

For any line in a coordinate plane, the rise between two of its points is the change in their y-coordinates and the run is the change in their x-coordinates. If points P and Q on line L have coordinates as shown in Figure 5-24, then going from P to Q,

Rise = y-coordinate of Q − y-coordinate of P $d - b$
Run = x-coordinate of Q − x-coordinate of P $c - a$

Thus, the slope of a line L is given by the following formula:

Slope of $L = \dfrac{d - b}{c - a}$

if (a, b) and (c, d) are any two points on L and $a \neq c$

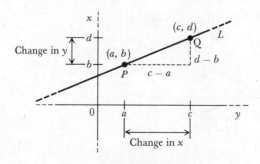

FIGURE 5-24

5-4: THE SLOPE OF A LINE

Problem 1 Find the slope of the line that passes through points $(2, 3)$ and $(5, 7)$.

SOLUTION:

Using the formula above,

$$\text{Slope of line} = \frac{7-3}{5-2} \qquad \text{Let } (a, b) \text{ be } (2, 3) \text{ and } (c, d) \text{ be } (5, 7); a = 2, b = 3, c = 5, \text{ and } d = 7 \text{ in the formula}$$

$$= \frac{4}{3}$$

Problem 2 Find the slope of the line having the equation

$$2y - 3 = x + 3$$

SOLUTION:

To use the slope formula, first you must find two points on the line. You can easily solve the equation for x:

$2y - 3 = x + 3$ *Given equation*

$2y - 6 = x$ *Add -3 to each side*

If $y = 5$, then

$2 \cdot 5 - 6 = x$

$x = 10 - 6$, or 4

Thus, $(4, 5)$ is a point on the line. If $y = 2$, then

$2 \cdot 2 - 6 = x$

$x = 4 - 6$, or -2

Thus, $(-2, 2)$ is another point on the line. Now use the formula:

$$\text{Slope of line} = \frac{2-5}{-2-4} \qquad \text{Let } (a, b) \text{ be } (4, 5) \text{ and } (c, d) \text{ be } (-2, 2); a = 4, b = 5, c = -2, \text{ and } d = 2 \text{ in the formula}$$

$$= \frac{-3}{-6}, \text{ or } \frac{1}{2}$$

If you had chosen (a, b) to be $(-2, 2)$ and (c, d) to be $(4, 5)$, then you would have gotten

$$\text{Slope of line} = \frac{5-2}{4-(-2)}$$

$$= \frac{3}{6}, \text{ or } \frac{1}{2}$$

The slope is the same, $\frac{1}{2}$.

Some lines have a negative slope, as the next two problems will demonstrate.

Problem 3 Find the slope of the line K passing through the points $(-4, 5)$ and $(2, -3)$.

SOLUTION:

By the slope formula,

Slope of $K = \dfrac{-3 - 5}{2 - (-4)}$ *Let (a, b) be $(-4, 5)$ and (c, d) be $(2, -3)$ in the formula*

$= \dfrac{-8}{6}$, or $-\dfrac{4}{3}$

Look at Figure 5-25 and imagine you are moving from left to right along a line in a coordinate plane. If you go *uphill*, the line has *positive* slope; if you go *downhill*, the line has *negative* slope.

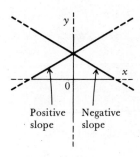

FIGURE 5-25

Problem 4 Find the slope of the line T passing through the points $(-1, 0)$ and $(-4, 4)$.

SOLUTION:

By the slope formula,

Slope of $T = \dfrac{4 - 0}{-4 - (-1)}$ *Let (a, b) be $(-4, 4)$ and (c, d) be $(-1, 0)$ in the formula*

$= \dfrac{4}{-4 + 1} = \dfrac{4}{-3}$

Thus, the slope of line T is $-\tfrac{4}{3}$.

Line K from Problem 3 and line T from Problem 4 have the same slope, $-\tfrac{4}{3}$. Does that mean that K and T are the same line? No, not necessarily. If the equations of K and T are equivalent, then lines

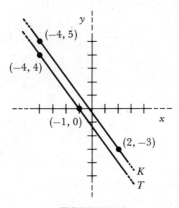

FIGURE 5-26

K and T are **coincident;** that is, they are the same line. However, two lines can have the same slope and yet not coincide; they can be **parallel.** Such is the case for lines K and T; they are two parallel lines, as shown in Figure 5-26.

Problem 5 Find an equation of the line that passes through the points $(3, 0)$ and $(6, -2)$.

SOLUTION:

By the slope formula,

$$\text{Slope of line} = \frac{-2 - 0}{6 - 3}$$

$$= \frac{-2}{3}$$

This slope can then be used in the **equation of the line $y = mx + b$,** where m is the slope, to obtain

$$y = -\frac{2}{3}x + b$$

By substituting either of the original points into this equation, you can solve for b. Using $(3, 0)$,

$y = -\frac{2}{3}x + b$ *Note: the point $(6, 2)$ would give the same solution for b!*

$0 = -\frac{2}{3}(3) + b$ $-2 = -\frac{2}{3}(6) + b$

$0 = -2 + b$ $-2 = -4 + b$

$2 = b$ $2 = b$

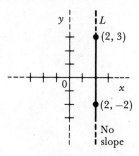

FIGURE 5-27

Replacing this value for b gives

$$y = -\frac{2}{3}x + 2$$

which is the equation of the line passing through the given points.

A vertical line has no slope. If you try to use the slope formula on the vertical line L of Figure 5-27, you get

Slope of $L = \dfrac{3 - (-2)}{2 - 2}$, or $\dfrac{5}{0}$ Let (a, b) be $(2, -2)$ and (c, d) be $(2, 3)$

Since division by zero is not possible, L has no slope. On the other hand, a horizontal line has zero slope, as indicated in Figure 5-28.

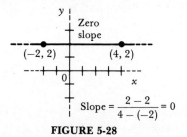

FIGURE 5-28

Problem 6 Assume that a company's profits y in dollars are related to the number x of items produced by a linear equation whose graph is a line L. If profits rise $300 for every 100 items produced, what is the slope of line L?

SOLUTION:

If (x, y) is any point on L, then increasing x by 100 produces an increase of 300 in y (see Figure 5-29). That is, the point $(x + 100, y + 300)$ is also on L. Thus, L has a slope of

$$\frac{300}{100}, \text{ or } 3$$

5-4: THE SLOPE OF A LINE

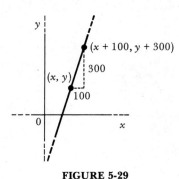

FIGURE 5-29

Quick Reinforcement

(a) Determine the slope of the line passing through the points $(-3, 1)$ and $(4, -2)$.

(b) Determine the slope of the graph of the linear equation $y - 3x + 4 = 0$.

Answers (a) $-\frac{3}{7}$ (b) 3

EXERCISES 5-4

Find the slope of the line passing through each pair of points.

1. (a) $(2, 1)$ and $(6, 3)$ (b) $(4, 2)$ and $(3, 1)$
2. (a) $(0, 2)$ and $(5, 3)$ (b) $(-1, -3)$ and $(1, 1)$
3. (a) $(-4, 1)$ and $(3, 2)$ (b) $(7, 1)$ and $(6, 3)$
4. (a) $(0, 5)$ and $(2, 3)$ (b) $(-9, -1)$ and $(-6, 0)$
5. (a) $(\frac{1}{2}, 1)$ and $(0, \frac{1}{4})$ (b) $(\frac{2}{3}, \frac{3}{4})$ and $(\frac{5}{3}, -\frac{1}{4})$
6. (a) $(1\frac{2}{5}, -4)$ and $(-\frac{3}{5}, 1)$ (b) $(-\frac{1}{2}, 2\frac{1}{4})$ and $(\frac{1}{2}, 2\frac{3}{4})$
7. (a) $(-7, -7)$ and $(-5, -5)$ (b) $(4, 4)$ and $(10, 10)$
8. (a) $(14, 9)$ and $(10, 7)$ (b) $(-20, -11)$ and $(-24, -12)$

Find the slope of the graph of each linear equation.

9. (a) $x + y = 3$ (b) $x - y = -2$
10. (a) $2x - y = 4$ (b) $3y + x = -2$
11. (a) $x = y + 5$ (b) $y = x - 6$
12. (a) $2y - 2x = 7$ (b) $2y + 5x = 12$
13. (a) $x - y + 1 = 0$ (b) $y + x - 9 = 0$
14. (a) $3y = 9x + 2$ (b) $3y - 6x = 5$
15. (a) $5x - y = 7$ (b) $7y - x = 1$
16. (a) $y = x$ (b) $3y = x$

Find the slope of each line below.

17. (a) 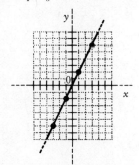 (b)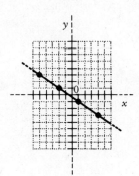

18. (a) 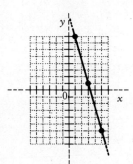 (b)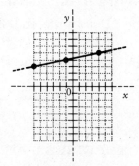

19. (a) 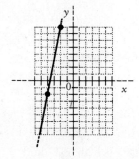 (b)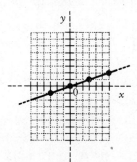

20. (a) 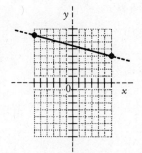 (b)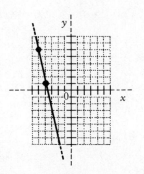

21. (a) Assume that a company's profits y in dollars are related to the number x of items produced by a linear equation whose graph is a line P. If profits rise $2000 for every 500 items produced, what is the slope of line P?

(b) An engine's revolutions in y revolutions per minute are related to the temperature of the engine in x degrees Celsius by a linear equation whose graph is a line E. If the engine's speed was increased by 350 revolutions per minute, generating an increase of 7°C, what is the slope of line E?

22. (a) A jogger's heartbeat in y beats per minute is related to his speed in x feet per second by a linear equation whose graph is line J. If his heartbeat increased 15 beats per minute when he increased his speed by 2 feet per second, what would be the slope of line J?

(b) A colony of bacteria of size y is related to its age x in hours by a linear equation whose graph is line B. If the bacteria count increases by 3200 every $\frac{1}{2}$ hour, what is the slope of line B?

5-5 EQUATIONS OF LINES

The graph of the linear equation

$$y = 2x + 5$$

is easily described. First, when $x = 0$, $y = 5$; that is, 5 is the y-intercept of the graph. Second, when $x = 1$, $y = 7$; that is, the point $(1, 7)$ is on the graph. In going from point $(0, 5)$ to point $(1, 7)$, there is a rise of 2 for a run of 1. Thus, the slope of the graph is $\frac{2}{1}$, or 2. For the equation above, the slope of its graph is 2, the coefficient of x, and the constant term is 5, the y-intercept.

Proceeding as above, we can make a general statement regarding the **slope-intercept form** of a linear equation.

The graph of the equation in two variables

$$y = mx + b$$

is a line with slope m and y-intercept b.

Thus, you can determine the graph of a line by looking at its equation in slope-intercept form.

Problem 1 Find the slope and y-intercept of the graph of the equation $y = 4x - 3$. Sketch its graph.

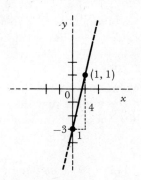

FIGURE 5-30

SOLUTION:

When you compare

$y = 4x - 3$

with

$y = mx + b$

you see that $m = 4$ and $b = -3$. Thus, the graph has slope 4 and y-intercept -3. If you rewrite the slope as $\frac{4}{1}$, then you see that the graph has a rise of 4 for a run of 1. Thus, starting from the point $(0, -3)$, a rise of 4 for a run of 1 takes you to the point $(1, 1)$ on the graph. You now have enough information to sketch the graph in Figure 5-30.

Not every linear equation is given in the form $y = mx + b$. However, a given linear equation can be transformed into slope-intercept form. Consider the following problem.

Problem 2 Find the slope and y-intercept of the linear equation $3x + 2y - 12 = 0$.

SOLUTION:

If you solve this equation for y, you get an equivalent equation in slope-intercept form.

$3x + 2y - 12 = 0$ *Given equation*

$2y = -3x + 12$ *Add $-3x + 12$ to each side*

$y = -\frac{3}{2}x + 6$ *Multiply each side by $\frac{1}{2}$*

5-5: EQUATIONS OF LINES

This last equation is equivalent to the given one. It is in slope-intercept form, so you can read off its slope, $-\frac{3}{2}$, and its y-intercept, 6. Thus, the given equation has slope $-\frac{3}{2}$ and y-intercept 6. You can sketch its graph by starting from the point $(0, 6)$ and making a rise of -3 for a run of 2 (or a rise of 3 for a run of -2). The resulting graph looks like the one in Figure 5-31.

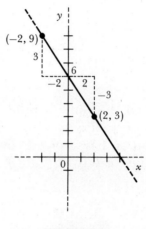

FIGURE 5-31

Given a line L, a linear equation whose graph is L is called an equation of L.

Problem 3 Find an equation of the line whose slope is $-\frac{3}{4}$ and whose y-intercept is $\frac{7}{8}$.

SOLUTION:

The slope-intercept form of such an equation is

$$y = -\frac{3}{4}x + \frac{7}{8}$$

You can clear fractions by multiplying each side by 8:

$$8y = -6x + 7$$

This is also an equation of the graph of the given line.

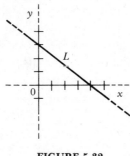

FIGURE 5-32

Problem 4 Find an equation of the line L of Figure 5-32.

SOLUTION:

You need only find the y-intercept and the slope of L to find its equation. The y-intercept is seen to be 3. Line L shows a fall of 3 for a run of 4; thus, L has slope $-\frac{3}{4}$. Therefore, the slope-intercept form of an equation of L is

$$y = -\frac{3}{4}x + 3$$

You can clear fractions by multiplying each side by 4:

$$4y = -3x + 12$$

This is an equation whose graph is L.

Problem 5 The profits y (in dollars) of a company are related by a linear equation to the number x of items sold. If there is a loss of \$140 when zero items are sold, and if there are profits of \$460 when 300 items are sold, what are the profits when 500 items are sold?

SOLUTION:

By what is given, $y = -140$ when $x = 0$, and $y = 460$ when $x = 300$. Therefore, $(0, -140)$ and $(300, 460)$ are points on the "profits" graph. Clearly, -140 is the y-intercept of the graph. The slope of the graph is given by

$$\frac{460 - (-140)}{300 - 0} = \frac{600}{300}, \text{ or } 2$$

Thus, the "profits" equation is

$$y = 2x - 140$$

When $x = 500$,

$y = 2 \cdot 500 - 140$, or 860

This means that there is a profit of $860 when 500 items are sold.

Quick Reinforcement

(a) Find the slope and y-intercept of the graph of the linear equation $y - 5x + 1 = 0$.

(b) Find an equation of the line with slope 3 and y-intercept -2.

(c) Find an equation of the line with x-intercept 2 and y-intercept 6.

Answers (a) Slope 5, y-intercept -1 (b) $y = 3x - 2$ (c) Points (2, 0) and (0, 6) are on the line $y = -3x + 6$.

EXERCISES 5-5

Find the slope and y-intercept of the line with the given equation. Graph it.

1. (a) $y = 5x + 4$ (b) $y = 7x - 1$
2. (a) $y = -2x + 6$ (b) $y = -x + 8$
3. (a) $y = \frac{2}{3}x - 2$ (b) $y = \frac{1}{2}x + 3$
4. (a) $y = -\frac{3}{4}x - 5$ (b) $y = -\frac{1}{3}x - 2$
5. (a) $y = 0.2x + 3$ (b) $y = -0.4x - 1$
6. (a) $3x - y + 6 = 0$ (b) $2x + y - 4 = 0$
7. (a) $7x + 5 + y = 0$ (b) $x - y + 9 = 0$
8. (a) $5x - 15 + 3y = 0$ (b) $6x + 4y + 16 = 0$
9. (a) $y - 6x = 0$ (b) $5y + 2x = 0$
10. (a) $2y - 5 = 4x - 5$ (b) $3x - 7y + 1 = 1$

Find an equation of the line having the given slope and y-intercept.

11. (a) $m = 3, b = 4$ (b) $m = -3, b = 2$
12. (a) $m = -1, b = 3$ (b) $m = 5, b = -1$
13. (a) $m = \frac{1}{2}, b = 0$ (b) $m = -\frac{2}{3}, b = 4$
14. (a) $m = \frac{5}{2}, b = -3$ (b) $m = 0, b = 5$
15. (a) $m = 0.4, b = 1.2$ (b) $m = 2.5, b = 3$

Find an equation that represents each of the given graphs.

16. (a) 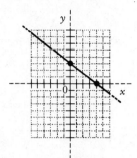 (b)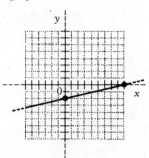

17. (a) 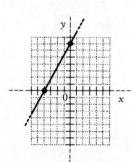 (b)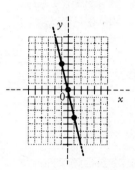

18. (a) 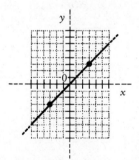 (b)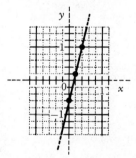

19. (a) 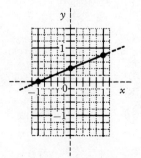 (b)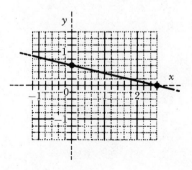

20. (a) 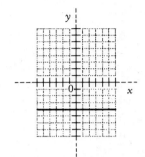 (b)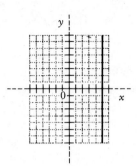

Find an equation of the line having the given intercepts.

21. (a) x-intercept $= 5$, y-intercept $= -1$
(b) x-intercept $= -3$, y-intercept $= 4$
22. (a) x-intercept $= 7$, y-intercept $= 2$
(b) x-intercept $= \frac{1}{2}$, y-intercept $= -1$
23. (a) x-intercept $= 3$, y-intercept $= 0$
(b) x-intercept $= 0$, y-intercept $= 5$
24. (a) x-intercept $= 0.3$, y-intercept $= 1$
(b) x-intercept $= 5.5$, y-intercept $= -0.5$

Find an equation of the line passing through the given points.

25. (a) $(5, 2)$ and $(3, 0)$ (b) $(0, 0)$ and $(-1, 4)$
26. (a) $(-2, 4)$ and $(1, 6)$ (b) $(3, 5)$ and $(7, 13)$
27. (a) $(1, 6)$ and $(3, 6)$ (b) $(-3, 2)$ and $(-3, 4)$
28. (a) $(16, -3)$ and $(8, -5)$ (b) $(1, 0)$ and $(5, 9)$
29. The Mad Hatter Company has developed a demand function relating the number of wigs they can sell, x, to the price, y. They can sell 500 wigs at $20 each, but if they make the price $70, they will not sell any wigs.
   (a) Write the demand equation for the Mad Hatter Company's wigs. Graph the equation.
   (b) If they priced the wigs at $30, how many would they sell? At $50? What would be their expected revenue at $20? $30? $50? $70?
   (Calculate revenue by multiplying price by number of expected sales.) Interpret your findings.
30. While doing a chemistry experiment, a student found that the volume in the test tube increased as the temperature rose and that the relationship was linear. At 0°C there were 32 cubic centimeters of chemical compound, while at 4°C there were 33 cubic centimeters of compound.
   (a) Write an equation showing the relationship between the temperature y and the volume x. Graph it.

(b) What would be the volume at $-2°C$? $6°C$?

31. Suppose the interest y you earn on savings is directly proportional to the amount x you invest. If you invest nothing, you earn nothing! If you invest $250, you earn $13.75 in one year.

   (a) Write an equation showing the relationship between earnings and investment (Hint: Your first point comes from the statement "invest nothing, earn nothing.") Graph the equation.

   (b) How much will $350 earn in one year? How much must you invest to earn $24.75 interest after one year?

 Calculator Activities

Find the slope and y-intercept of the given equation and graph it.

1. (a) $55y - 22x = 451$ (b) $0.34y + 1.51 = 2.4$
2. (a) $0.54x + 0.36y = 1.62$ (b) $235x - 47y = 423$

Answer the following questions using the equation

$$y - 3.25x + 1545.50 = 0$$

which relates profit y in thousands of dollars with the number of items x that are produced.

3. (a) What is the initial profit?
   (b) How much will your profits be if you produce 2500 items?
4. (a) What is your break-even analysis (that is, $y = 0$)?
   (b) How much must you produce to realize $10,000 in profits?
5. (a) Sketch the graph.
   (b) What would be the approximate profit on one item if 3000 items were produced?

5-6 NONLINEAR GRAPHS

So far, we have focused on how to graph a linear equation. To make such a graph, you need to plot only two points and connect them with a line. For another type of an equation, a **nonlinear equation,** you must plot several points to get a clear picture of the graph. For any particular equation, the procedure is to make a table of values, plot the corresponding points on a grid, and connect the points systematically with a smooth curve. If you sketch the curve carefully, you should be able to get an idea of what the actual graph looks like.

Problem 1 Sketch the graph of the equation $y = x^2$.

SOLUTION:

A table of values for x between -3 and 3 is given below:

x	-3	-2.5	-2	-1.5	-1	-0.5	0	0.5	1	1.5	2	2.5	3
y	9	6.25	4	2.25	1	0.25	0	0.25	1	2.25	4	6.25	9

For large values of x, y is much larger; for example, when $x = 10$, $y = 100$. It is hard to choose a scale that will accommodate small numbers and also such large ones! The points of the table are plotted in Figure 5-33 and are connected with a smooth curve in Figure 5-34. The resulting curve is called a **parabola,** and it is discussed further in Chapter 12.

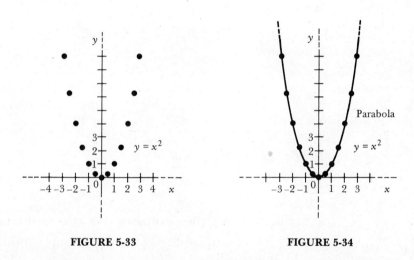

FIGURE 5-33   FIGURE 5-34

Problem 2 Sketch the graph of the equation $y = 2^x$.

SOLUTION:

In this equation, the variable x is an exponent. A table of values for x between -3 and 3 is given below:

x	-3	-2	-1	0	1	2	3
y	$\frac{1}{8}$	$\frac{1}{4}$	$\frac{1}{2}$	1	2	4	8

$2^{-3} = \dfrac{1}{2^3} = \dfrac{1}{8}$, $2^3 = 8$, *and so on*

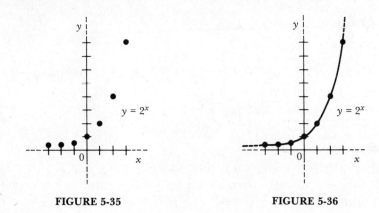

FIGURE 5-35   FIGURE 5-36

The points of the table are plotted in Figure 5-35 and connected with a smooth curve in Figure 5-36. Graphs of this type are called **exponential curves.**

The following problem is a hypothetical application of nonlinear graphs to economics.

Problem 3 The graph of Figure 5-37 is a parabola, the same kind of curve (upside down) as in Figure 5-34. It represents the relationship between profit y (in thousands of dollars) and items produced x (in hundreds) for a certain manufacturing company. Use the graph to answer the following questions.

(a) What are the coordinates of point P and what information do they convey?

(b) How many items must the company produce to break even?

(c) What is their maximum profit? How many items do they have to produce to realize this profit?

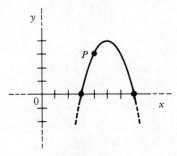

FIGURE 5-37

SOLUTION:

(a) The coordinates of P are evidently $(4, 3)$. This means that when the production rate is 400 items, the profit is $3000.

(b) To break even is to have profit $y = 0$. That is, if they underproduce with 300 items or overproduce with 700 items, their profit is 0.

(c) A maximum profit is attained when the curve reaches its highest point. This occurs when $x = 5$, giving $y = 3$. That is, a production rate of 500 items yields a maximum profit of $3000.

Quick Reinforcement

(a) Sketch the graph of the equation $y = \frac{1}{2}x^2$ between $x = -4$ and $x = 4$.

(b) If B represents the number of bacteria (in hundreds) present in a culture at time t (in seconds), what information is conveyed by point P in the figure?

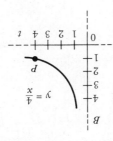
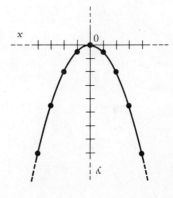

Answers (a)

(b) P has coordinates $(4, 1)$; therefore, at 4 seconds, there are 100 bacteria present.

EXERCISES 5-6

Graph each of the following equations by making a table of values and plotting the points.

 1. (a) $y = -x^2$ (b) $y = -2x^2$
2. (a) $y = x^2 + 1$ (b) $y = x^2 - 1$
3. (a) $y = -x^2 + 2$ (b) $y = -x^2 - 3$

4. (a) $y = 3^x$ (b) $y = 3^{2x}$
5. (a) $y = |x|$ (b) $y = |-x|$
6. (a) $y = x^3$ (b) $y = -x^3$
7. (a) $y = 2x^3 + 1$ (b) $y = 4x^3 - 2$
8. (a) $y = -3x^3$ (b) $y = -2x^3 - 5$
9. (a) $y = \dfrac{1}{x}$ (b) $y = \dfrac{1}{x^2}$
10. (a) $y = \dfrac{1}{x} + 3$ (b) $y = \dfrac{1}{x} - 2$
11. (a) $y = 4 - \dfrac{1}{x^2}$ (b) $y = 3 + \dfrac{1}{x^2}$

12. A garden supply store has plotted the graph showing the relationship between the price per pound of earthworms and their total revenue from the earthworms. Use their graph (below) to answer the following questions.
    (a) At what price per pound will they attain the maximum revenue?
    (b) If they decided on a given day to set the price at point P on the graph, what price have they chosen, and what will be their revenue?

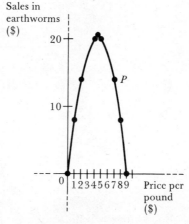

13. If $1 is deposited in a bank at 6% annual interest compounded daily, the amount of money in dollars, A, in the account after t years is given by the following graph.

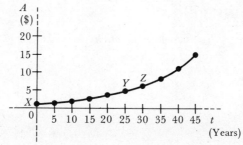

(a) Interpret the meaning of the point X on the graph.

(b) Approximately what would be the value of $1 at this interest rate after 25 years? After 30 years?

14. The cost of operating a car at given speeds is shown in the graph below.

   (a) What speed is the most economical, and what would it cost, approximately, to operate a car at that speed for one hour?

   (b) What would it cost to travel 150 miles at an average speed of 50 miles per hour?

KEY TERMS

Ordered pair
Grid
Grid points
Origin
x-axis
y-axis
Coordinate system
Coordinate plane
Coordinates
x-coordinate
y-coordinate
Quadrants

Graph
Linear equation in two variables
x-intercept
y-intercept
Slope
Coincident
Parallel
Slope-intercept form
Nonlinear equation
Parabola
Exponential curves

REVIEW EXERCISES

Find the solution set of the following equations by solving for x. Find three ordered pairs for each.

1. (a) $2x + 10y - 3 = 0$ (b) $5y - x + \frac{1}{2} = 0$
2. (a) $3x - 9y^2 = 12$ (b) $x + 2y^2 - 7 = 0$

Find the solution set of the following equations by solving for y. Find three ordered pairs for each.

3. (a) $5x - y + 12 = 0$ (b) $2x + 8y = 3$
4. (a) $2x^2 + y = 1$ (b) $xy = 5$
5. The relationship of gestation period G to longevity L is $20L - 3G = 1$.
   (a) Solve for L.
   (b) Find three ordered pairs in the solution set and interpret their meaning.

Plot the following points on a grid.

6. (a) $(-4, 6)$ and $(3, 2)$ (b) $(5, 1)$ and $(-2, -2)$
7. (a) $(0, -2)$ and $(-3, -5)$ (b) $(3, 0)$ and $(9, -1)$

Find the coordinates of the following points.

8. (a) 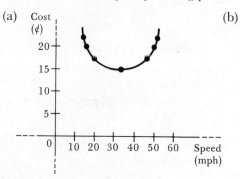 (b)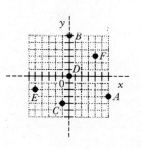

Graph each set of ordered pairs.

9. (a) $\{(1, 0), (2, 3), (3, 0)\}$
(b) $\{(-2, 0), (-1, 1), (0, 0), (1, 1), (2, 0)\}$

Find the set of ordered pairs having the following figures as graphs.

10. (a) 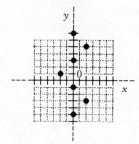 (b)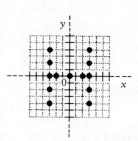

11. The points on the graph below represent the sales in thousands of dollars earned by Louise Chin as an encyclopedia salesperson over a ten-year period.
  (a) Find the coordinates of points $A, B,$ and C and interpret their meaning.
  (b) Find the coordinates of points D and E. If Louise earns 22% of her sales, how much did she earn in 1979 and 1980?

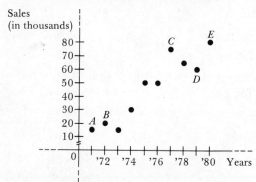

Sketch the graph of each equation, using the given values for x or y.

12. (a) $x + 2 = y$, with $x = -2, 0,$ and 2
    (b) $0.5x + 2.5y = 1$, with $x = 7, -3,$ and -8
13. (a) $\frac{1}{4}y + 3x = 2$, with $y = 12, -4,$ and 0
    (b) $5y - 7 = x$, with $y = 2, 0.2,$ and 0

Sketch the graph of each equation by choosing values for x or y. Plot three points on each graph.

14. (a) $x - y = 7$ (b) $y + x - 1 = 0$
15. (a) $2x - \frac{1}{4}y = 1$ (b) $0.2y + 0.5x = 1.3$

Sketch the graph of each equation by determining the x- and y-intercepts.

16. (a) $2x - 3y = 18$ (b) $x + 5y = 15$
17. (a) $4x - 7y + 42 = 0$ (b) $\frac{1}{2}y - \frac{1}{4}x + \frac{3}{8} = 0$
18. The college bookstore has determined that its total profit y from the sale of school pennants is $0.10 per pennant for x pennants, given by the equation $y = 0.1x$.
    (a) Choose three values for x and find corresponding values for y. Sketch the graph.
    (b) By reading your graph, how many pennants must be sold to bring in a profit of $30 on pennants?

Find the slope of the line passing through each pair of points.

19. (a) $(4, -2)$ and $(1, 0)$ (b) $(0, -1)$ and $(2, 2)$
20. (a) $(4, 11)$ and $(-1, 6)$ (b) $(2\frac{1}{2}, 7)$ and $(3\frac{1}{2}, 9)$

Find the slope of the graph of each linear equation.

21. (a) $x - y = 6$ (b) $x + y = -2$
22. (a) $4x + 2 = y$ (b) $6x - 3y = 1$

Find the slope of each line drawn below.

23. (a) 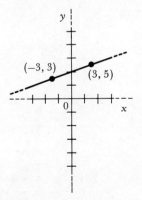 (b)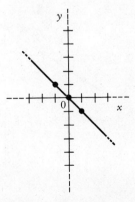

24. (a) 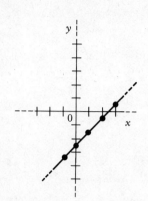 (b)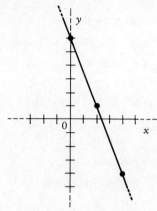

Find the slope and y-intercept of the line with the given equation. Graph it.

25. (a) $y = 7x + 3$ (b) $y = -3x - 1$
26. (a) $x - 2y = 6$ (b) $y + 3x = 4$

Find an equation of the line having the given slope and y-intercept.

27. (a) $m = -5, b = 2$ (b) $m = 6, b = 1$
28. (a) $m = \frac{1}{4}, b = 3$ (b) $m = -\frac{1}{2}, b = \frac{3}{4}$

Find an equation that represents each of the given graphs.

29. (a) 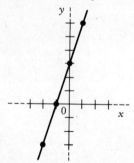 (b)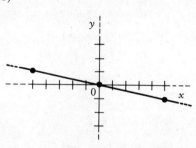

30. (a) 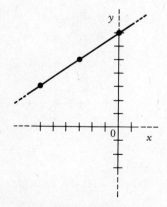 (b)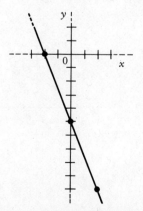

Find an equation of the line passing through the given intercepts.

31. (a) x-intercept $= -2$, y-intercept $= 0$
   (b) x-intercept $= 4$, y-intercept $= 3$
32. (a) x-intercept $= \frac{2}{3}$, y-intercept $= 2$
   (b) x-intercept $= -5$, y-intercept $= 0.4$

Find an equation of the line having the given points.

33. (a) $(7, 3)$ and $(10, 9)$ (b) $(5, 1)$ and $(-3, 9)$
34. (a) $(4, -2)$ and $(3, -1)$ (b) $(-1, 5)$ and $(6, 3)$
35. The Revolutionary Home Computer Company has developed a demand function relating the number of model HAL home computers they can sell, x, to the possible pricing, y. They find they can sell 2500 model HALs at a price of \$450; but if they raise the price to \$650, they will not sell any HALs.
   (a) Write a demand equation for the model HAL computers and graph it.
   (b) If the company priced the HALs at \$400, how many would they expect to sell? Which price would bring in a greater total revenue, the \$400 price, or the \$450 price?

Graph each of the following equations by making a table of values and plotting the points.

36. (a) $y = 3x^2$ (b) $y = -3x^2$
37. (a) $y = 4^x$ (b) $y = 4^{2x}$
38. (a) $y = 3x^3 + 1$ (b) $y = \dfrac{1}{x^2} + 2$

39. A recording company plotted a graph (below) showing the relationship between their profit, in tens of thousands of dollars, and the number of records they cut, in thousands.
   (a) What record volume will bring in the maximum profit per record?
   (b) If the company decided to produce a new record and market 4000 of them, what profit could they expect?

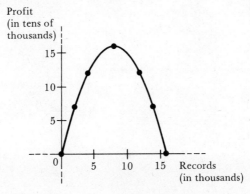

186 CHAPTER 5: EQUATIONS IN TWO VARIABLES AND THEIR GRAPHS

40. A psychological experiment was done related the number of three-syllable words that could be memorized in given lengths of time. As can be seen from the graph, after a certain period of time, the subject cannot substantially increase the number of words memorized when given extra time.
   (a) After how many minutes has the subject memorized seventeen words?
   (b) Between which two points is the increase in learning the greatest? The least?

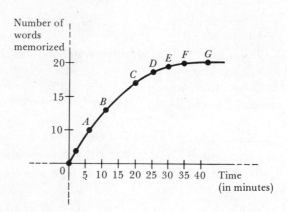

Calculator Activities

1. A manufacturer of high-quality cassette tapes for stereo tape recorders and home computers has found that the price p of his high-quality tape is related to its demand d by the linear equation

   $$27{,}000p + 0.6d - 102{,}000 = 0$$

   (a) If p is plotted on the x-axis and d on the y-axis, find the x- and y-intercepts and sketch the graph.
   (b) With a price of $3.75, approximately what would be the demand?
2. Find an equation of the line passing through the given points. Use the slope-intercept equations.
   (a) $(510, 41)$ and $(39, -486)$ (b) $(-2.11, 4)$ and $(-4, -3.21)$
3. Find the slope and y-intercept of the line with the given equation. Graph it.
   (a) $0.31y - 0.5x = -1.5$ (b) $-21x - 152y = 484$
4. If you have a slope of 456 and a y-intercept of 1050, determine what y is if $x = 2049$.

6 Systems of Linear Equations

Brenda Hamilton is doing a chemistry experiment that will require an 8% saline solution (8% salt, 92% water). She has 1.5 liters of a 9% saline solution and a quantity of a 6% solution. How much of the 6% solution must she mix with the 9% solution to get her desired 8% solution?

6-1 SYSTEMS OF TWO LINEAR EQUATIONS IN TWO VARIABLES

We begin this section with the following problem.

Problem 1 John bought two paperbacks at a cost of $3.75. One book was quite a bit more expensive than the other. In fact, if we doubled the amount paid for the cheaper book, we would still need to add $0.15 to get the amount paid for the more expensive book. How much did John pay for each book?

SOLUTION:

Use two different variables to represent the cost of each book.

Let x = cost of cheaper book, in dollars
Let y = cost of more expensive book, in dollars

Then

$x + y = 3.75$ *The two books cost $3.75*

$2x + 0.15 = y$ *Double the cost of the cheaper book, then add $0.15 to get the cost of the more expensive book*

The problem can be solved if you can find values of the variables that are solutions of *both* linear equations. One way to proceed is as follows.

$x + (2x + 0.15) = 3.75$ *Replace y in the 1st equation by its value from the second equation*

$x + 2x + 0.15 = 3.75$

$3x + 0.15 = 3.75$

$3x = 3.60$ *Add -0.15 to each side*

$x = 1.20$ *Multiply each side by $\frac{1}{3}$*

Thus, the cheaper book costs $1.20 and the more expensive book costs $3.75 − $1.20, or $2.55.

√ CHECK:

$2x + 0.15 = y$

$2 \cdot (1.20) + 0.15 = 2.55$

$2.40 + 0.15 = 2.55$ √

This is a problem whose mathematical formulation involves two linear equations in two variables,

$$\begin{cases} x + y = 3.75 \\ 2x + 0.15 = y \end{cases}$$

We call such a set of equations a **system of equations.** A *solution* of such a system consists of values of the variables that make both equations true. Using ordered pairs, you found the solution of the system above to be $(1.20, 2.55)$, that is, $x = 1.20$, $y = 2.55$.

The following problem shows how to use graphs to solve a system of linear equations in two variables. It also illustrates why there is usually only one solution.

Problem 2 Solve the following system graphically.

$$\begin{cases} 2x + y = -1 \\ -3x + 2y = 5 \end{cases}$$

6-1: SYSTEMS OF TWO LINEAR EQUATIONS IN TWO VARIABLES

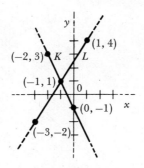

FIGURE 6-1

SOLUTION:

As shown in Figure 6-1, equation $2x + y = -1$ has line K as its graph. It is sketched using the two solutions $(-2, 3)$ and $(0, -1)$ of this equation. Similarly, $(-3, -2)$ and $(1, 4)$ are two solutions of the second equation. Thus, line L is the graph of equation $-3x + 2y = 5$.

Each point on line K represents a solution of the first equation, and each point on line L represents a solution of the second equation. Hence, the **point of intersection** of K and L represents a solution of both equations. By inspection of the figure, $(-1, 1)$ is the solution of the system. It is also the *only* solution, since two non-parallel lines intersect in exactly one point.

✓ CHECK:

Let $x = -1$, $y = 1$

$$2x + y = -1 \qquad\qquad -3x + 2y = 5$$
$$2 \cdot (-1) + 1 = -1 \qquad\qquad (-3) \cdot (-1) + 2 \cdot 1 = 5$$
$$-2 + 1 = -1 \;\checkmark \qquad\qquad 3 + 2 = 5 \;\checkmark$$

Problem 3 Solve the following system graphically.

$$\begin{cases} 2x + 3y = 14 \\ y = x - 2 \end{cases}$$

SOLUTION:

The graph of

$$2x + 3y = 14$$

Let $y = 0$ to get $x = 7$

Let $x = 0$ to get $y = \dfrac{14}{3}$

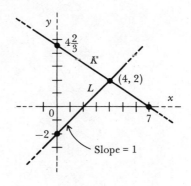

FIGURE 6-2

has x-intercept 7 and y-intercept $\frac{14}{3}$. It is the line K of Figure 6-2. The graph of $y = x - 2$ has slope 1 and y-intercept -2. It is the line L of Figure 6-2. Apparently, the lines intersect at point $(4, 2)$. Thus, $(4, 2)$ is the solution of the system.

√ CHECK:

Let $x = 4, y = 2$

$$2x + 3y = 14 \qquad y = x - 2$$
$$2 \cdot 4 + 3 \cdot 2 = 14 \qquad 2 = 4 - 2 \;\checkmark$$
$$8 + 6 = 14 \;\checkmark$$

Given a system of two linear equations, there is a corresponding system of lines, K and L, the graphs of the two given equations. There are three ways K and L can be related.

1. K and L can *intersect* in one point. Then the system has one solution. For example, the system of Problem 2 has one solution, $(-1, 1)$.

2. K and L can *coincide*. Then every point on K (or L) is a solution of the system, and the system has an infinite number of solutions. For example, the system

$$\begin{cases} x - y = 1 \\ 2y = 2x - 2 \end{cases}$$

is such that the graph of each equation has slope 1 and y-intercept -1. Thus, $K = L$, as shown in Figure 6-3.

3. K and L can be *parallel* but not coincident. Then the system has no solution. For example, the system

6-1: SYSTEMS OF TWO LINEAR EQUATIONS IN TWO VARIABLES

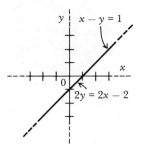

FIGURE 6-3

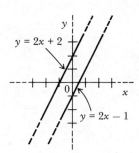

FIGURE 6-4

$$\begin{cases} y = 2x + 2 \\ y = 2x - 1 \end{cases}$$

has K and L parallel (same slope 2) but not coincident (different y-intercepts). This is shown in Figure 6-4.

Graphing can be effective in determining whether you have one, none, or infinite solutions for the system. However, it is not always practical to solve a system of linear equations by graphing. If the solution involves fractions or large numbers, it might be very difficult to read the exact coordinates of the point of intersection. That is why we must develop other methods of solving systems of linear equations.

Quick Reinforcement

Solve the system

$$\begin{cases} x - y = 1 \\ x + y = 2 \end{cases}$$

by carefully drawing the graph of each equation.

Answer $\left(\frac{3}{2}, \frac{1}{2} \right)$

EXERCISES 6-1

Graph each system and describe the nature of its solution. If the solution is one point, try to specify it, and check the solution in the given system.

1. (a) $\begin{cases} y + 2x + 3 = 0 \\ x - y = 0 \end{cases}$ (b) $\begin{cases} 2x + 3y - 2 = 0 \\ x + y = 0 \end{cases}$

2. (a) $\begin{cases} x + y = 12 \\ y - x = 4 \end{cases}$ (b) $\begin{cases} 5x + 4y = 7 \\ 2x - 3y = 12 \end{cases}$

3. (a) $\begin{cases} 2x - y = 4 \\ y = 2x + 3 \end{cases}$ (b) $\begin{cases} 3x + 2y = -12 \\ x - 2y + 20 = 0 \end{cases}$

4. (a) $\begin{cases} 3x - 4y - 8 = 0 \\ 4x + 5y + 10 = 0 \end{cases}$ (b) $\begin{cases} 3x + y = 2 \\ 6x = 4 - 2y \end{cases}$

5. (a) $\begin{cases} 4x - 10y = 30 \\ 3x = -4y - 12 \end{cases}$ (b) $\begin{cases} 5x + y - 7 = 0 \\ 2x + 2y + 2 = 0 \end{cases}$

6. (a) $\begin{cases} x = 3y \\ y = \frac{1}{2}x \end{cases}$ (b) $\begin{cases} 2x = 20 - 5y \\ y = 10 - x \end{cases}$

7. (a) $\begin{cases} 3x - 9y = 3 \\ 3y = x - 1 \end{cases}$ (b) $\begin{cases} 3x - 3y = 1 \\ y = x + 5 \end{cases}$

8. (a) $\begin{cases} y - 4x = 7 \\ y + x = 2 \end{cases}$ (b) $\begin{cases} 4y + 5x + 20 = 0 \\ 8y = 5x + 20 \end{cases}$

9. (a) $\begin{cases} 5x + 5y = 15 \\ x = y - 4 \end{cases}$ (b) $\begin{cases} 3x + 3y = 1 \\ y = x + 5 \end{cases}$

10. (a) $\begin{cases} 2y = 7x - 6 \\ 4y + x + 12 = 0 \end{cases}$ (b) $\begin{cases} 5y - 3x = 23 \\ y - 2x + 8 = 0 \end{cases}$

Use two variables, x and y, to set up a system of equations for each of the following problems. Solve graphically.

11. (a) On July 1, the men's department of the Hillsdale Emporium ordered 3 wallets and 5 dozen handkerchiefs for $54. On August 1, they ordered 5 wallets and 2 dozen handkerchiefs for $52, at the same unit prices as the July 1 order. Find the cost of 1 wallet and 1 dozen handkerchiefs.

    (b) To make a mixture of "gorp" for quick energy for a Sierra Club hike, the leader bought 3 pounds of raisins and 2 pounds of assorted nuts for $3.86. The following month, at the same price per pound (inflation not yet having caught up with raisins and nuts), he bought 5 pounds of raisins and 3 pounds of nuts for $6.15. Find the price of 1 pound of raisins and 1 pound of nuts. (Notice that a graphical solution is not always very easy to read. Try changing axis units if you have problems reading your solution.)

12. (a) At a local charity crafts fair, 4 hand-knit sweaters and 3 large stuffed animals cost $63, while 3 hand-knit sweaters and 4 stuffed animals cost $56. Find the price of 1 sweater and 1 stuffed animal.

    (b) Two brothers were teasing a friend who had asked them how old they were. The older brother said, "I'll tell you how old we are, but only in a riddle. The sum of our ages is 22 years, and the difference of our ages is 4 years." Since their friend was fairly good at algebra, he was able to show them, by graphs, how old they were. Can you?

6-2 SUBSTITUTION METHOD OF SOLVING A SYSTEM

Here is another approach to solving a system such as

$$\begin{cases} 2x + 3y = 14 \\ y = x - 2 \end{cases}$$

Any values of x and y that make both equations of the above system true also make both equations of the new system below true:

$$\begin{cases} 2x + 3(x - 2) = 14 \\ y = x - 2 \end{cases}$$ *Substitute $x - 2$ for y in the 1st equation*

The first equation of the new system involves only one variable, x. Solve that equation:

$2x + 3(x - 2) = 14$ *1st equation*
$2x + 3x - 6 = 14$
$5x = 20$
$x = 4$

Return to the second equation:

$y = x - 2$ *2nd equation*
$y = 4 - 2$ *Replace x by 4*
$y = 2$

Thus, $x = 4$ and $y = 2$ is the solution of the given system. This agrees with the graphical solution from Section 6-1.

Given any system of two linear equations in x and y, you can find an equivalent system, that is, a system having the same solution, in which one equation is solved for y in terms of x or solved for x in terms of y. Then you can solve the resulting system as above.

This method of solving a system is called the **substitution method.** It is the method that was suggested at the beginning of Section 6-1. Further examples follow.

Problem 1 Solve the following system by the substitution method.

$$\begin{cases} 3x - 5y = 2 \\ 3x + y = 32 \end{cases}$$

SOLUTION:

$$\begin{cases} 3x - 5y = 2 \\ 3x + y = 32 \end{cases}$$ *Given system*

194 CHAPTER 6: SYSTEMS OF LINEAR EQUATIONS

$$\begin{cases} 3x - 5y = 2 \\ y = 32 - 3x \end{cases} \quad \text{Solve 2nd equation for } y$$

Substitute the value of y from the second equation into the first equation to obtain an equation in one variable, x.

$$3x - 5(32 - 3x) = 2$$
$$3x - 160 + 15x = 2$$
$$18x = 162$$
$$x = 9$$

To find the value of y, give x the value 9 in the equation $y = 32 - 3x$.

$$y = 32 - 3x$$
$$y = 32 - 3 \cdot 9$$
$$y = 32 - 27$$
$$y = 5$$

Thus, the solution of the given system is $(9, 5)$.

√ CHECK:

Let $x = 9, y = 5$

$3x - 5y = 2$	$3x + y = 32$
$3 \cdot 9 - 5 \cdot 5 = 2$	$3 \cdot 9 + 5 = 32$
$27 - 25 = 2$ √	$27 + 5 = 32$ √

Problem 2 Solve the following system by the substitution method.

$$\begin{cases} -2x + 3y = 8 \\ 5x - 4y = 1 \end{cases}$$

SOLUTION:

Solve one of the equations for y in terms of x, or x in terms of y. (This time, solve the first equation for x in terms of y.)

$$-2x + 3y = 8 \qquad \textit{1st equation}$$
$$-2x = 8 - 3y \qquad \textit{Add } -3y \textit{ to each side}$$
$$x = -4 + \frac{3}{2}y \qquad \textit{Multiply each side by } -\frac{1}{2}$$

The new system

$$\begin{cases} x = -4 + \frac{3}{2}y \\ 5x - 4y = 1 \end{cases}$$

6-2: SUBSTITUTION METHOD OF SOLVING A SYSTEM

is equivalent to the given system. Now replace x in the second equation by its value from the first equation above. The resulting equation involves only one variable, y.

$$5\left(-4 + \frac{3}{2}y\right) - 4y = 1$$

$$5 \cdot \frac{3}{2}y = \frac{15}{2}y$$

$$-20 + \frac{15}{2}y - 4y = 1$$

$$\frac{15}{2}y - 4y = \left(\frac{15}{2} - 4\right)y$$

$$-20 + \frac{7}{2}y = 1$$

$$= \left(\frac{15}{2} - \frac{8}{2}\right)y = \frac{7}{2}y$$

$$\frac{7}{2}y = 21$$

Multiply each side by $\frac{2}{7}$: $21 \cdot \frac{2}{7} = 6$

$$y = 6$$

Substituting 6 for y in the first equation,

$$x = -4 + \frac{3}{2} \cdot 6$$

$$x = -4 + 9$$

$$x = 5$$

Thus, the solution is $(5, 6)$.

√ CHECK:

Let $x = 5$, $y = 6$

$$-2x + 3y = 8 \qquad\qquad 5x - 4y = 1$$
$$-2 \cdot 5 + 3 \cdot 6 = 8 \qquad\qquad 5 \cdot 5 - 4 \cdot 6 = 1$$
$$-10 + 18 = 8 \quad \checkmark \qquad\qquad 25 - 24 = 1 \quad \checkmark$$

Problem 3 Solve the following system of equations.

$$\begin{cases} 4x + 2y = 6 \\ y = -2x + 3 \end{cases}$$

SOLUTION:

Proceeding as before,

$$4x + 2(-2x + 3) = 6 \qquad \textit{Replace } y \textit{ in the 1st equation by } -2x + 3 \textit{ from the 2nd equation}$$

CHAPTER 6: SYSTEMS OF LINEAR EQUATIONS

$4x - 4x + 6 = 6$ Solve resulting equation for x

$0x + 6 = 6$ True for all values of x; $0x = 0$

Thus, for every value of x, $(x, -2x + 3)$ is a solution of the system. In other words,

$\{(x, -2x + 3) \mid x \text{ any real number}\}$

is the solution set of the system. This means that the graphs of the two given equations are coincident.

Quick Reinforcement

Solve the system

$$\begin{cases} x + 2y = 0 \\ 2x - y = -5 \end{cases}$$

by the substitution method.

Answer: $(-2, 1)$

EXERCISES 6-2

Solve each system by the substitution method. Check your solution.

1. (a) $\begin{cases} 2x + 3y = 4 \\ x = 2 \end{cases}$ (b) $\begin{cases} x + 4y = 6 \\ x = -6 \end{cases}$

2. (a) $\begin{cases} y = -3 \\ 2x + y = 5 \end{cases}$ (b) $\begin{cases} y = 5 \\ 3x + 2y + 8 = 0 \end{cases}$

3. (a) $\begin{cases} 3x - 4y = 6 \\ 2x + y = 4 \end{cases}$ (b) $\begin{cases} 3x - 2y = 14 \\ 2x + y = 0 \end{cases}$

4. (a) $\begin{cases} 2x + y = 5 \\ x + 3y = 0 \end{cases}$ (b) $\begin{cases} x + 3y = 1 \\ 2x + y = -3 \end{cases}$

5. (a) $\begin{cases} 2x + 3y = 1 \\ 3x - y = 7 \end{cases}$ (b) $\begin{cases} 5x - 4y = 1 \\ 3x - 6y = 6 \end{cases}$

6. (a) $\begin{cases} 5x + 3y = 16 \\ x = y \end{cases}$ (b) $\begin{cases} y = 3x \\ 2x + 3y = 22 \end{cases}$

7. (a) $\begin{cases} 3x - y + 14 = 0 \\ x = -2y \end{cases}$ (b) $\begin{cases} x = 5y + 21 \\ 3x + 4y = -13 \end{cases}$

8. (a) $\begin{cases} 0.5x - 0.6y = 1.6 \\ 1.5x + 2y = 1 \end{cases}$ (b) $\begin{cases} 1.2x - 0.4y = -1.4 \\ 1.5x + 0.9y + 5.7 = 0 \end{cases}$

9. (a) $\begin{cases} 2.4x + 1.8y + 8.4 = 0 \\ 0.2x = 0.2y - 1.4 \end{cases}$ (b) $\begin{cases} 0.5y - x = 2.5 \\ 1.4x + 2.1y = -2.1 \end{cases}$

10. (a) $\begin{cases} 11x + 2y = 1 \\ 9x - 3y = 24 \end{cases}$ (b) $\begin{cases} 8x - y = 29 \\ 2x = 11 - y \end{cases}$

11. (a) $\begin{cases} \frac{3}{2}x + y = 2\frac{1}{2} \\ y = 2x + 1 \end{cases}$ (b) $\begin{cases} 5x + 6y = 31 \\ y = x + \frac{2}{3} \end{cases}$

12. (a) $\begin{cases} x - 5y + 6 = 0 \\ y - 4x + 3 = 0 \end{cases}$ (b) $\begin{cases} 4x + 6y - 2 = 0 \\ 6x - 9y - 15 = 0 \end{cases}$

13. (a) $\begin{cases} 5x - 3y = 7 \\ 7x + 12y + 1 = 0 \end{cases}$ (b) $\begin{cases} 5x + 2y + 19 = 0 \\ y - 3 = 0 \end{cases}$

14. (a) $\begin{cases} \frac{1}{4}x = 3y \\ \frac{3}{4}x + 6y = \frac{15}{4} \end{cases}$ (b) $\begin{cases} 0.6x + 0.4y = 0.4x + 0.4 \\ 1.6x + 2.4y = 1.6x - 1.6 \end{cases}$

15. (a) $\begin{cases} x = \frac{1}{2}y - 3 \\ 2x + 5y = 10 \end{cases}$ (b) $\begin{cases} 3x + y + 8 = 0 \\ 3y - 5x = 4 \end{cases}$

16. (a) $\begin{cases} 2x - 3y - 1 = -3x \\ 4y + 2 = 7x \end{cases}$ (b) $\begin{cases} 3x - 8 = 2y - 3 - 5 \\ y - 2x - 1 = -8 - 4x \end{cases}$

Calculator Activities

Solve each system by the substitution method. Check your solution.

1. (a) $\begin{cases} 59x - 24y = 180 \\ y + 15x = 12 \end{cases}$ (b) $\begin{cases} 1.21x - 0.5y = 3.9 \\ x + 0.02y = 1 \end{cases}$

2. (a) $\begin{cases} x - 0.01y = 2 \\ 2.3x - 0.023y = 4.6 \end{cases}$ (b) $\begin{cases} y - 29x = 18 \\ 58x - 2y + 10 = 0 \end{cases}$

3. (a) $\begin{cases} 1289x - 435y = 902 \\ y - 341x = 0 \end{cases}$ (b) $\begin{cases} 2.04x + 3.58y - 1.66 = 0 \\ x + 6.26y = 5.90 \end{cases}$

4. (a) $\begin{cases} 91.3x + 45.7y - 102.2 = 0 \\ 2.7x - 16.2y = 29.7 \end{cases}$ (b) $\begin{cases} 375x - 750y = 225 \\ 35y = 210x + 455 \end{cases}$

6-3 ADDITION METHOD OF SOLVING A SYSTEM

As you have seen, the substitution method of solving a system can lead to calculations involving fractions. Now we will discuss a related method that bypasses such calculations. Basic to this new method is the following property. If

$$A = B$$

and

$$C = D$$

are true equations, then so is

$$A + C = B + D$$

true. For example,

$$3 = 5 - 2$$

and

$$8 + 4 = 12$$

are true. Hence,

$$3 + (8 + 4) = (5 - 2) + 12$$

is true.

If you start with a system of equations of the form

$$\begin{cases} A = B \\ C = D \end{cases}$$

you can obtain an equivalent system by keeping one equation of the system and taking the sum of the two equations as the second equation:

$$\begin{cases} A = B \\ A + C = B + D \end{cases}$$

For example, the system

(1) $\quad \begin{cases} 2x + y = 7 \\ x + 3y = 6 \end{cases} \quad \begin{matrix} A = B \\ C = D \end{matrix}$

is equivalent to

(2) $\quad \begin{cases} 2x + y = 7 \\ 3x + 4y = 13 \end{cases} \quad \begin{matrix} A = B \\ A + C = B + D \end{matrix}$

or to

(3) $\quad \begin{cases} x + 3y = 6 \\ 3x + 4y = 13 \end{cases} \quad \begin{matrix} C = D \\ A + C = B + D \end{matrix}$

Any solution of system (1) is also a solution of systems (2) and (3) by our remarks above. Conversely, any solution of (2) or (3) is also a solution of system (1). In fact, $(3, 1)$ is a solution of system (1):

$$\begin{cases} 2x + y = 7 \\ x + 3y = 6 \end{cases} \quad \begin{matrix} 2 \cdot 3 + 1 = 7 \\ 3 + 3 \cdot 1 = 6 \end{matrix}$$

Therefore, it is a solution of system (2):

$$\begin{cases} 2x + y = 7 \\ 3x + 4y = 13 \end{cases} \quad \begin{matrix} 2 \cdot 3 + 1 = 7 \\ 3 \cdot 3 + 4 \cdot 1 = 13 \end{matrix}$$

Of course, $(3, 1)$ is also a solution of system (3).

6-3: ADDITION METHOD OF SOLVING A SYSTEM

Starting with any system of two linear equations in two variables, you can form an equivalent system as above by using one of the given equations together with their sum. If the new system is easier to solve than the given one, then that is the one you solve. In this way, you will have also solved the given system. This method of solving a system is called the **addition method.**

Problem 1 Solve the following system by the addition method.

$$\begin{cases} -3x + 2y = -10 \\ 3x - 4y = 8 \end{cases}$$

SOLUTION:

First, find the sum of the two given equations.

$$-3x + 2y = -10$$
$$3x - 4y = 8$$ \hfill *Given equations*

$$(-3 + 3)x + (2 - 4)y = -10 + 8$$ \hfill *Add sides*
$$0x - 2y = -2 \qquad 0x = 0$$
$$-2y = -2$$ \hfill *Their sum*

Hence, the given system is equivalent to the system

$$\begin{cases} -3x + 2y = -10 \\ -2y = -2 \end{cases}$$ \qquad *One of the given equations* / *Their sum*

In turn, this system is equivalent to

$$\begin{cases} -3x + 2y = -10 \\ y = 1 \end{cases}$$ \qquad *Multiply each side of the 2nd equation by* $-\frac{1}{2}$

This system is easily solved. From the second equation, $y = 1$. Substituting 1 for y in the first equation,

$$-3x + 2 \cdot 1 = -10$$
$$-3x = -12$$
$$x = 4$$

Thus, the system has the solution $(4, 1)$.

√ CHECK:

Let $x = 4$, $y = 1$

$$-3x + 2y = -10 \qquad\qquad 3x - 4y = 8$$
$$-3 \cdot 4 + 2 \cdot 1 = -10 \qquad\qquad 3 \cdot 4 - 4 \cdot 1 = 8$$
$$-12 + 2 = -10 \quad √ \qquad\qquad 12 - 4 = 8 \quad √$$

CHAPTER 6: SYSTEMS OF LINEAR EQUATIONS

To compare the substitution and addition methods, let us solve the system of Problem 1 from Section 6-2 again, this time by the addition method.

Problem 2 Solve the system below by the addition method.

$$\begin{cases} 3x - 5y = 2 \\ 3x + y = 32 \end{cases}$$

SOLUTION:

The whole point to adding the equations in the previous problem was to eliminate one of the variables. Adding equations in this problem won't eliminate either the terms involving x or the terms involving y. However, if you first multiply each side of one equation by -1 and then add the equations, you will eliminate the x-term.

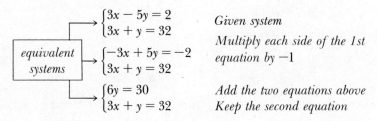

If $6y = 30$, then $y = 5$. Now give y the value 5 in the second equation, $3x + y = 32$.

$$3x + 5 = 32$$
$$3x = 27$$
$$x = 9$$

The solution of the given system is $(9, 5)$, which agrees with our earlier solution. Which method is better? Perhaps the addition method is simpler, because it avoids unnecessary fractions.

Problem 3 Solve the following system by the addition method.

$$\begin{cases} 4x + 3y = 6 \\ 2x + 5y = -4 \end{cases}$$

SOLUTION:

Adding equations will not eliminate one of the variables. However, if you first multiply each side of one equation by a number such that the terms involving one of the variables become negatives of each other, then you can add equations to eliminate a variable.

6-3: ADDITION METHOD OF SOLVING A SYSTEM

$\begin{cases} 4x + 3y = 6 \\ 2x + 5y = -4 \end{cases}$ *Make the terms involving x negatives of each other*

$\begin{cases} 4x + 3y = 6 \\ -4x - 10y = 8 \end{cases}$ *Multiply the 2nd equation above by -2*

Adding equations,

$\begin{cases} 4x + 3y = 6 \\ -7y = 14 \end{cases}$ *1st equation above*
Sum of equations above

From the second equation, $y = -2$. From the first equation,

$$4x + 3 \cdot (-2) = 6$$
$$4x - 6 = 6$$
$$4x = 12$$
$$x = 3$$

The solution is $(3, -2)$.

✓ CHECK:

Let $x = 3$, $y = -2$

$\quad\quad 4x + 3y = 6 \quad\quad\quad\quad 2x + 5y = -4$
$4 \cdot 3 + 3 \cdot (-2) = 6 \quad\quad 2 \cdot 3 + 5 \cdot (-2) = -4$
$\quad\quad\quad 12 - 6 = 6 \;\checkmark \quad\quad\quad\quad 6 - 10 = -4 \;\checkmark$

Problem 4 Solve the system below by the addition method.

$\begin{cases} 2x - 3y = 8 \\ 3x + 2y = 25 \end{cases}$

SOLUTION:

If you multiply the first equation by 2 and the second equation by 3, then the terms involving y will become negatives of each other and you can add the two equations.

$\begin{cases} 2x - 3y = 8 \\ 3x + 2y = 25 \end{cases}$ *Given system*

equivalent $\to$ $\begin{cases} 4x - 6y = 16 \\ 9x + 6y = 75 \end{cases}$ *Multiply the 1st equation by 2*
Multiply the 2nd equation by 3

$\begin{cases} 4x - 6y = 16 \\ 13x = 91 \end{cases}$ *1st equation above*
Sum of equations above

By the second equation above, $x = 7$. Solving the first equation for y,

$$4 \cdot 7 - 6y = 16$$
$$28 - 6y = 16$$
$$-6y = -12$$
$$y = 2$$

The solution is $(7, 2)$.

CHECK:

✓ Let $x = 7$, $y = 2$

$$2x - 3y = 8 \qquad\qquad 3x + 2y = 25$$
$$2 \cdot 7 - 3 \cdot 2 = 8 \qquad\qquad 3 \cdot 7 + 2 \cdot 2 = 25$$
$$14 - 6 = 8 \quad ✓ \qquad\qquad 21 + 4 = 25 \quad ✓$$

Problem 5 Solve the following system by the addition method.

$$\begin{cases} \dfrac{3}{2}x + \dfrac{2}{3}y = -\dfrac{2}{3} \\ \dfrac{5}{2}x - \dfrac{4}{3}y = 5 \end{cases}$$

SOLUTION:

You can eliminate denominators by multiplying each equation by the LCM of its denominators (LCD). Each equation has LCD = 6, so multiply each equation by 6. This gives the following equivalent system.

$$\begin{cases} 9x + 4y = -4 \\ 15x - 8y = 30 \end{cases} \qquad \begin{matrix} A = B \\ C = D \end{matrix}$$

Solve:

$$\begin{cases} 18x + 8y = -8 \\ 15x - 8y = 30 \end{cases} \qquad \begin{matrix} 2A = 2B \\ C = D \end{matrix}$$

$$\begin{cases} 18x + 8y = -8 \\ 33x = 22 \end{cases} \qquad \begin{matrix} 2A = 2B \\ 2A + C = 2B + D \end{matrix}$$

Divide the last equation by 33:

$$x = \frac{2}{3}$$

Solve for y in the first equation of the above system:

$$18 \cdot \frac{2}{3} + 8y = -8$$

$$12 + 8y = -8$$

$$8y = -20$$

$$y = -\frac{5}{2}$$

Thus, the solution of the given system is $(\frac{2}{3}, -\frac{5}{2})$.

CHECK:

✓ Let $x = \frac{2}{3}$, $y = -\frac{5}{2}$

$$\frac{3}{2}x + \frac{2}{3}y = -\frac{2}{3} \qquad\qquad \frac{5}{2}x - \frac{4}{3}y = 5$$

$$\frac{3}{2} \cdot \frac{2}{3} + \frac{2}{3} \cdot \left(-\frac{5}{2}\right) = -\frac{2}{3} \qquad\qquad \frac{5}{2} \cdot \frac{2}{3} - \frac{4}{3} \cdot \left(-\frac{5}{2}\right) = 5$$

$$1 - \frac{5}{3} = -\frac{2}{3} \ \checkmark \qquad\qquad \frac{5}{3} + \frac{10}{3} = 5$$

$$\frac{15}{3} = 5 \ \checkmark$$

The following problem shows what happens when the system has no solution.

Problem 6 Solve the system

$$\begin{cases} 3x - 2y = 1 \\ -6x + 4y = -8 \end{cases}$$

SOLUTION:

The following systems are equivalent:

$\begin{cases} 3x - 2y = 1 \\ -6x + 4y = -8 \end{cases}$ *Given system*

$\begin{cases} 6x - 4y = 2 \\ -6x + 4y = -8 \end{cases}$ *Multiply the 1st equation by 2*

$\begin{cases} 6x - 4y = 2 \\ 0 = -6 \end{cases}$ *1st equation*
Sum of two equations above

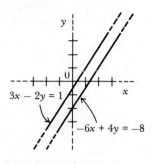

FIGURE 6-5

The equation $0 = -6$ is false. Therefore, there are no values of x and y that make both equations true. In other words, the given system has no solution and, as shown by Figure 6-5, the graphs of the two equations must be parallel lines.

Quick Reinforcement

Use the addition method to solve the system

$$\begin{cases} 2x - 3y = 1 \\ 5x - 2y = -1 \end{cases}$$

Answer $\left(-\frac{5}{11}, -\frac{7}{11}\right)$

EXERCISES 6-3

Solve each system by the addition method. Check your solution.

1. (a) $\begin{cases} 3x - y = -5 \\ x - y = 7 \end{cases}$ (b) $\begin{cases} 4x - y = 3 \\ x + y = 7 \end{cases}$

2. (a) $\begin{cases} 3x + y = 5 \\ x - y = 3 \end{cases}$ (b) $\begin{cases} 3x - 4y = 6 \\ 2x + y = 4 \end{cases}$

3. (a) $\begin{cases} x + 4y = 34 \\ y = 4x \end{cases}$ (b) $\begin{cases} x + 4y = -15 \\ 3x - 2y = -3 \end{cases}$

4. (a) $\begin{cases} x + 2y = 3 \\ x - 4y = -9 \end{cases}$ (b) $\begin{cases} 3x - y = 2 \\ x - y = 6 \end{cases}$

5. (a) $\begin{cases} 4x - 2y = -16 \\ 2x - 4y = -14 \end{cases}$ (b) $\begin{cases} x - y = 6 \\ 3x - 4y = 23 \end{cases}$

6. (a) $\begin{cases} 3x + 2y = 3 \\ 2x - 3y = -11 \end{cases}$ (b) $\begin{cases} 2x + 2y = 28 \\ 4x + 3y = 31 \end{cases}$

7. (a) $\begin{cases} 3x + 5y = 2 \\ 15x - 5y = 4 \end{cases}$ (b) $\begin{cases} 6x - 9y = -45 \\ 5x + 3y = 15 \end{cases}$

8. (a) $\begin{cases} 1.5x + y = 3.5 \\ x + 1.5y = 4 \end{cases}$ (b) $\begin{cases} 1.8x - 1.5y = 2.4 \\ 0.2x + 0.4y = 1.4 \end{cases}$

9. (a) $\begin{cases} \dfrac{x}{3} + \dfrac{y}{2} = 2 \\ 2x + 3y = 12 \end{cases}$ (b) $\begin{cases} \dfrac{x}{2} - \dfrac{y}{3} = \dfrac{5}{6} \\ \dfrac{x}{5} - \dfrac{y}{4} = \dfrac{1}{10} \end{cases}$

10. (a) $\begin{cases} 2x - 2.4y = 6.4 \\ 1.5x + 2y = 1 \end{cases}$ (b) $\begin{cases} 1.4x + 2.1y = 7 \\ 2x + 2.5y = 5 - 0.5y \end{cases}$

11. (a) $\begin{cases} 4x + y = 13 \\ 6x + 12y = 72 \end{cases}$ (b) $\begin{cases} 2x + 2y = 26 \\ 7x - y = 67 \end{cases}$

12. (a) $\begin{cases} \dfrac{3x}{7} + \dfrac{2y}{3} = 5 \\ x - y = 4 \end{cases}$ (b) $\begin{cases} 2x + 3y = 31 \\ 4x - 3y = 17 \end{cases}$

13. (a) $\begin{cases} \tfrac{1}{2}x + \tfrac{1}{3}y = 14 \\ \tfrac{1}{3}x + \tfrac{1}{2}y = 11 \end{cases}$ (b) $\begin{cases} x + \tfrac{1}{2}y = 8 \\ \tfrac{1}{2}x + y = 7 \end{cases}$

14. (a) $\begin{cases} 3x - 8y + 3y + 2 = x + 5y + 16 \\ 6x - 2y = 2x + 28 \end{cases}$ (b) $\begin{cases} 20x + 42y = 180 \\ 6x + 13y = 10 - 4x - 8y \end{cases}$

15. (a) $\begin{cases} 5x - 2y = 16 + 4x - 10 \\ 4x + 3y = 60 + 2x + y \end{cases}$ (b) $\begin{cases} 3y - 2x = 7 + 4y - x \\ 6x + 2y + 8 = 22 \end{cases}$

16. (a) $\begin{cases} \dfrac{x-4}{8} - \dfrac{y-3}{3} = 0 \\ 2x - \dfrac{3y-6}{4} = 21 \end{cases}$ (b) $\begin{cases} \dfrac{x-y}{5} + \dfrac{x+y}{19} = 4 \\ x - y = 10 \end{cases}$

Solve each system by any method. Check your solutions.

17. (a) $\begin{cases} 3x + 5y - 15 = 0 \\ 6x + 10y = -30 \end{cases}$ (b) $\begin{cases} 3x - 5y = 15 \\ x - \tfrac{5}{3}y = 5 \end{cases}$

18. (a) $\begin{cases} \tfrac{1}{2}x - y = -3 \\ -x + 2y = 6 \end{cases}$ (b) $\begin{cases} 2x + 3y = 2y - 2 \\ 3x + 2y = 2x + 2 \end{cases}$

19. (a) $\begin{cases} 2m - 3n = 1 - 3m \\ 4n = 7m - 2 \end{cases}$ (b) $\begin{cases} 0.2p - 0.5q - 0.07 = 0 \\ -0.3q + 0.8p - 0.79 = 0 \end{cases}$

20. (a) $\begin{cases} 0.2a + 0.5b = 0.54 \\ 0.3b - 0.6a - 0.18 = 0 \end{cases}$ (b) $\begin{cases} \tfrac{7}{8}x + \tfrac{1}{4}y = 3\tfrac{1}{8} \\ \tfrac{2}{3}x - \tfrac{1}{3}y - 1\tfrac{1}{3} = 0 \end{cases}$

21. (a) $\begin{cases} 275x + 385y = 7652 \\ 27.5x + 5.07y = 15.06 \end{cases}$ (b) $\begin{cases} 7200x + 8100y = 79{,}200 \\ 10{,}000x - 4000y = 61{,}200 \end{cases}$

22. (a) $\begin{cases} 12x + 16y = 3(4x - 2y) \\ 5x = 7y \end{cases}$ (b) $\begin{cases} 2(3x - 5y) - 5x = 5 - 6y \\ 4y = 6x - 7 \end{cases}$

Calculator Activities

Solve the following systems by the addition method. Check your solution.

1. (a) $\begin{cases} 48x - 29y = 11 \\ 15x - 31y = 1 \end{cases}$ (b) $\begin{cases} 2.01x - 0.7y = 1 \\ 0.105x + 0.01y = 2.3 \end{cases}$

2. (a) $\begin{cases} 7.32x + 4.45y = 9.54 \\ 3.05x - 9.30y = 11.22 \end{cases}$ (b) $\begin{cases} 935x + 740y = 1030 \\ 1125x - 335y = 845 \end{cases}$

Solve the following systems by either the substitution or addition method. Check your solution.

3. (a) $\begin{cases} 112y - 27x = 12 \\ 54x - 224y = -24 \end{cases}$ (b) $\begin{cases} 0.05x - 0.07y = 0.02(30) \\ x - y = 500 \end{cases}$

4. (a) $\begin{cases} x + y = 244 \\ 0.35x - 1.24y = -1.02 \end{cases}$ (b) $\begin{cases} 195x + 143y = 286 \\ 30x + 47y = 74 \end{cases}$

6-4 APPLIED PROBLEMS

In some applications of mathematics, the problems encountered can be solved by solving systems of linear equations in two variables. Some hypothetical examples follow:

Problem 1 The Atlas Corporation sells two types of doorknobs, the expensive, gold-plated kind for $10 and the cheaper, chromium-plated variety for $7. During one month, Atlas Corporation sold 520 doorknobs for $4600. For inventory purposes, how many of each type did Atlas Corporation sell?

SOLUTION:

Introduce a variable for the number of each type sold.

x = number of gold-plated doorknobs sold

y = number of chromium-plated doorknobs sold

Atlas Corporation sold 520 doorknobs, so

$x + y = 520$

Atlas Corporation received

$\begin{array}{rl} 10x & \text{dollars for gold-plated knobs} \\ +\ \underline{7y} & \text{dollars for chromium-plated knobs} \\ 4600 & \text{total dollars received} \end{array}$

That is,

$10x + 7y = 4600$

The problem is solved by solving the system of equations below:

$$\begin{cases} x + y = 520 \\ 10x + 7y = 4600 \end{cases}$$

$$\begin{cases} -10x - 10y = -5200 \\ 10x + 7y = 4600 \end{cases} \quad \text{Multiply each side of the 1st equation by } -10$$

$$\begin{cases} -3y = -600 \\ 10x + 7y = 4600 \end{cases} \quad \text{Add equations to get rid of x-terms}$$

Thus, $y = 200$ and

$$10x + 7 \cdot 200 = 4600$$
$$10x = 3200$$
$$x = 320$$

Atlas Corporation sold 320 gold-plated doorknobs and 200 chromium-plated doorknobs.

√ CHECK:

$$320 + 200 = 520 \quad \checkmark \qquad 10 \cdot 320 + 7 \cdot 200 = 4600$$
$$3200 + 1400 = 4600 \quad \checkmark$$

The flowchart in Figure 6-6 can help you solve this kind of applied problem. It is modeled after the one in Figure 3-1.

Problem 2 A road construction crew consists of Caterpillar operators paid $80 per day and laborers paid $40 per day. The daily payroll is $1520. There are 2 laborers doing odd jobs and 4 laborers assigned to work with each Cat operator. Describe the makeup of the crew.

SOLUTION:

Use the flowchart.

There are Cat operators and laborers
Each Cat operator is paid
$80 per day
Each laborer is paid $40 per day
Daily payroll is $1520
2 laborers doing odd jobs
4 laborers working with
each Cat operator

Bits of information

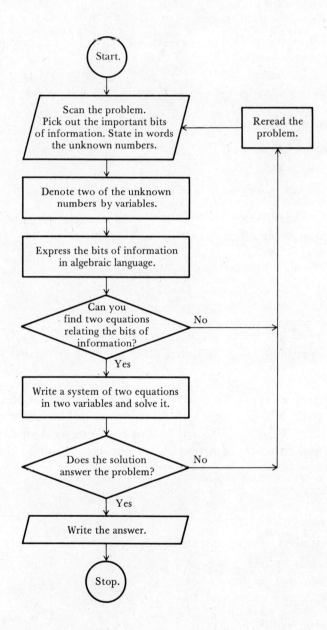

FIGURE 6-6

$x =$ number of Cat operators *Variables for the*
$y =$ number of laborers *two unknowns*

$80x\ \ $ dollars paid to Cat operators each day
$\underline{+\ 40y\ \ }$ dollars paid to laborers each day
$1520\ \ $ dollars paid each day

$80x + 40y = 1520$ *1st equation*

$2\ \ $ laborers doing odd jobs
$\underline{+\ 4x\ \ }$ laborers working with Cat operators
$2 + 4x\ \ $ total number of laborers

$y = 2 + 4x$ *2nd equation*

$\begin{cases} 80x + 40y = 1520 \\ y = 2 + 4x \end{cases}$ *System of equations relating bits of information*

Solve by substitution.

$80x + 40(2 + 4x) = 1520$ *Replace y by $2 + 4x$, from the 2nd equation*
$80x + 80 + 160x = 1520$
$240x = 1520 - 80$
$240x = 1440$
$x = 6$

Then, going back to the second equation,

$y = 2 + 4 \cdot 6$
$ = 26$

Thus, there are 6 Cat operators and 26 laborers.

√ CHECK:

$26 = 2 + 4 \cdot 6$ √ $80 \cdot 6 + 40 \cdot 26 = 1520$
$480 + 1040 = 1520$ √

Problem 3 A rectangular field is constructed so that twice its length equals three times its width. Also, three times its length exceeds four times its width by 27 meters. How much fencing must you buy to completely enclose the field?

SOLUTION:

Assume the field is as pictured in Figure 6-7.

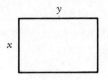

FIGURE 6-7

$y =$ length of the field, in meters $x =$ width of the field, in meters	*Variables*

Twice length equals three times width
Three times length exceeds four *Bits of information*
times width by 27 meters

$2y =$ twice length
$3x =$ three times width
$3y =$ three times length *Algebraic language*
$4x + 27 = 27$ meters more than
four times width

$$\begin{cases} 2y = 3x \\ 3y = 4x + 27 \end{cases} \quad \textit{System of equations}$$

Solve for an equivalent system.

$$\begin{cases} y = \frac{3}{2}x & \textit{Multiply 1st equation above by } \frac{1}{2} \\ 3y = 4x + 27 & \textit{2nd equation above} \end{cases}$$

Substitute $y = \frac{3}{2}x$ in the second equation.

$$3 \cdot \frac{3}{2}x = 4x + 27$$

$$\frac{9}{2}x = 4x + 27$$

$$\frac{1}{2}x = 27 \qquad \textit{Add } -4x \textit{ to each side; } \frac{9}{2}x - 4x = \left(\frac{9}{2} - 4\right)x = \frac{1}{2}x$$

$$x = 54$$

Then substitute $x = 54$ in the first equation.

$$y = \frac{3}{2} \cdot 54 = 3 \cdot 27 = 81$$

Thus, the field is 54 meters by 81 meters. Its perimeter is

$$2x + 2y = 2 \cdot 54 + 2 \cdot 81$$
$$= 108 + 162$$
$$= 270$$

You must buy 270 meters of fencing to enclose the field.

Problem 4 It takes a jet $3\frac{1}{2}$ hours to go one way on a trip of 3150 kilometers and $4\frac{1}{2}$ hours on the return trip against the wind. What is the average speed of the wind and of the jet in still air?

SOLUTION:

Jet takes $3\frac{1}{2}$ hours on a trip with wind

Jet takes $4\frac{1}{2}$ hours on return trip against wind   *Bits of information*

Trip is 3150 kilometers each way

x = average speed of wind, km/hr

y = average speed of jet in still air, km/hr   *Variables*

Rate × Time = Distance

Going (with wind)	$y + x$	$\frac{7}{2}$	3150
Returning (against wind)	$y - x$	$\frac{9}{2}$	3150

Thus,

$$\begin{cases} \frac{7}{2}(y + x) = 3150 \\ \frac{9}{2}(y - x) = 3150 \end{cases}$$ *Equations relating variables*

$$\begin{cases} y + x = \frac{2}{7}(3150) \\ y - x = \frac{2}{9}(3150) \end{cases}$$ *Multiply each side of 1st equation by $\frac{2}{7}$; each side of 2nd equation by $\frac{2}{9}$*

$$\begin{cases} y + x = 900 \\ y - x = 700 \end{cases}$$

$$2y = 1600$$ *Add equations*

$$y = 800$$

Substitute $y = 800$ in the first equation.

$$x + 800 = 900$$
$$x = 100$$

The jet has an average speed of 800 km/hr in still air. The wind has an average speed of 100 km/hr.

√ CHECK:

$$\frac{7}{2}(900) = 3150 \quad \checkmark$$

$$\frac{9}{2}(700) = 3150 \quad \checkmark$$

Quick Reinforcement

How many each of nickels and dimes must you have to total 25 coins with a value of $1.75?

Answer: 15 nickels and 10 dimes

EXERCISES 6-4

Solve each of the following problems using two variables.

1. (a) Joshua noticed that his algebra class had 31 students. He didn't want to bother counting the men and women, but saw that the number of men exceeded the number of women by 5. How many men were in the class?

   (b) Marcy emptied her pocket of change before making a long-distance telephone call, to see how much she had and for how long she would be able to talk on the telephone. She counted twenty coins consisting of nickels and dimes. How many of each did she have if their total value was $1.60?

2. (a) A temporary employment agency charges $8 per hour for a moderately experienced keypunch operator and $5 per hour for an entry-level clerk typist. With several employees on vacation, a local firm called to have a keypunch operator and clerk-typist fill in from the agency. The bill from the agency for their services came to $88. The clerk-typist worked for two more hours than the keypunch operator. What was the total time spent by each of the temporary employees, and how much did each earn?

(b) While shopping for some new clothes before his wedding, Jaime figured that the cost of two new coats and three hats was $310 but that if he spent only $205 he could have four hats with one coat. What would be his cost if he simplified his needs and bought only one hat and one coat?

3. (a) In a small ecosystem, the number of prey y must exceed the number of predators x by 55, while together they must total 115. How many of each should be in the ecosystem?

(b) To install a new solar-heated swimming pool, the construction company hired four skilled and eight unskilled workers and paid them $300. For a similar pool, the company hired seven skilled and six unskilled workers and paid them $375. What were the earnings of each type of worker?

4. (a) In planning to paint a miniature portrait, an artist originally chose a rectangular canvas with a perimeter of 54 centimeters. He decided, however, that it wasn't quite the right shape for what he wanted. He doubled the width and decreased the length by 4 centimeters and ended up with a new perimeter of 58 centimeters. What were the dimensions of his original canvas?

(b) In preparation for a backyard wildlife sanctuary, a rectangular area was laid out with the length 2 meters greater than the width. This did not work out very well, however, and the decision was made to increase the width by 7 meters and the length by 2 meters. This new, larger sanctuary had a perimeter that was 64 meters more than three times the original width. What will be the dimensions of the final wildlife sanctuary?

5. (a) An airplane pilot making a turn-around flight between Chicago and Washington, D.C., found that on the flight out to Washington, with tail winds all the way, he could make the 2000-kilometer trip in 4 hours. However, the return trip, with head winds, took him 5 hours. What was the speed of the wind? How fast would the airplane have been going if there had been no wind at all?

(b) The college Outing Club took a day for a canoe trip and picnic. The canoe trip downstream took $\frac{3}{4}$ of an hour to travel the planned 9 kilometers. After the picnic, the return trip upstream took an hour. How fast was the Outing Club paddling the canoe, assuming there had been no current? What was the speed of the current?

6. (a) Wishing to log some flying time, you have rented an airplane for two hours. You decide to fly due west until you have to turn around to be back at the airport at the end of your allotted time. The cruising speed of the airplane is 160 km/hr in still air. If there is a 40 km/hr wind blowing from the west, how long should you head west before you turn around, and how long will it take you to get back?

(b) A salmon fishing boat put out to sea early in the morning traveling with the tide. It took 20 minutes to cover the 6 kilometers to get to the captain's favorite fishing grounds. The return trip in the afternoon, however, was against the tide and took 36 minutes. At what speed was the captain maintaining the boat, and what was the speed of the tide, in kilometers per hour?

7. (a) A dairy has a quantity of milk containing 3% butterfat and 15 gallons of cream containing 30% butterfat. How many gallons of milk should be mixed with the cream to obtain a mixture containing 9% butterfat? (Don't forget to convert the percents to decimals!)

(b) The Mendoza family deposited $10,000 in a savings bank, some at an annual interest rate of 5% and the rest at an annual rate of 6%. If the interest at the end of one year amounts to $560, how much money was deposited at each interest rate?

8. (a) The radiator of a car holds 17 quarts of water. If it now contains 15% antifreeze, how many quarts must be replaced by antifreeze to give the car a 60% solution in its radiator?

(b) Sterling silver contains 92.5% pure silver. How many grams of pure silver and sterling silver must be mixed to obtain 100 grams of a 94% silver alloy?

9. (a) A total of 3640 votes were cast in a local city council election with two nonpartisan candidates running. Three times the number of votes received by the first candidate was 720 votes more than twice the number received by the second candidate. What was the vote count for each candidate?

(b) A tire manufacturing company produces two grades of tires. The lower grade requires five pounds of rubber and takes three hours of labor, giving a total production cost of $26.50 per tire. The higher grade of tire requires four pounds of rubber with five hours of labor and has a total production cost of $32.00. How much does the company outlay for a pound of rubber, and what is its hourly labor cost?

10. (a) In economics, a product reaches a state of equilibrium when the supply S equals the demand D. If we let x be the quantity of the product and if the supply function for a given product is $S = \frac{1}{2}x$ and the demand function is $D = 9 - x$, then for what quantity will equilibrium exist $(S = D)$?

(b) In business, the break-even point occurs where the equation representing the revenue intersects the equation representing the costs. If a firm sells a product for $10 a unit, then their revenue (R) equation would be $R = 10y$, where y is the number of units sold. If the cost equation has fixed costs of $100 and variable costs of $5 per unit, then the Cost (C) equation would be $C = 5y + 100$. The break-even point would be where cost equals revenue, or $C = R$. Find the break-even point.

11. (a) The cost function of a firm selling a reasonably low-priced calculator is $C = 8.3x + 84$, where x is the number of calculators sold. If the calculators sell for $12.50 each, how many calculators must be sold to achieve the break-even point? (See Problem 10b above.)

(b) A coffee company mixes two kinds of coffee to form a blend that they will sell for $0.88 per pound. They mix coffee normally costing $0.85 per pound with coffee costing $0.90 per pound to get ten pounds of the blend. How many pounds of each kind of coffee do they need to make the ten pounds of blend?

12. (a) When she was ordering gift packages of meat and cheese from a Wisconsin mail-order company, Sarah noticed that there was a fixed fee for shipping and handling up through ten pounds. For shipments over ten pounds there was an additional per pound cost. Sarah was ordering an assortment of gifts to be sent to herself (which she would distribute at holiday time) totaling 42 pounds, which would cost her $3.45 in shipping and handling fees. She was also ordering 30 pounds of merchandise to be sent back East to her family. This was going to cost an additional $2.49. What were the fixed fee and additional per-pound fee that she used to compute her shipping and handling charges?

(b) Two neighboring cities, Whoville and Howville, were informed by their state environmental agency to clean up their air! Whoville, which has only light manufacturing, must have one-half the pollution count of Howville, which has a large amount of heavy industry within its borders. Together, their pollution count can be only 1550 units. Determine the allowable pollution count for Whoville and Howville.

Nostalgia Problems

1. (a) A man bought at one time 3 bushels of wheat and 5 bushels of rye for 38 shillings; and at another time 6 bushels of wheat and 3 bushels of rye for 48 shillings. What was the price of a bushel of each?

(b) A man spent 30 cents for apples and pears, buying his apples at the rate of 4 for a cent and his pears at the rate of 5 for a cent. He afterward let a friend have half of his apples and one-third of his pears for 13 cents, at the same rate. How many did he buy of each sort? (Hint: Let $x =$ number of apples and $y =$ number of pears; then $x/4 =$ cost of one apple and $y/5 =$ cost of one pear.)

2. (a) A merchant has sugar at 9 cents and 13 cents a pound, and he wishes to make a mixture of 100 pounds that shall be worth 12 cents a pound; how many pounds of each quality must he take?

(b) A person has a saddle worth $50, and two horses. When he saddles the poorer horse, the horse and saddle are worth twice as much as the better horse; but when he saddles the better horse, the horse and saddle are together worth three times the other. What is the value of each horse?

Calculator Activities

1. (a) A fast-food chain pays its inexperienced high-school help $2.95 per hour and its more experienced managers $4.15 per hour. Cathy, the kitchen manager, and her younger brother Tomas (who was still in high school), chief floor-washer, both worked part-time at the same fast-food outlet store. One week, with Tomas working two more hours than Cathy, their combined earnings were $92.70. How many hours did they each spend on their respective jobs?

   (b) The Hillview Tennis Club was planning on adding some new courts for its expanding membership. Originally they planned to pour concrete in a rectangular area where the width would be 2.5 meters less than the length, but after rechecking their figures, they realized that this area would not be adequate. However, adding 7.2 meters to the width and 1.8 meters to the length would give them the appropriate area. They found that this new rectangular space would have a perimeter 65.8 meters more than 3 times the original width. What would be the dimensions of the new tennis court area?

2. (a) Gus and Amy Petropolis have managed to save $12,500 toward a down payment on a house they will buy in the not-too-distant future. They put part of the money in a passbook account at $5\frac{1}{4}$% so they can place a deposit easily when they find their house, and the rest at $7\frac{1}{2}$% in a certificate of deposit. If their annual interest is $722.50, how much did they put in the certificate of deposit?

   (b) The Santiana Company plans to manufacture a new low-priced transistor radio, which they would like to market at $10.25. They have determined that their fixed costs will be $148.20 and that their variable costs will be $4.60 per radio. They are trying to determine what their break-even point will be. Can you compute this for them? (Recall that break-even is when revenue, which is number of units times price per unit, equals cost, which is fixed costs plus variable costs times number of units. See Exercises 6-4, Problem 10b.)

KEY TERMS

System of equations
Point of intersection
Substitution method
Addition method

REVIEW EXERCISES

Graph each system and describe the nature of its solution. If the solution is one point, attempt to identify it, and then check the solution in the original system of equations.

1. (a) $\begin{cases} 4x - y = 5 \\ 2x + 2y = 20 \end{cases}$ (b) $\begin{cases} x - y = 4 \\ 4x - 2y = 36 \end{cases}$

2. (a) $\begin{cases} 4y = x + 4 \\ y + x = \frac{3}{4} \end{cases}$ (b) $\begin{cases} 0.1y = 0.2x - 0.8 \\ y + 1.5x + 1 = 0 \end{cases}$

Use two variables, x and y, to set up a system of equations for each of the following problems. Solve graphically.

3. (a) Two sisters were playing with some other children when a friend of the older one asked how old her little sister was. The older sister, who was addicted to mathematics puzzles, answered, "If you add up our ages, you will get a twelve, but if you subtract them, you will have a four." Since her friend also enjoyed such puzzles, she quickly found their ages. Can you?

   (b) While on a shopping trip, Lazlo bought some new elegant clothes to wear in his nightclub act. The store he preferred to patronize was having a special sale where all coats were one price and all slacks were another price. He bought four coats and five pairs of slacks and spent $266. Very pleased with his purchases, he went back the next day and bought an additional two coats and eight pairs of slacks, costing him $276. What did he pay for each coat and each pair of slacks?

Solve each system by the substitution method. Check your solution.

4. (a) $\begin{cases} y - x = 4 \\ y + 4x = 9 \end{cases}$ (b) $\begin{cases} x + 3y = 10 \\ x - y = 6 \end{cases}$

5. (a) $\begin{cases} 4x + 3y + 13 = 0 \\ \frac{x}{3} = \frac{y}{3} + 3 \end{cases}$ (b) $\begin{cases} 2.4x = 2.4y + 3.0 \\ x + 4y + 2.5 = 0 \end{cases}$

Solve each system by the addition method. Check your solution.

6. (a) $\begin{cases} 4x + 2y = -4 \\ 5x - 3y = -5 \end{cases}$ (b) $\begin{cases} 5x - 2y = 9 \\ 3x - 4y = -3 \end{cases}$

7. (a) $\begin{cases} 2x + 3y = 38 \\ 6x + 5y = 82 \end{cases}$ (b) $\begin{cases} 4x + 6y = 46 \\ 5x - 2y = 10 \end{cases}$

8. (a) $\begin{cases} 1.2x + 0.3y = 30.6 \\ 0.25x + y = 12.00 \end{cases}$ (b) $\begin{cases} 2.5y + 1.5x = 46.5 \\ 0.6y + 0.8x = 16.0 \end{cases}$

Solve by either substitution or addition. Check your solution.

9. (a) $\begin{cases} 3x + 5y = 3 \\ 2x - y + 11 = 0 \end{cases}$ (b) $\begin{cases} 3x + y = 9 \\ 2x - 3y = 6 \end{cases}$

10. (a) $\begin{cases} \dfrac{4x+3y}{9}+y=4 \\ x+\dfrac{8x-2y}{5}=7 \end{cases}$ (b) $\begin{cases} \dfrac{x}{4}+\dfrac{5y}{6}=14 \\ \dfrac{x}{8}+\dfrac{y}{6}=4 \end{cases}$

Solve each of the following problems using two variables.

11. (a) While building some graduated bookshelves, Lee Yin decided to take an eight-meter board and cut it into two pieces, with the first piece one meter less than twice the length of the second. What will be the lengths of Lee Yin's two bookshelves?

    (b) On a certain day with moderate winds, a pilot figured out that she could fly her airplane at 540 km/hr when she was flying with the wind, but that her speed would be only 400 km/hr when she was flying against the wind. What would her airplane speed be if she were flying in still air, and what was the wind speed?

12. (a) Judy bought a yard of double knit material and three yards of orlon material for a total of $25. Very pleased with the results of her sewing, she bought an additional four yards of the doubleknit and five yards of orlon for a total of $64. What did she pay per yard for each material?

    (b) Two joggers begin running on a quarter-mile track at the same place. One runs at 8 min/mile while the other clips along at 6 min/mile. How far will the slower runner have run when the faster runner catches up?

13. (a) Two species of bacteria are to be kept in one test tube, where they will feed on two resources. Each day the test tube is supplied with 14,000 units of the first resource and 8000 units of the second. Each species consumes 2 units a day of the first resource. One species consumes one unit a day of the second resource, while the other consumes three units a day of the second resource. What is the population of each species if they are to coexist in equilibrium and consume all of the supplied resources?

    (b) Twenty pounds of a given ore P combined with thirty pounds of another ore R produce 26.5 pounds of silver, while five pounds of ore P combined with ten pounds of ore R produce 7.75 pounds of silver. Find the percentage of silver in each ore.

Nostalgia Problems

1. (a) A gentleman being asked the ages of his two sons, answered, "If to the sum of their ages 18 be added, the result will be double the age of the elder; but if 6 be taken from the difference of their ages, the remainder will equal the age of the younger." What were their ages?

    (b) A market-woman bought eggs to the amount of 65 cents, some at the rate of 2 for a cent and some at the rate of 3 for 2 cents. She afterward sold them all for 120 cents, thereby gaining a half a cent on each egg. How many of each kind did she buy?

7 Inequalities in Two Variables

The Alcar Rent-A-Car Corporation is planning its annual advertising campaign. The marketing committee is meeting to discuss the allocation of at most $300,000 for national media advertising. Each one-page ad in a national magazine costs approximately $15,000, while a minute of television time is $45,000. Alcar's advertising department recommends at least 3 magazine ads and at least 5 television commercials. What are the possible combinations of ads that the marketing committee can recommend that will not exceed the budget, but will satisfy the advertising department?

7-1 LINEAR INEQUALITIES IN TWO VARIABLES

If in any linear equation in two variables you replace the equals sign by an inequality sign, the resulting expression is called a **linear inequality in two variables.** Thus

$$y = 3x - 2$$

is an example of a linear equation in the variables x and y, and

$$y < 3x - 2 \quad \text{Other inequality signs are } >, \leq, \text{ and } \geq$$

is an example of a linear inequality in two variables.

Several types of real-world problems give rise to linear inequalities in two variables. For example, if the number y of predators is to be no more than twice the number x of prey, then the prey-predator relationship is described by the linear inequality in two variables

$y \leq 2x$

A **solution** of a linear inequality in two variables is any pair of values of the variables that make the inequality true. For example, $(5, 2)$ is a solution of

$y < 3x - 2$

because it gives us

$2 < 3 \cdot 5 - 2 \qquad 2 < 13$ *is true*

The ordered pair $(1, 7)$ is not a solution because it gives us

$7 < 3 \cdot 1 - 2 \qquad 7 < 1$ *is false*

Similarly, $(15, 20)$ is a solution of the predator-prey inequality $y \leq 2x$, while $(40, 90)$ is not. (You can have 15 prey and 20 predators but not 40 prey and 90 predators.)

The set of all solutions of a linear inequality in two variables is called the **solution set** of the inequality. Two linear inequalities having the same solution set are said to be *equivalent*.

You can use the addition and multiplication properties for inequalities (see Chapter 4) to solve linear inequalities in two variables as well as those in one variable.

Problem 1 Solve the inequality $2x + y < 3x - 2$.

SOLUTION:

Solve the inequality for y; that is, find an equivalent inequality having y alone on one side.

equivalent		
	$2x + y < 3x - 2$	*Given inequality*
	$-2x + 2x + y < -2x + 3x - 2$	*Add $-2x$ to each side*
	$0 + y < (-2 + 3)x - 2$	
	$y < x - 2$	$(-2 + 3)x = 1x$, or simply x

Thus, (x, y) is a solution provided y is less than $x - 2$. In set notation,

$\{(x, y) \mid y < x - 2\}$ *The set of all ordered pairs (x, y) of numbers such that y is less than $x - 2$*

is the solution set of the inequality $2x + y < 3x - 2$. A few solutions are listed below.

Let $x = 10$. Then any y less than $10 - 2$, or 8, will work. Thus, $(10, 0)$ and $(10, 7)$ are solutions.

Let $x = 3$. Then any y less than $3 - 2$, or 1, will work. Thus, $(3, \frac{1}{2})$ and $(3, -5)$ are solutions.

Let $x = -4$. Then any y less than $-4 - 2$, or -6, will work. Thus, $(-4, -7)$ and $(-4, -20)$ are solutions.

Problem 2 Solve the inequality $x + 2y \geq 3 - x$.

SOLUTION:

Solve for y:

$x + 2y \geq 3 - x$	*Given inequality*
$-x + x + 2y \geq 3 - x - x$	*Add $-x$ to each side*
$2y \geq 3 - 2x$	
$y \geq \dfrac{3}{2} - x$	*Multiply each side by $\frac{1}{2}$*

The solution set is

$\{(x, y) \mid y \geq \dfrac{3}{2} - x\}$ *The set of all ordered pairs (x, y) such that y is greater than or equal to $\frac{3}{2} - x$*

Let $x = 0$. Then $y \geq \frac{3}{2}$. Thus, $(0, \frac{3}{2})$, $(0, 2)$, and $(0, 10)$ are solutions.
Let $x = \frac{3}{2}$. Then $y \geq 0$. Thus, $(\frac{3}{2}, 0)$, $(\frac{3}{2}, 3)$, and $(\frac{3}{2}, 100)$ are solutions.
Let $x = 10$. Then $y \geq -\frac{17}{2}$. Thus, $(10, -\frac{17}{2})$, $(10, -5)$, and $(10, 5)$ are solutions.

Problem 3 Solve the inequality $x \leq 2$.

SOLUTION:

You can consider $x \leq 2$ a linear inequality in two variables because $x \leq 2$ is equivalent to $x + 0y \leq 2$. Thus, (x, y) is a solution for all values of x less than or equal to 2 and all values of y. Thus, its solution set is

$\{(x, y) \mid x \leq 2, y \text{ any value}\}$

Some solutions are $(1, 17)$, $(2, -5)$, $(0, 6)$, and $(-77, 88)$.

Problem 4 Solve the inequality $\frac{1}{2}x - \frac{3}{5}y \leq 2$.

SOLUTION:

$\frac{1}{2}x - \frac{3}{5}y \leq 2$ Given inequality

$10\left(\frac{1}{2}x - \frac{3}{5}y\right) \leq 10 \cdot 2$ Multiply each side by 10 to clear fractions

$5x - 6y \leq 20$

$-6y \leq 20 - 5x$ Add $-5x$ to each side

$y \geq -\frac{20}{6} + \frac{5}{6}x$ Multiply each side by $-\frac{1}{6}$ and reverse the inequality sign

$y \geq -\frac{10}{3} + \frac{5}{6}x$ Reduce all terms

The solution set is

$$\left\{(x, y) \mid y \geq -\frac{10}{3} + \frac{5}{6}x\right\}$$

If $x = 0$, then $y \geq -\frac{10}{3}$. Thus, $(0, -3)$ and $(0, 0)$ are solutions. If $x = 4$, then $y \geq 0$. Thus, $(4, 6)$ and $(4, 10)$ are solutions.

Quick Reinforcement

Solve each inequality for y.

(a) $2x - 3y \leq 1$

(b) $y - x > 2x + 1$

Answers: (a) $y \geq \frac{2}{3}x - \frac{1}{3}$ (b) $y > 3x + 1$

EXERCISES 7-1

Solve each inequality by first solving for y, then writing the complete solution set, and finally listing three solutions.

1. (a) $3x + y > 5$ (b) $4x - 2y < 6$
2. (a) $5x - 7y \geq 5 - 2y$ (b) $3x + 6 > 2y - 3$
3. (a) $2(x + 3y) < 8$ (b) $3(y - 2) \leq x + 3$
4. (a) $\frac{2}{3}x - 5 + \frac{y}{3} > 0$ (b) $\frac{1}{2}y - \frac{3}{4}x < \frac{3}{8}$
5. (a) $1.4y + 0.7x < 2.1x + 1.4y$ (b) $7x \geq 0$
6. (a) $4y + 6 < 0$ (b) $0.4(2x - 1) < 0.5(y + 7)$

7. (a) $3y - 5x > 8 + 2x$ (b) $2(y + x) \leq 3y - 6(x - 2)$

8. (a) $\dfrac{11x}{9} - \dfrac{2y + 1}{3} > 0$ (b) $\dfrac{7y}{8} + \dfrac{2x - 3}{2} < 1$

9. (a) $10x + 10y - 5 < 0$ (b) $5(2x + 3y) + 1 > 6$

10. (a) $2(2x + y - 5) \leq 3(4x - 3y)$ (b) $4(6y - 3x + 2) \geq 6(x - y + 1)$

Write the inequality that abbreviates each of the following sentences. Label your variables, where necessary. Solve each inequality for one of the variables, then write the complete solution set and list three solutions.

11. (a) The revenue R should exceed costs C by at least $4090.
    (b) The costs C must be less than the revenue R by at least $12,400.
12. (a) The total population of the two sister cities is no more than 500,000.
    (b) The maximum capacity of the two county jails is 84 people.
13. (a) With inflation, your return I on investment P should be at least 12%.
    (b) To keep up with the cost of living, the combined salaries of a husband and wife needed to be at least $2200 per month.
14. (a) The amounts of two types of bacteria in the culture should differ by no more than 50.
    (b) The chemistry laboratory assistant knew that they needed at least 135 petri dishes in the two standard sizes.

 Calculator Activities

Solve each inequality for y, then write the complete solution set and list three solutions.

1. (a) $298y - 415 > 39x + 1896$ (b) $0.0213 + 2.9x < y - 4.998$
2. (a) $55(48 - 12y) < 489 + x$
   (b) $1.02(5.8 - 2.11y) > -0.0918 + 0.1x$
3. (a) $2.034(y + x) \geq 3.239y - x$ (b) $672(y - 233) \leq x + 321$
4. (a) $0.05x - 0.07y \leq 0.02 + 0.13y$ (b) $297x + 534 \geq 688y - 201$

7-2 GRAPHS OF INEQUALITIES IN TWO VARIABLES

Because the solution set of an inequality in two variables is a set of ordered pairs, it has a graph in a coordinate plane. This graph is called the **graph of the given inequality.**

The line $y = x$ cuts diagonally across the coordinate plane, as shown in Figure 7-1. It is made up of points (x, x) for which x is any real number. The inequality $y > x$ has as its graph all points (x, y) for which $y > x$. As you see in the figure, these points are *above* the line $y = x$. Similarly, the graph of the inequality $y < x$ consists of all points *below* the line $y = x$.

CHAPTER 7: INEQUALITIES IN TWO VARIABLES

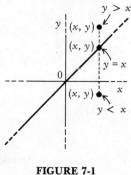

FIGURE 7-1

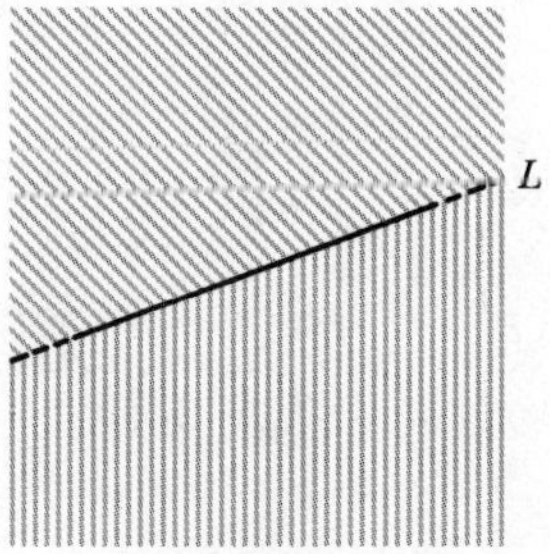

FIGURE 7-2

Every line L in a plane divides the plane into two sections, the shaded regions of Figure 7-2. Each section is called a **half-plane,** and line L is called the **edge** of each half-plane.

The line $y = x$ in Figure 7-3 divides the plane into two half-planes. The half-plane above the line is the graph of the inequality $y > x$; the half-plane below the line is the graph of the inequality $y < x$. The line $y = x$ is dashed to show it does not belong in either half-plane.

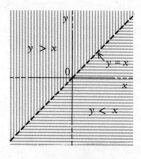

FIGURE 7-3

Problem 1 Graph the inequality $2x + y < 3x - 2$.

SOLUTION:

From Section 7-1, the solution set of this inequality is

$$\{(x, y) \mid y < x - 2\}$$

The graph of this set consists of all points (x, y) *below* the line L with equation

$y = x - 2$ If $y < x - 2$, then the point (x, y) is below the point $(x, x - 2)$ on L

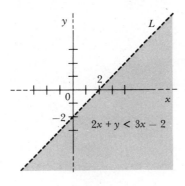

FIGURE 7-4

It is the shaded half-plane of Figure 7-4, not including its edge L. Note that the edge has slope 1 and y-intercept -2.

Each linear inequality has an associated linear equation. Thus, in the problem above, inequality

$$2x + y < 3x - 2$$

has the associated equation

$$2x + y = 3x - 2$$

The graph of the equation is a line L. The graph of the inequality is one of the two half-planes having edge L. You can decide which half-plane is the graph of the inequality by testing any point not on L to see if it is a solution of the inequality. An easy point to test is the origin $(0, 0)$, which is not on L. Is it a solution of the inequality?

$2x + y < 3x - 2$	*Given inequality*
$2 \cdot 0 + 0 < 3 \cdot 0 - 2$	*Let $x = 0, y = 0$*
$0 + 0 < 0 - 2$	
$0 < -2$	*False*

Thus, $(0, 0)$ is not a solution of the inequality. You conclude that the solution of the inequality is the half-plane not containing the origin $(0, 0)$ as shown in Figure 7-4.

Problem 2 Graph the inequality $3x - 2y - 1 \leq 5 - x$.

SOLUTION:

The associated line L has the equation

226 CHAPTER 7: INEQUALITIES IN TWO VARIABLES

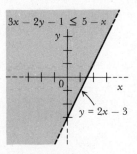

FIGURE 7-5

$$3x - 2y - 1 = 5 - x$$
$$-2y = 6 - 4x$$
$$y = 2x - 3 \quad \text{Slope 2, y-intercept } -3$$

Line L is graphed in Figure 7-5. Is $(0, 0)$ a solution of the inequality?

$$3 \cdot 0 - 2 \cdot 0 - 1 \leq 5 - 0 \quad \text{Let } x = 0, y = 0$$
$$-1 \leq 5 \quad \text{True}$$

Therefore, the solution of the given inequality is the half-plane of the figure that contains the origin. Of course, the half-plane also contains its edge L (as indicated by a solid line).

Problem 3 Graph the inequality $2x + 5 \leq 0$.

SOLUTION:

Assume that $2x + 5 \leq 0$ is an inequality in two variables, say of the form

$$2x + 0y + 5 \leq 0$$

You cannot solve this inequality for y, so you solve it for x instead.

$$2x + 5 \leq 0$$
$$2x \leq -5$$
$$x \leq -\frac{5}{2}$$

Its solution set is

$$\{(x, y) \mid x \leq -\frac{5}{2}, y \text{ any real number}\}$$

To graph this set, first graph the equation

$$x = -\frac{5}{2}$$

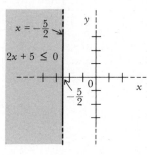
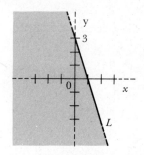

FIGURE 7-6   FIGURE 7-7

It is the vertical line L of Figure 7-6. The graph of the inequality

$x \leq -\dfrac{5}{2}$ *The values of x are equal to or to the left of $-\tfrac{5}{2}$*

is the shaded half-plane on and to the left of line L.

Problem 4 Find an inequality whose graph is the half-plane of Figure 7-7.

SOLUTION:

The edge L has y-intercept 3 and slope -3. Hence, its equation is

$y = -3x + 3$

The half-plane contains all points *on* (L is a solid line) and *below* line L. The inequality

$y \leq -3x + 3$ $\leq$ *because it is* on (=) *and* below (<) *line L*

has the given half-plane as its graph.

Quick Reinforcement

Graph the inequality $2y - 4x \geq 2$.

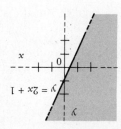

Answer

EXERCISES 7-2

Graph each inequality.

1. (a) $x < 2$ (b) $y > 5$
2. (a) $y > -3$ (b) $x < -1$
3. (a) $x \leq \frac{1}{2}$ (b) $y \geq -\frac{1}{4}$
4. (a) $5y > 3$ (b) $-4x < 16$
5. (a) $y \geq 3x + 1$ (b) $y < \frac{1}{2}x - 2$
6. (a) $2y < 8x + 10$ (b) $3y > 2x - 9$
7. (a) $11x > -33$ (b) $3y \geq 2$
8. (a) $3x - 2y < x + 5$ (b) $2x + 3y < -x + 4y - 3$
9. (a) $y - 5 < x + 7$ (b) $3(x + 4) < 2y + 15$
10. (a) $1.8x > 0.6y + 1.2$ (b) $0.5y > 0.5x - 2.5$
11. (a) $\frac{x-3}{3} \leq \frac{y+x}{2}$ (b) $\frac{x}{3} + \frac{2}{9} \geq 0$
12. (a) $7x + 2y > 7(x - y + 2)$ (b) $11x + 2(y - 3) < 2x - 6$
13. (a) $\frac{4}{7}x - y > \frac{3}{7}x + 3$ (b) $\frac{9x}{7} < y$
14. (a) $7.8x - 20.8 < 13y - 20.8$ (b) $2.5x + 5y > 7.5 - x$
15. (a) $3(2x - 4y) + 12 > 0$ (b) $4(x - 5y) < 40$

Write the inequality that abbreviates each of the following sentences, and graph it.

16. (a) The amount of profit on an item y is greater than four times its cost x.
    (b) The sum of the ages of the two parents is less than 62.

17. (a) The amount of interest earned on the first savings account at an interest rate of 8% plus the amount of interest earned on the second savings account at an interest rate of 7% is greater than $700.
    (b) The maximum perimeter of a back yard is 120 meters.

18. (a) The manufacturing cost x of an item plus the amount of profit desired y must not exceed the demand price of $14.
    (b) To earn an A on a final exam, the sum of the scores on the multiple-choice section x and the essay section y must be at least an 87.

19. (a) To stay comfortably aloft, a pilot knew that the speed y of his airplane minus the speed x of the head wind must be at least 160 km/hr.
    (b) While shopping for meat and vegetables, Melissa Khalid knew that the amount x spent on meat plus the amount y spent on vegetables had better not exceed the $22 she had in her wallet!

7-2: GRAPHS OF INEQUALITIES IN TWO VARIABLES

Find the inequalities whose graphs are the shaded half-planes below.

20. (a) 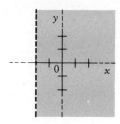 (b)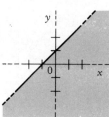

21. (a) 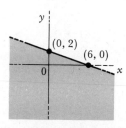 (b)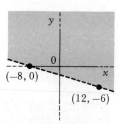

22. (a) 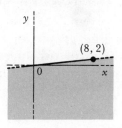 (b)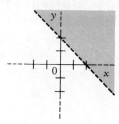

23. (a) 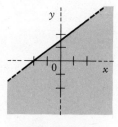 (b)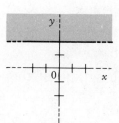

24. (a) 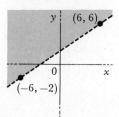 (b)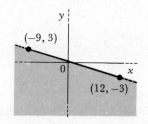

230 CHAPTER 7: INEQUALITIES IN TWO VARIABLES

25. (a) 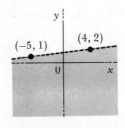 (b)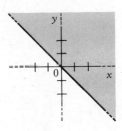

Calculator Activities

Graph each inequality.

1. (a) $4891 - y \leq 2x + 4789$ (b) $3.21 + 0.2y < x - 1.89$
2. (a) $-0.015 + 4.6x \geq 2.3(y - 1.2)$ (b) $15(21 - 2y) > 30x + 155$

7-3 SYSTEMS OF LINEAR INEQUALITIES

A system of inequalities in two variables, like a system of equations, has as its solution the set of all ordered pairs of numbers that are solutions of every inequality of the system. The graph of a system of inequalities is the intersection of the graphs of the individual inequalities of the system. For a system of two linear inequalities in two variables, the graph of the system will be the intersection of two half-planes.

Problem 1 Graph the following system.

$$\begin{cases} y > x \\ y < -x \end{cases}$$

SOLUTION:

The graph of the inequality $y > x$ is the shaded half-plane *above* the line

$y = x$ Line K of Figure 7-8

The graph of $y < -x$ is the half-plane *below* the line

$y = -x$ Line L of the figure

The graph of the system is the intersection of these two half-planes. It is the doubly shaded region (omitting edges) of the figure. In set notation, the solution set is

$\{(x, y) \mid y > x\} \cap \{(x, y) \mid y < -x\}$

7-3: SYSTEMS OF LINEAR INEQUALITIES

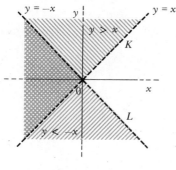

FIGURE 7-8   FIGURE 7-9

Problem 2 Graph the system

$$\begin{cases} 2x + y \geq 1 \\ 2y - x \leq 2 \end{cases}$$

SOLUTION:

Solve each inequality for y:

$2x + y \geq 1$	1st inequality
$y \geq -2x + 1$	Add $-2x$ to each side
$2y - x \leq 2$	2nd inequality
$2y \leq x + 2$	Add x to each side
$y \leq \frac{1}{2}x + 1$	Multiply each side by $\frac{1}{2}$

Therefore, the given system is equivalent to the new system:

$$\begin{cases} y \geq -2x + 1 \\ y \leq \frac{1}{2}x + 1 \end{cases}$$

The lines

$y = -2x + 1$ Line K of Figure 7-9, slope -2 and y-intercept 1

and

$y = \frac{1}{2}x + 1$ Line L of Figure 7-9, slope $\frac{1}{2}$ and y-intercept 1

are edges of the graphs of the two inequalities of the system. Thus, the graph of the inequality $y \geq -2x + 1$ is the shaded half-plane on and above line K, and the graph of the inequality $y \leq \frac{1}{2}x + 1$ is the shaded half-plane on and below line L. The intersection of these two half-planes is the graph of the given system. It is the doubly shaded region (containing edges) of the figure. In set notation, the solution set is

$$\{(x, y) \mid y \geq -2x + 1\} \cap \{(x, y) \mid y \leq \tfrac{1}{2}x + 1\}$$

Problem 3 Graph the system

$$\begin{cases} x \geq 1 \\ x \leq 3 \end{cases}$$

SOLUTION:

As Figure 7-10 shows, the graph of $x \geq 1$ is the shaded half-plane to the right of the vertical line $x = 1$; the graph of $x \leq 3$ is the shaded half-plane to the left of the vertical line $x = 3$. Hence, the graph of the system is the doubly shaded region between the two parallel lines $x = 1$ and $x = 3$, including the lines.

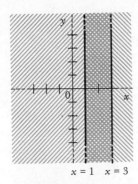

FIGURE 7-10

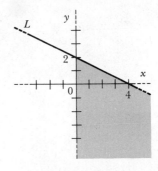

FIGURE 7-11

Problem 4 Find a system of inequalities whose graph is the shaded region of Figure 7-11.

SOLUTION:

Line L has slope $-\tfrac{1}{2}$ and y-intercept 2. Hence, its equation is

$$y = -\tfrac{1}{2}x + 2 \qquad y = mx + b,\ \text{slope } m,\ y\text{-intercept } b$$

The half-plane below L (including L) is the graph of the inequality

$$y \leq -\tfrac{1}{2}x + 2$$

The shaded region of the figure is *below L* and *to the right* of the *y*-axis. The equation of the *y*-axis is

$$x = 0$$

Hence, the half-plane to the right of the *y*-axis (including the *y*-axis) is the graph of the inequality

$$x \geq 0$$

Therefore, the shaded region of the figure is the graph of the following system of inequalities:

$$\begin{cases} y \leq -\frac{1}{2}x + 2 \\ x \geq 0 \end{cases}$$

Another type of problem involving two inequalities requires that you find all solutions of *either* one *or* the other inequality. If this solution set is S, then S is the *union* of the solution sets of the two inequalities.

Problem 5 Let S be the set of all solutions of either

$$2x - y \leq 5 \quad \text{or} \quad y + x \leq 1$$

Graph the set S.

SOLUTION:

The graph of $2x - y \leq 5$ is the half-plane on and above the line

$$y = 2x - 5 \quad \text{Line K of Figure 7-12, slope 2 and y-intercept } -5$$

The graph of $y + x \leq 1$ is the half-plane on and below the line

$$y = -x + 1 \quad \text{Line L of Figure 7-12, slope } -1 \text{ and y-intercept } 1$$

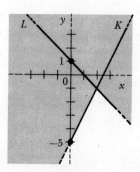

FIGURE 7-12

The graph of set S is the union of the two half-planes above. It is the total shaded region indicated in the figure. In set notation, the solution set is

$$\{(x, y) \mid 2x - y \leq 5\} \cup \{(x, y) \mid y + x \leq 1\}$$

There are many applications of systems of linear inequalities to economics and other disciplines. In fact, there is an entire area of mathematics called linear programming that is devoted to a study of systems of linear inequalities. Often, more than two inequalities are involved. The following problem is a simple example of such a system.

Problem 6 Mr. McGilly, a local farmer, raises only pigs and geese. He wants to raise no more than 16 animals. It costs him $3 to raise a pig and $4 to raise a goose. His total budget for raising the animals is $60. What combinations of pigs and geese can he raise?

SOLUTION:

Let x = number of pigs he raises

Let y = number of geese he raises

It costs $3x$ dollars to raise x pigs and $4y$ dollars to raise y geese. By what is given,

$$\begin{cases} x \geq 0 & \textit{He cannot raise a negative number of pigs or geese} \\ y \geq 0 & \\ x + y \leq 16 & \textit{He raises at most 16 animals} \\ 3x + 4y \leq 60 & \textit{He spends at most \$60} \end{cases}$$

The graph of the above system of linear inequalities is the shaded region of Figure 7-13. Any ordered pair

(a, b)   *a, b integers*

in this region is a solution of the problem. For example, some solutions are

(6, 4)   *6 pigs, 4 geese*
(0, 15)   *0 pigs, 15 geese*
(10, 5)   *10 pigs, 5 geese*
(4, 12)   *4 pigs, 12 geese*
(16, 0)   *16 pigs, 0 geese*

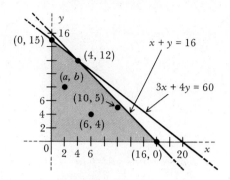

FIGURE 7-13

Quick Reinforcement

Graph the system of inequalities

$$\begin{cases} y + x < 4 \\ 2x - y < -1 \end{cases}$$

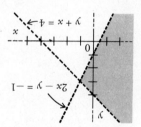

Answer

EXERCISES 7-3

Graph the following systems of inequalities.

💡 1. (a) $\begin{cases} y - x > 4 \\ y + 4x \leq 9 \end{cases}$ (b) $\begin{cases} 4x + 12y \leq 5 \\ x + 2y > 1 \end{cases}$

2. (a) $\begin{cases} y - x \leq 4 \\ y + 6x > 7 \end{cases}$ (b) $\begin{cases} 3x - y < 10 \\ 2x + y > 0 \end{cases}$

3. (a) $\begin{cases} 6x \leq 5y \\ 4x > 7y \end{cases}$ (b) $\begin{cases} x + 5y < 6 \\ x - 6y \geq 12 \end{cases}$

4. (a) $\begin{cases} 2x + 3y > 7 \\ 8x + 10y > 26 \end{cases}$ (b) $\begin{cases} 5x - 3y < 15 \\ 2y - x > 14 \end{cases}$

5. (a) $\begin{cases} \frac{2}{7}x - \frac{1}{2}y > \frac{3}{2} \\ \frac{2}{5}x + y \leq 5 \end{cases}$ (b) $\begin{cases} \frac{2x+y}{3} < 1 \\ -x \leq 4 \end{cases}$

6. (a) $\begin{cases} 1.6(4x - y) < 8 \\ \frac{-(x+2y)}{2} \geq 2 \end{cases}$ (b) $\begin{cases} 0.1y \leq 1 \\ 2x - 1.5y > 4.5 \end{cases}$

7. (a) $\begin{cases} -\frac{1}{2}x > 1 \\ \frac{2}{5}x + y + 1 \leq 0 \end{cases}$ (b) $\begin{cases} 3x - y + 6 < 0 \\ 4y - 3x + 8 > 0 \end{cases}$

Find the union of the graphs of each pair of inequalities.

8. (a) $\begin{cases} 2x - 4y + 5 > -3 \\ -y \leq 5x - 7 \end{cases}$ (b) $\begin{cases} 6x - 5y - 4 < 6 \\ x - 2 \geq 0 \end{cases}$

9. (a) $\begin{cases} 4x < 2y + 5 \\ 10 + 2y > 0 \end{cases}$ (b) $\begin{cases} 2(y - x) + 4 > 0 \\ 3x - 7y \leq 28 \end{cases}$

10. (a) $\begin{cases} 2.4 < 0.8x + 0.4y \\ -2.5 > 0.5x - 2.5y \end{cases}$ (b) $\begin{cases} 2.7x + 1.2y \leq 4.8 \\ -3.5x + 4.2y > 8.4 \end{cases}$

Find the systems of inequalities whose graphs are the shaded regions below.

11. (a) 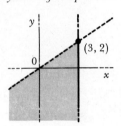 (b)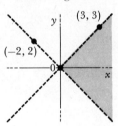

12. (a) 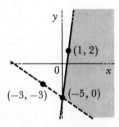 (b)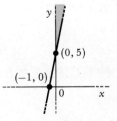

13. (a) 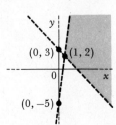 (b)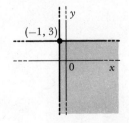

14. (a)   (b)

15. (a)   (b)

16. (a)   (b)

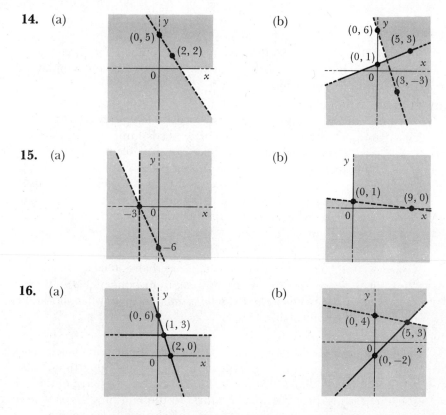

Write the system of inequalities that represents each of the following situations. Graph the system of inequalities.

17. (a) While shopping for holiday gifts, Ann Romana decided she would buy no more than fifteen presents. If she has $20 to spend and the small gifts she is looking at cost $1 and $2 apiece, what combinations of presents can she buy? (Note: Although we only start by setting up two inequalities with two unknowns, the very nature of this practical application leads us to conclude that she is not going to buy less than zero of either gift. We will see that most practical applications limit us to the first quadrant, thus adding two more inequalities to our system, $x \geq 0$, and $y \geq 0$. See Problem 6 in this section.)

(b) While shopping for school supplies, a student found that the cost of two pencils and one pen was at least $4.00, but the cost of eight pencils and ten pens did not exceed $8.00. What possible combinations of pens and pencils could the student purchase? (Note: He is not going to purchase a piece of a pen or pencil! So although the solution area has an infinite number of points, here you would only be interested in the points that will give positive integers for both variables.)

18. (a) A pizza parlor can make a maximum of 40 pizzas in an hour if they really work hard. If their maximum costs per hour can be no more than $150 and it costs $5 to make a special combination pizza and $2 to make a cheese-only pizza, what would be desirable combinations for them to make during an hour?

(b) A candy manufacturer makes two different kinds of candy: Chocolate Dreams and Chewy Chocolates. A Chocolate Dream needs 2 units of chocolate per candy, a Chewy Chocolate needs 1 unit of chocolate per candy, and there are 100 units of chocolate available per quarter hour of machine time. Also, a Chocolate Dream needs 1 unit of caramel per candy, while a Chewy Chocolate needs 3 units of caramel per candy, and there are at most 150 units of caramel available per quarter hour of machine time. What possible combinations of candies can be made every quarter hour? (Hint: Let $x =$ number of Chocolate Dreams and $y =$ number of Chewy Chocolates.)

19. (a) Grace McCambell wishes to invest no more than $30,000, some in the bank (at $5\frac{1}{4}\%$ interest per year) and some in bonds (at 9% interest per year). If she needs at least $2000 annual income, what allocations of her funds are available to her?

(b) A manufacturer of fishing poles produces two models — a fresh-water fishing pole and a surf-casting fishing pole. Both poles need work done in the Fabricating Department and Finishing Department. The fresh-water pole needs 4 hours in the Fabricating Department, while the surf-casting pole requires 6 hours. In the Finishing Department, they both require one hour. If the Fabricating Department has a maximum of 96 work-hours available per week and the Finishing Department has at most 20 work-hours available per week, what possible combinations of poles can they manufacture in a given week?

20. (a) Because of a large increase in demand for their fishing poles, the company in problem 19b has hired more workers to increase their output. If the Fabricating Department now has a maximum of 180 work-hours per week and the Finishing Department now has at most 35 work-hours available per week and the other information remains the same, what combinations of poles can they now manufacture in one week?

(b) A fish and game agency wishes to stock a lake with trout and bass. The determining factors on the number of fish to stock are the availability of food supplies for the newly stocked fish. It has been determined that the average trout will consume 0.4 kilogram of minnows and 0.2 kilogram of crabs and other crustaceans, while the bass will eat 0.2 kilogram of minnows and 0.2 kilogram of crabs and crustaceans. The lake can maintain at most 100 kilograms of minnows and 90 kilograms of crabs and other crustaceans on a daily basis. How many of each type of fish might the agency stock in the lake?

21. (a) The number of predators and preys within an ecosystem must be such that either their total number is less than 500 or the number of prey is at least 200 more than the number of predators.

(b) A company's income y and costs x must satisfy one of the following: either their difference is at least 10,000 or the income is greater than twice the costs.

Calculator Activities

Graph the following systems of inequalities.

1. (a) $\begin{cases} 4891 - y \leq 2x + 4789 \\ 3.21 + 0.2y < x - 1.089 \end{cases}$ (b) $\begin{cases} 15(21 - 2y) \geq 30x + 155 \\ -0.015 + 4.6x > 2.3(y - 1.2) \end{cases}$

2. (a) $\begin{cases} 160x + 128y - 640 \leq 0 \\ -56x + 84y - 168 \leq 0 \end{cases}$ (b) $\begin{cases} 0.036x + 0.012y \geq 0.048 \\ 0.270x + 0.405y \leq 1.08 \end{cases}$

Find a system of linear inequalities that solves each of the following problems. Graph the systems.

3. (a) Your company's costs are $2.75x + 442.50$, where x denotes the number of items produced. Your profits y must be at least $1296.75 more than your costs, but your costs will be at least $859.29. Graph all possible profits and numbers of items produced.

(b) The fish and game agency of problem 20b wishes to stock a different lake with bullheads and rainbow trout. Each rainbow trout consumes 0.318 kilogram of minnows and 0.156 kilogram of crabs each day, while each bullhead consumes 0.214 kilogram of minnows and 0.144 kilogram of crabs each day. If this lake can maintain no more than 110 kilograms of minnows and 86 kilograms of crabs, how many of each type of fish can be stocked in the lake? Assuming the local fishermen like to have a relatively equal chance on catching one type of fish or the other, give several possible combinations (not parts of fish, remember!) that would be desirable.

KEY TERMS

Linear inequality in two variables
Half-planes
Edge

REVIEW EXERCISES

Solve each inequality by first solving for y, then writing the complete solution set, and finally listing three solutions.

1. (a) $5x - 7y + 3 > 0$ (b) $y + 8x - 7 < 0$

2. (a) $\frac{7}{4}x + \frac{x-y}{2} \leq 2$ (b) $1.8(2x - 3y) + 0.3y > 3.6$

Write the inequality that abbreviates each of the following sentences, solve for one of the variables, and list three solutions.

3. (a) Revenue R minus the costs C should be at least $6012.
   (b) The birth rate is at least 1.4 times the death rate.

4. (a) The sum of the two children's allowances in a five-year period never exceeded $2.75.
   (b) Since Mario and Julio were using the same library card and the library rule said no one may check out more than 8 books at a time, the sum of their checked-out books was not greater than the library maximum.

Graph each inequality.

5. (a) $3y < 4x - 9$ (b) $7x - 2y > 6$
6. (a) $4x + 3y - 1 > 2(y + 4)$ (b) $3(x - 2y) \leq 5(x + 4y)$
7. (a) $4x - y \geq \frac{5 + 7y}{3}$ (b) $1.8x - 4.5(y + 3) > 8.1$
8. (a) $2.8x < 8.4$ (b) $\frac{x}{7} - y + \frac{2}{7} < 2x - y$

Write the inequality that abbreviates each of the following sentences, then graph the inequality.

9. (a) The cost of an item is no more than 75% of the selling price.
   (b) The number of pages in the algebra manuscript that are text together with the number that are exercises must not exceed 485.

10. (a) The maximum perimeter of the mural for the reception room wall is 14 meters.
    (b) The minimum amount of space that the child-care facility needed for its indoor and outdoor use was 200 square meters.

Find the inequalities whose graphs are the shaded half-planes below.

11. (a) 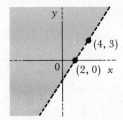 (b)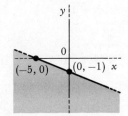

12. (a) 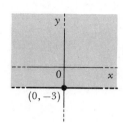 (b)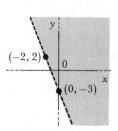

Graph the following system of inequalities.

13. (a) $\begin{cases} 2x - y > 4 \\ 3y < 4x - 7 \end{cases}$ (b) $\begin{cases} x + 9y < 3 \\ 4y - 5x + 3 > 0 \end{cases}$

14. (a) $\begin{cases} 2(3x + y) < 2 \\ \dfrac{1}{2}x - \dfrac{x - y}{6} > \dfrac{1}{3} \end{cases}$ (b) $\begin{cases} \dfrac{5x - 10y}{4} + \dfrac{1}{4} > 1 \\ \dfrac{4y}{3} - (2x + 1) - 1 < 0 \end{cases}$

15. (a) $\begin{cases} 4x + 1.6(y + 3) \le 0 \\ 0.6x \le 2.1 \end{cases}$ (b) $\begin{cases} 5.6y + 1.4 > 2.1(2x - 1) \\ 0.5y > 2.5 \end{cases}$

16. Describe the union of the graphs of the following pairs of inequalities.
   (a) $y > 1$
   $3(5x + 2y) > 12$
   (b) $x \le 7$
   $7 - 6(y + 2x) \le 0$

Find the systems of inequalities whose graphs are the shaded regions below.

17. (a) 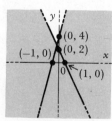 (b)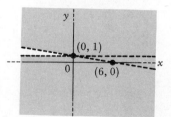

18. (a) 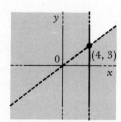 (b)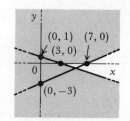

Write the set of inequalities that represents each of the following situations. Graph the system of inequalities.

19. (a) A group of students were going to buy soda for a party. They planned on buying no more than 25 six-packs made up of two different flavors of soda. They had a maximum of $30 to spend, and the orange soda cost $2 per six-pack, while the cherry-cola was $2.50 per six-pack. What possible combinations of purchases of the two flavors of soda can the students make?

    (b) Six swizzles plus 9 swozzles cost less than or equal to $18, or the difference in price between a swozzle and a swizzle must be at least $1. What possible combinations of prices might be available for swizzles and swozzles?

20. (a) A restaurant has been reserved for a large office party. A maximum of 84 persons are expected to attend. Two main courses will be available: prime rib and chicken Kiev. Experience indicates that the prime rib is selected at least twice as often as the chicken Kiev. What possible combinations of dinners should the restaurant plan on having available?

    (b) The Westford Winery is planning on producing no more than 5000 bottles of its popular Cabernet wine this year. Their plan is to make at least three times as many liter bottles as half-liter bottles. What possible combinations of liter and half-liter bottles can they produce?

Calculator Activities

Solve each inequality for y, then write the complete solution set and list three solutions.

1. (a) $908x - 483y < 778x + 640$ (b) $1.03(2x - 0.5y) \geq 5.22x + 2.13$
2. (a) $0.19y - 3.81x + 4.44 > 0$ (b) $335x - 34(26y + 13) < 96x$
3. Graph each inequality.
   (a) $550y - 330x + 1650 \leq 0$ (b) $0.195(x + 4) > 0.130y + 0.975$
4. Graph the following systems of inequalities.
   (a) $\begin{cases} 648y - 144x + 972 > 0 \\ 42(5x + 17) - 56y < 98 \end{cases}$ (b) $\begin{cases} 2.08x + 3(2.6y - 3.64) \leq 1.04 \\ 3.11y - 4.665x - 10.885 > 0 \end{cases}$

Find a system of linear inequalities that solves each of the following problems. Graph the systems.

5. (a) A carton of Valentine boxed candy arrived at the novelty store with a maximum of 144 boxes of candy inside. The boxes came in two sizes—the $\frac{1}{4}$-pound size and the $\frac{1}{2}$-pound size. If the delivery person said that the carton weighed no more than 51 pounds, what possible combinations of Valentine candy were delivered?

(b) You have a chance at a sales job in one of two companies. While the work in both jobs is similar, their manner of payment is somewhat different. The first company guarantees that your earnings will be at least 8% of your sales, while the second company says that your earnings will be at least $51 per week plus 5% commission on sales. You also know that it is unlikely that your sales will be any more than $4000 in any one week. What possible salaries could you earn for each company with varying sales? (Let x be your weekly sales for each company, maximum of $4000, and y be your corresponding weekly earnings.)

Polynomials

Kathy Parker is collecting information for a well-known department store's market survey of male consumers. The data she collects will be compiled with the data from all the forms she's filled out and will be used to evaluate complex formulas for determining how the department store will allocate its advertising and displays budget.

8-1 THE LANGUAGE OF POLYNOMIALS

Figure 8-1 is made up of two squares and a rectangle. Its area is

$$x^2 + 3x + 9$$

The more complex Figure 8-2, made up of two squares and three rectangles, has an area of

$$x^2 + xy + y^2 + 4x + 3y$$

The height attained by a ball t seconds after it is thrown vertically upward with a velocity of 20 meters per second is

$$20t - 5t^2$$

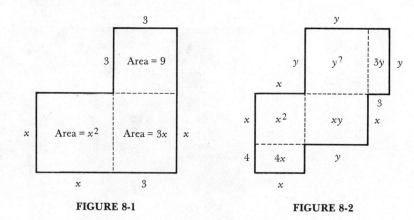

FIGURE 8-1   FIGURE 8-2

The x, y, and t appearing in these algebraic expressions are variables. Such algebraic expressions are examples of polynomials.

Some other examples of polynomials are given below.

$2x + 3.2$ Sum of two terms, $2x$ and 3.2

$3a^2 + (-4a) + 7$ Sum of three terms, $3a^2$, $-4a$, and 7

$4x^2 + \frac{7}{2}xy + y^2 + 5x$ Sum of four terms, $4x^2$, $\frac{7}{2}xy$, y^2, and $5x$

$6a^2x$ One term, $6a^2x$

Polynomials such as

$3a^2 + (-4a) + 1$ and $6x^2 + (-3.1y^3)$

may be written as

$3a^2 - 4a + 1$ and $6x^2 - 3.1y^3$ Always, $a + (-b) = a - b$

A *polynomial* is either a term or a sum of terms. Each term is either a number or a product of a number and positive integral powers of one or more variables.

A term consisting of a number alone is called a **constant term**. For example, the polynomial

$x^2 + 3x + 9$

has the constant term 9. A polynomial such as

$x^2 + 7xy + y^2$

does not appear to have a constant term, but it can be considered as having a constant term of zero.

Every nonconstant term consists of a product of a number and powers of variables. For example, the term $4x^2$ is the same as

$4 \cdot x^2$ Number 4, variable x to the 2nd power

and the term

$-3xy^2$

is the same as

$(-3) \cdot x \cdot y^2$ Number -3, variables x,y with x to the 1st power and y to the 2nd power

The number portion of a term is called the **coefficient.** Thus, 4 is the coefficient of x^2 in the term $4x^2$, and -3 is the coefficient of xy^2 in the term $-3xy^2$. For the polynomial

$3a^2 - 4ab + 5b^2$ The sign of a term may be taken to be a part of the coefficient; thus, $-4ab = (-4)ab$

3 is the coefficient of a^2, -4 is the coefficient of ab, and 5 is the coefficient of b^2.

The terms x^2 and y^2 of the polynomial

$x^2 - 7xy + y^2$

do not appear to have coefficients. However, since

$x^2 = 1 \cdot x^2$

and

$y^2 = 1 \cdot y^2$

you can assign x^2 and y^2 the coefficient 1. Thus, every nonconstant term of a polynomial has a coefficient. In the polynomial

$\frac{3}{8}a^3 - a^2 + \frac{2}{9}a + 3$ ■* $-a^2 = (-1) \cdot a^2$

a^2 has the coefficient -1.

While

$3x^2 + 7xy - \frac{4}{y^2}$

is a sum of terms, it is not a polynomial. The term $-\frac{4}{y^2}$ involves *division* by a variable, which is not allowed in a polynomial.

---

*A ■ is used between a step and its running commentary whenever there might be confusion between them—such as here, where both are equations.

A polynomial having only one term is a **monomial,** having two terms a **binomial,** and having three terms a **trinomial.** For example,

$\quad\quad\quad 7x^2 \quad$ and $\quad 3u \quad\quad\quad$ *Monomials*
$\quad\quad 3x + 4y \quad$ and $\quad 2a + 7 \quad\quad$ *Binomials*
$5.11x^2 - 3x + 4.2 \quad$ and $\quad a^2 - 2ab + b^2 \quad$ *Trinomials*

The terms of a polynomial can be arranged in any order. For example, the polynomials

$3x + 7 + 4x^2 \quad$ *Terms $3x$, 7, and $4x^2$*

and

$4x^2 + 3x + 7 \quad$ *Terms $4x^2$, $3x$, and 7*

have the same terms and hence are equal. If a polynomial contains only one variable, as in the example above, then the terms can always be arranged so that the first term contains the highest power of the variable, the second term the second highest power, the third term the third highest power, and so on, with the constant term last. A polynomial arranged this way is expressed in **standard form.** The highest power of the variable in a polynomial is called the **degree** of the polynomial.

POLYNOMIAL	STANDARD FORM	DEGREE
$3x + 7 + 4x^2$	$4x^2 + 3x + 7$	2
$3 + 2a^2 - 4a + a^3$	$a^3 + 2a^2 - 4a + 3$	3
$2y - y^2 + 7y^4$	$7y^4 - y^2 + 2y$	4

Monomials containing the same variables and the same powers of each variable are called **like monomials.** For example, the pairs

$\quad\quad 7x^2 \quad$ and $\quad 5x^2 \quad\quad x^2$ *in both*
$\quad\quad 3xy^2 \quad$ and $\quad 2xy^2 \quad\quad xy^2$ *in both*
$\quad\quad 9a \quad$ and $\quad -12a \quad\quad a$ *in both*
$-2a^2b^3 \quad$ and $\quad 8a^2b^3 \quad\quad a^2b^3$ *in both*

consist of like monomials. On the other hand,

$\quad 2xy^2 \quad$ and $\quad 4x^2y$

are not like monomials because $2xy^2$ contains the first power of x and the second power of y, whereas $4x^2y$ contains the second power of x and the first power of y.

The distributive property can be used to find sums or differences of like monomials. For example,

$7x^2 + 5x^2 = (7 + 5)x^2$, or $12x^2$

$3xy^2 - 2xy^2 = 3xy^2 + (-2)xy^2 = [3 + (-2)]xy^2$, or $xy^2 \quad 1xy^2 = xy^2$

$-\frac{2}{7}a^2b^3 + \frac{8}{7}a^2b^3 = \left(-\frac{2}{7} + \frac{8}{7}\right)a^2b^3$, or $\frac{6}{7}a^2b^3$

$9a - (-12a) = 9a + 12a = (9 + 12)a$, or $21a$

You can see that the sum or difference of two like monomials is again a monomial.

If a polynomial contains terms that are like monomials, then the polynomial can be simplified by combining the like terms. For example, the polynomial

$2x^2 + 3x - 7x + 5$ *Like terms $3x$ and $-7x$ can be added:*
$3x + (-7x) = (3 - 7)x$, or $-4x$

can be simplified to

$2x^2 - 4x + 5$

Also,

$3a^2 + 12ab + 7ab + 3b^2 = 3a^2 + (12 + 7)ab + 3b^2$
$= 3a^2 + 19ab + 3b^2$

Every polynomial can be evaluated by giving values to its variables. For example, consider the polynomial

$x^2 + xy + y^2 + 4x + 3y$

which represents the area of a field having the shape pictured in Figure 8-2. If $x = 10$ meters and $y = 6$ meters, then the area of the field is

$10^2 + 10 \cdot 6 + 6^2 + 4 \cdot 10 + 3 \cdot 6 =$ *First evaluate*
$100 + 60 + 36 + 40 + 18$, or 254 square meters   *monomials, then add terms*

The polynomial

$20t - 5t^2$

given earlier tells the height of the ball t seconds after it is thrown upward. When $t = 0$, height is

$20 \cdot 0 - 5 \cdot 0^2 = 0 - 0$, or 0 meters   *Ball is on ground*

When $t = 1$, height is

$20 \cdot 1 - 5 \cdot 1^2 = 20 - 5 \cdot 1$, or 15 meters   Ball is 15 *meters above ground*

When $t = 2$, height is

$20 \cdot 2 - 5 \cdot 2^2 = 40 - 20$, or 20 meters   Ball is 20 *meters above ground*

Quick Reinforcement

Answer true or false about the given polynomial:

$$\frac{3}{2}x^3 - 2x^2 + x + 5x + 1$$

(a) The coefficient of x^2 is 2.
(b) The coefficient of x^3 is 3.
(c) The polynomial has degree 2.
(d) The polynomial has no constant term.
(e) When $x = 1$, the value of the polynomial is $6\frac{1}{2}$.
(f) The polynomial simplifies to $\frac{3}{2}x^3 - 2x^2 + 6x + 1$.

Answers (a) False (b) False (c) False (d) False (e) True (f) True

---

EXERCISES 8-1

Tell whether or not each expression is a polynomial. If it is a polynomial, tell whether it is a monomial, binomial, or trinomial.

💡 1. (a) $2x^2 + 4$ (b) $x^2 - 4x$

2. (a) $a^2 - 4a + \dfrac{7}{a}$ (b) $b^3 - 3b + \dfrac{4}{b^2}$

3. (a) $-y^2$ (b) $x^3 + 2x^2 + 3\left(\dfrac{1}{x}\right)$

4. (a) $\frac{1}{2}z^2 + 2zy + \frac{1}{4}y^2$ (b) $4.2p^2 + 0.5p - 1.1$

5. (a) $\dfrac{5}{x^2 + y}$ (b) $2x - y^5$

6. (a) $\dfrac{1}{b^3} + \dfrac{1}{b^2} + \dfrac{1}{b}$ (b) r^{12}

7. (a) $\dfrac{2r^2}{3} - \dfrac{r}{5} - \dfrac{2}{9}$ (b) $5.3z^5 + 4z^3 - 3.2$

8. (a) $m^2 - n^2$ (b) $2 - \dfrac{5p}{q}$

9. (a) $\dfrac{2}{5}a^2 - \dfrac{1}{5}a + \dfrac{4}{5a}$ (b) $1700x^5 - x^2 + 1$

10. (a) $0.034y + 0.221y^4$ (b) $1225x^9$

List the coefficients and constant term of each polynomial.

11. (a) $6x + 5$ (b) $4x - 7$
12. (a) $0.12a$ (b) $-\frac{1}{4}x^2 + \frac{2}{3}x$
13. (a) $2y^2 + y - 7$ (b) $x^2 - 5x + 2$
14. (a) $0.14z - 0.5$ (b) $-5y + 7y^2 - 17y^3$
15. (a) $234a^3 - 2a + 445$ (b) $2.36p^2 + 3.11p - 5.82$
16. (a) $x^5 + 2x + 1$ (b) $12y + 4$
17. (a) $b^3 + b^2 + 5b$ (b) $-x^2 - x$
18. (a) $26y^3 - 14y^2 + 2$ (b) $110z$
19. (a) $2p^5 - p^3$ (b) $5x^2 - 3x + 4$
20. (a) $9.04x - 10.22$ (b) $\frac{3}{7}y + \frac{6}{7}$

Express each polynomial in standard form.

21. (a) $x^3 + 5 - 7x$ (b) $3 + x - 4x^2$
22. (a) $5.7 + 2.3x$ (b) $0.126y + 1.005y^2 - 0.702$
23. (a) $3x^2 - 7x^3 + 2x$ (b) $14 + x^4 - x^2 + 5x^3$
24. (a) $\frac{1}{9}a + \frac{2}{3}a^4 + \frac{5}{9} + \frac{1}{3}a^3$ (b) $\frac{1}{2}x^3 - \frac{1}{4}x + 3$
25. (a) $y^3 - 6 + 3y - y^2$ (b) $15 + x^5$
26. (a) $0.9 - x^3 + 1.2x$ (b) $3.5y + 3.2y^3 + 3.7y^2 - 7.8$

Simplify each polynomial by combining like terms. Express the polynomial in standard form if it involves only one variable.

27. (a) $4x + 13x$ (b) $7x + 9x$
28. (a) $3xy^2 - 2x^2y + 7xy^2$ (b) $4ab + 7a - 6ab$
29. (a) $4x^2 + 2x + 3x - x^2 + 7$ (b) $x^3 + 3x^2 + x^2 - x^3$
30. (a) $\frac{2}{5}x^4y + \frac{1}{2}xy^4 + \frac{3}{5}x^4y - \frac{1}{2}xy^4$ (b) $\frac{6}{7}a - \frac{3}{4}a^2 + 1\frac{1}{4}a^2 + 1\frac{1}{7}a - \frac{5}{7}$
31. (a) $4.2x^3 - 2.4x^2 + 3.5x^3 + 1.2x^2 - 5$
    (b) $0.08x + 0.15 + 0.27x - 0.30 + 0.26x^2$
32. (a) $3xy + \frac{1}{3}xy^2 + \frac{3}{4}x^2y - \frac{2}{3}xy^2 + 5xy + \frac{1}{4}x^2y - 2$
    (b) $14ab^2 + 32ab^2 - 12a^2b + 45ab^2 + 9ab - 64a^2b$

Evaluate each of the following polynomials at the given values.

33. (a) $x^2 - 3x + 2$, for $x = 3$ (b) $y^3 - 4$, for $y = 2$
34. (a) $m^2 - 2m + 1$, for $m = \frac{1}{2}$ (b) $\frac{1}{3}a^3 - \frac{2}{3}a + 1$, for $a = 3$
35. (a) $4.5 - 3.2x^3$, for $x = 2$ (b) $3.1 + 6.2x$, for $x = 1.1$

36. A car has traveled a certain distance in t hours. The distance is given in miles, according to the polynomial $10t^2$.
    (a) Find the distance when $t = 2$ hours.
    (b) Find the distance when $t = 5$ hours.
37. A firm estimates that it will sell a certain number of units of a product after spending a thousands of dollars on advertising. The number of units sold is found by evaluating the polynomial $-a^2 + 300a + 6$.
    (a) How many units will be sold if they spend $3000 on advertising? (Let $a = 3$.)
    (b) How many units will be sold if they spend $5500 on advertising?
38. The speed of blood in an artery is given by the polynomial $1.28 - 10{,}000x^2$, where x is the distance in centimeters from the center of the artery.
    (a) Find the velocity of the blood (in centimeters per second) when $x = 0.01$ centimeters.
    (b) Find the velocity of the blood at the center of the artery (when $x = 0$).
39. The total cost in thousands of dollars for manufacturing x thousand record albums is shown by the polynomial $x^2 - 8x + 2$.
    (a) Find the total cost of manufacturing 12 thousand albums.
    (b) Find the total cost of manufacturing 13,500 albums.
40. The population of a city has grown from 100,000 people in 1968 according to the polynomial $2000t^2 + 100{,}000$, where t is the number of years since 1968.
    (a) Find the population 10 years after 1968.
    (b) Find the population in 1980.

Calculator Activities

Evaluate each of the following polynomials for the indicated values of the variables.

1. (a) $t^2 - 3t + 4$, for $t = 88$ (b) $-2x^3 + 0.5x - 3$, for $x = 0.29$
2. (a) $-21ab^2 + 14ab - 100$, for $a = 12$ and $b = 2.1$
   (b) $4xyz + z^2y - 2.8x - \frac{1}{2}y$, for $x = -12$, $y = 1$, and $z = -3.11$
3. If an object has height $64 - 16t^2$ feet after t seconds, then how high is it in
   (a) 2.0 seconds?   (b) 3.2 seconds?
4. The temperature (in degrees Fahrenheit) of a person during an illness is given by $-0.1t^2 + 1.2t + 98.6$ after time t, measured in days.
   (a) Find the person's temperature (in degrees Fahrenheit) at $t = 1.5$ days.
   (b) Find the person's temperature at $t = 1.9$ days.
5. For a given medication of y cubic centimeters (cc), there is a resulting change in blood pressure given by $-0.05y^2 + 0.003y^3$.
   (a) What is the blood pressure when $y = 20$ cc?
   (b) Find the blood pressure when $y = 32$ cc.

8-2 SUMS AND DIFFERENCES OF POLYNOMIALS

The commutative and associative properties let you add the terms of an indicated sum in any order; the answer will always be the same. For example,

$$3x + 4 + 5x^2 = (3x + 4) + 5x^2 \quad \text{By definition, } a + b + c = (a + b) + c$$
$$= 5x^2 + (3x + 4) \quad \text{By commutative property}$$
$$= (5x^2 + 3x) + 4 \quad \text{By associative property}$$
$$= 5x^2 + 3x + 4 \quad \text{By definition, again}$$

Thus,

$$3x + 4 + 5x^2 = 5x^2 + 3x + 4$$

The sum of two polynomials is also a polynomial consisting of the sum of the terms of both polynomials. It can be simplified by adding like terms and using the distributive property. For example,

$$(3x^2 + 5x + 9) + (5x^2 + 8x + 1) = (3x^2 + 5x^2) + (5x + 8x) + (9 + 1)$$

Adding like terms

$$= (3 + 5)x^2 + (5 + 8)x + 10$$

Using distributive property

$$= 8x^2 + 13x + 10$$

Thus,

$$(3x^2 + 5x + 9) + (5x^2 + 8x + 1) = 8x^2 + 13x + 10$$

Now find the sum of $7a^2 - 12ab$ and $4ab + \frac{3}{4}b^2$.

$$(7a^2 - 12ab) + \left(4ab + \frac{3}{4}b^2\right) = 7a^2 + (-12ab + 4ab) + \frac{3}{4}b^2$$

Adding like terms

$$= 7a^2 + (-12 + 4)ab + \frac{3}{4}b^2$$

Using distributive property

$$= 7a^2 + (-8)ab + \frac{3}{4}b^2$$

Thus,

$$(7a^2 - 12ab) + \left(4ab + \frac{3}{4}b^2\right) = 7a^2 - 8ab + \frac{3}{4}b^2$$

You can call the above method of adding two polynomials the *horizontal method*. There is also the *vertical method*. In this method, you line up like terms vertically. For example, add $\frac{7}{2}x^2 + 9x + 21$ to $-3x^2 + 11x + 9$.

$$+ \begin{array}{r} \frac{7}{2}x^2 + 9x + 21 \\ -3x^2 + 11x + 9 \\ \hline \frac{1}{2}x^2 + 20x + 30 \end{array}$$

with boxes: $\boxed{\frac{7}{2}x^2 - 3x^2}$, $\boxed{9x + 11x}$, $\boxed{21 + 9}$

Thus,

$$\left(\frac{7}{2}x^2 + 9x + 21\right) + (-3x^2 + 11x + 9) = \frac{1}{2}x^2 + 20x + 30$$

Now add $3r^2 + 7t^2$ to $5r^2 - 3rt - 15t^2$. Remember, $5r^2 - 3rt - 15t^2 = 5r^2 + (-3rt) + (-15t^2)$.

$$+ \begin{array}{r} 3r^2 + 0rt + 7t^2 \\ 5r^2 - 3rt - 15t^2 \\ \hline 8r^2 - 3rt - 8t^2 \end{array}$$

You can fill in "missing" terms with a 0 coefficient; $0rt = 0 \cdot rt = 0$

with boxes: $\boxed{3r^2 + 5r^2}$, $\boxed{0rt - 3rt}$, $\boxed{7t^2 - 15t^2}$

Thus,

$$(3r^2 + 7t^2) + (5r^2 - 3rt - 15t^2) = 8r^2 - 3rt - 8t^2$$

Just as you can add two polynomials by adding like terms, so can you subtract two polynomials by subtracting like terms. The difference is also a polynomial. For example,

$$(7x^2 + 3x + 5) - (4x^2 + 8x - 2) = (7x^2 - 4x^2) + (3x - 8x) + [5 - (-2)]$$

Subtract like terms
$$= (7 - 4)x^2 + (3 - 8)x + (5 + 2)$$
$$= 3x^2 + (-5x) + 7$$

Thus, using the horizontal method,

$$(7x^2 + 3x + 5) - (4x^2 + 8x - 2) = 3x^2 - 5x + 7$$

Find $\left(\frac{3}{2}x^2 + 6xy\right) - (2xy + 12y^2)$ using the vertical method.

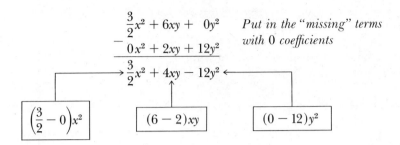

Thus,

$$\left(\frac{3}{2}x^2 + 6xy\right) - (2xy + 12y^2) = \frac{3}{2}x^2 + 4xy - 12y^2$$

You can also subtract by changing signs and then adding, since

$$a - b = a + (-b) = a + (-1)b$$

Thus,

$$\begin{aligned}(3x^2 - 4x + 5) - (7x^2 + 1.1x - 6) &= (3x^2 - 4x + 5) + (-1)(7x^2 + 1.1x - 6)\\ &= (3x^2 - 4x + 5) + (-7x^2 - 1.1x + 6)\\ &= (3x^2 - 7x^2) + (-4x - 1.1x) + (5 + 6)\\ &= -4x^2 - 5.1x + 11\end{aligned}$$

Hence,

$$(3x^2 - 4x + 5) - (7x^2 + 1.1x - 6) = -4x^2 - 5.1x + 11$$

Using the vertical method,

$$\begin{array}{r}5a^2 - 7ab + 0.3b^2\\ -\underline{2a^2 - 3ab + 0.2b^2}\\ 3a^2 - 4ab + 0.1b^2\end{array} \quad \text{is the same as} \quad \begin{array}{r}5a^2 - 7ab + 0.3b^2\\ +\underline{-2a^2 + 3ab - 0.2b^2}\\ 3a^2 - 4ab + 0.1b^2\end{array}$$

As another example, find $(5x^2 - 8xy + y^2) - (3x^2 - 8xy - 2y^2)$. This difference equals each of the following polynomials:

$$\begin{aligned}(5x^2 - 8xy + y^2) + (-1)(3x^2 - 8xy - 2y^2)\\ = (5x^2 - 8xy + y^2) + (-3x^2 + 8xy + 2y^2)\\ = (5 - 3)x^2 + (-8 + 8)xy + (1 + 2)y^2\\ = 2x^2 + 3y^2\end{aligned}$$

That is,

$$(5x^2 - 8xy + y^2) - (3x^2 - 8xy - 2y^2) = 2x^2 + 3y^2$$

Quick Reinforcement

Perform the indicated operations.

(a) $(2x^2 - 3x + 1) - (4x^3 - 2x^2 - 5)$

(b) $(3ab^2 - 2ab + 4b) + (-4a^2b - 5ab - 3a)$

Answers (a) $-4x^3 + 4x^2 - 3x + 6$ (b) $3ab^2 - 4a^2b - 7ab + 4b - 3a$

EXERCISES 8-2

In each of the following exercises, find the indicated sum or difference using the horizontal method.

💡 1. (a) $(6x^2 + 3x - 1) + (2x^2 + x + 4)$
  (b) $(4x^2 - 5x + 8) + (-3x^2 + 11x - 4)$
2. (a) $(-5x^2 + x + 3) + (x^2 + 4x - 1)$
  (b) $(3x^2 + 4xy + y^2) + (5x^2 + xy - 2y^2)$
3. (a) $(-\frac{1}{4}x^4 - \frac{1}{2}x^3 + 2x) + (\frac{3}{4}x^3 + x^2 + \frac{1}{2}x)$
  (b) $(\frac{2}{3}y^3 + \frac{1}{6}y^2 - \frac{5}{6}) + (\frac{2}{3}y^3 - \frac{1}{3}y^2 + \frac{1}{3})$
4. (a) $(1.2a^3 - 3.5a^2 + 8.2a) - (0.6a^3 - 2.5a^2 + 4.2a)$
  (b) $(3.5x^2y^2 - 7.6xy + 8.5) - (3.2x^2y^2 + 2.4xy - 1.5)$
5. (a) $(y^2 + 5) - (2y^2 - 5y + 6)$ (b) $(x^3 + 2x + 3) - (4x^3 + 2x^2 + 5)$
6. (a) $(4w^2 + 2wz + z^2) - (w^2 + 5z^2)$ (b) $(6wz + 2z^2) - (7w^2 + 2wz - z^2)$
7. (a) $(0.1x^2 + 0.9x - 0.5) + (0.1x^3 + 0.2x^2 + 0.4x)$
  (b) $(2.3y^4 + y^3 + 4.2y^2) + (7.5y^3 - 8.7y^2 + y + 4.3)$
8. (a) $(11a^2 + 2a - 6) + (a^2 + 5a - 7)$
  (b) $(5b^3 - 7b^2) - (11b^3 + 8b + 2)$
9. (a) $(-t^3 + 2t^2 + 1) - (3t^3 + 7t) + (2t^2 + 5t + 3)$
  (b) $(17x + 3) + (5x^2 + 7x + 9) + (-24x - 12)$
10. (a) $(5x^3 - 7x^2 + 11) + (-x^3 + 4x^2 - 6) - (6x^3 + 3x^2 + 2x - 5)$
  (b) $(1.1m^2 + 0.2m - 3.4) - (-0.1m^2 + 0.5m - 1.4) + (0.8m^2 + 2.0)$

In each of the following exercises, find the indicated sum or difference using the vertical method.

💡 11. (a) $(x^2 - 4x + 1) + (2x^2 - 7x + 9)$
  (b) $(3x^2 - 5x + 7) + (13x^2 + 14x - 17)$
12. (a) $(1.5y^2 - 1.3y + 2.1) - (1.1y^2 + 1.8)$
  (b) $(1025a^2 + 986a + 1134) + (1328a^2 - 442a + 109)$
13. (a) $(11m^3 + 2m - 3) - (5m^3 + 4m^2 + 8)$
  (b) $(62x^3 + 7) - (15x^2 + 2x + 9)$

8-2: SUMS AND DIFFERENCES OF POLYNOMIALS

14. (a) $(\frac{5}{12}b^2 - \frac{1}{3}b + \frac{3}{4}) + (\frac{2}{3}b^2 + \frac{1}{3}b - \frac{1}{4})$
    (b) $(\frac{1}{2}m^4 - \frac{1}{4}m^2 + \frac{1}{8}) + (m^3 + \frac{1}{2}m^2 + \frac{3}{8})$
15. (a) $(-y^3 + 3y^2 + y - 5) + (3y^3 - 2y^2 + 8)$
    (b) $(1.06x^2 + 3.20x - 5.11) - (2.50x^2 - 3.25)$
16. (a) $(136y^4 - 297y^2 + 105y - 332) - (302y^4 + 728y^3 - 72y + 631)$
    (b) $(-459m^3 + 619m^2 + 903) + (782m^3 - 540m + 225)$

In each of the following exercises, perform the operations as indicated.

17. (a) $[(3x^2 + 5x - 1) + (5x^2 - 4x + 3)] - (9x^2 + x - 7)$
    (b) $(5a^2 - 7ab + 9b^2) - [(10ab + 5b^2) + (11a^2 - 9ab)]$
18. (a) $(4.2x^2 + 3.9xy + 5.0y^2) + [(6.3x^2 - 7.1y^2) + (0.1x^2 + 6.2xy)]$
    (b) $[(5280m + 8246n - 3421mn) - (3492m - 4439mn)] - (4452m - 5210n)$
19. (a) $[(2p^3 - 4p + 1) + (3p^2 + 7p + 5)] + (9p^3 - 10p^2 - 12)$
    (b) $[(\frac{1}{2}x^2 - \frac{1}{4}xy + y^2) - (\frac{3}{4}x^2 + \frac{1}{2}xy - \frac{1}{2}y^2)] + (x^2 + 1\frac{1}{4}xy + \frac{1}{4}y^2)$
20. (a) $[(5a - 4b + 9c) + (2a - 7b + 11c)] - (a - 10b + 3c)$
    (b) $(8w^2 - 3wx - 10x^2) - [(4w^2 - x^2) - (7w^2 + 8wx - 2x^2)]$
21. (a) Add $4x^2 + 7 - 3x$, $5x - 2$, and $6x^2 + 1$.
    (b) Add $9y^2 + 2$, $3y + 5$, and $7y^2 + 8y - 4$.
22. (a) Subtract $3a^2 + 5b^2$ from $9a^2 - 5ab - 2b^2$.
    (b) Find the difference between $7x^2 + 4xy$ and $2x^2 - 6y^2$.
23. (a) Subtract $9p^2 + 4p - 2$ from the sum of $p^2 - 6$ and $5p - 2$.
    (b) Add $9m^3 + 6m^2 - 3$ to the difference between $6m^3 - 1$ and $5m^2 + 2m$.
24. (a) From the sum of $\frac{2}{3}x^2 - 3x + 2$ and $\frac{1}{3}x^2 + \frac{1}{2}x - \frac{1}{4}$ subtract $x^2 - \frac{1}{4}x + \frac{1}{4}$.
    (b) To the sum of $5.2x^2 + 3.1y^2$ and $3.7xy - 4.2y^2$ add $8.1x^2 + 9.1xy - 6.4y^2$.
25. (a) Find the sum of $4.28ab - 4.33b^2$, $9.01a^2 + 2.85ab - 2.02b^2$, and $3.30a^2 - 5.92ab + 9.48b^2$.
    (b) Subtract $342m^4 - 6m^3 + 583m + 21$ from $922m^4 + 651m^3 - 76m^2 - 90$.

Calculator Activities

Combine the following polynomials.

1. (a) $(485x^3 - 2910x^2 + 512x) + (1481x^4 - 2015x^3 - 989x - 5)$
   (b) $(9.213xy^2 - 31xy + 0.029x - 0.533) - (4.210y^2 - 0.398xy^2 + 4.888x)$
2. (a) $(0.94a^2 - 2.10ab + 3.36b^2) + [(1.12a^2 + 7.08b^2) + (0.65ab - 8.85b^2)]$
   (b) $(2907m^3 + 4399m^2 - 439m + 5569)$
       $- [(903m^2 - 4432m - 6510) + (2208m^3 + 9962)]$
3. (a) From $59x^5 + 1.8x^4 + 415x^3 - 6.49$ subtract the sum of $451.3x^4 + 9x^3 - 6.48x^2$ and $784x^4 - 0.39x^2 + 0.51$.
   (b) If a company produces x items, receives $3.29x$ in revenue, and has $2.88x + 2500$ in costs, write an expression for the company's profit.

258 CHAPTER 8: POLYNOMIALS

8-3 PRODUCTS OF MONOMIALS

The product of two monomials is another monomial whose coefficient is the product of the coefficients of the two given monomials. The variable part of the product is the product of the variable parts of the two given monomials. For example,

$$3x^2 \cdot 5x^3 = 3 \cdot 5 \cdot x^2 \cdot x^3 \qquad \text{3 · 5, or 15, is the coefficient of the product}$$
$$= 15x^5 \qquad x^2 \cdot x^3 = (x \cdot x) \cdot (x \cdot x \cdot x), \text{ or } x^5 \text{ is the variable part of the product}$$

Thus,

$$3x^2 \cdot 5x^3 = 15x^5$$

As another example, let us find $\frac{2}{3}x^3y^4 \cdot (-5xy^3)$:

$$\frac{2}{3}x^3y^4 \cdot (-5xy^3) = \frac{2}{3} \cdot (-5) \cdot (x^3 \cdot x) \cdot (y^4 \cdot y^3) \qquad \text{Combine coefficients, } x\text{'s, and } y\text{'s}$$
$$= \frac{-10}{3}x^4y^7$$

$x^3 \cdot x = (x \cdot x \cdot x) \cdot x$, or x^4; $y^4 \cdot y^3 = (y \cdot y \cdot y \cdot y) \cdot (y \cdot y \cdot y)$, or y^7

Thus,

$$\frac{2}{3}x^3y^4 \cdot (-5xy^3) = \frac{-10}{3}x^4y^7$$

Finding the product of two monomials often depends on our finding the products of powers of variables. As you have just seen,

$$x^2 \cdot x^3 = x^5 \qquad x^2 \cdot x^3 = x^{2+3}$$

and

$$y^4 \cdot y^3 = y^7 \qquad y^4 \cdot y^3 = y^{4+3}$$

These results illustrate the following property of exponents.

For any positive integers m and n,

$$x^m \cdot x^n = x^{m+n} \qquad \textit{1st law of exponents}$$

Problem 1 Find these products.

(a) $\frac{7}{4}x^7 \cdot 4x^9$ (b) $-\frac{2}{9}x^2y^3 \cdot -3x^4y^5$ (c) $a^{10}b^5 \cdot 8a^{13}b^{11}$

8-3: PRODUCTS OF MONOMIALS

SOLUTION:

(a) $\frac{7}{4}x^7 \cdot 4x^9 = \left(\frac{7}{4} \cdot 4\right) \cdot (x^7 \cdot x^9)$

$\qquad\qquad = 7x^{16}$ *Use 1st law of exponents.* $x^7 \cdot x^9 = x^{7+9}$

Thus,

$\frac{7}{4}x^7 \cdot 4x^9 = 7x^{16}$

(b) $-\frac{2}{9}x^2y^3 \cdot -3x^4y^5 = \left(-\frac{2}{9}\right) \cdot (-3)x^{2+4}y^{3+5}$ *Use 1st law of exponents*

Thus,

$-\frac{2}{9}x^2y^3 \cdot -3x^4y^5 = \frac{2}{3}x^6y^8$ $\qquad \left(-\frac{2}{9}\right) \cdot (-3) = \frac{6}{9} = \frac{2}{3}$

(c) $a^{10}b^5 \cdot 8a^{13}b^{11} = (1 \cdot 8)a^{10+13}b^{5+11}$ *The coefficient of $a^{10}b^5$ is 1; use 1st law of exponents*

Thus,

$a^{10}b^5 \cdot 8a^{13}b^{11} = 8a^{23}b^{16}$

The first law of exponents can be extended to a product of three or more monomials. For example,

$x^3 \cdot x^2 \cdot x^4 = x^{3+2+4}$

$\qquad\qquad = x^9$

Also,

$2a^2b \cdot 3a^3b^3 \cdot 5a^5b^7 = (2 \cdot 3 \cdot 5)(a^2 \cdot a^3 \cdot a^5)(b^1 \cdot b^3 \cdot b^7)$

$\qquad\qquad\qquad\qquad = 30a^{2+3+5}b^{1+3+7}$

Thus,

$2a^2b \cdot 3a^3b^3 \cdot 5a^5b^7 = 30a^{10}b^{11}$

Any power of a monomial is also a monomial. For example,

$(2x)^3 = 2x \cdot 2x \cdot 2x$

$\qquad = (2 \cdot 2 \cdot 2) \cdot (x \cdot x \cdot x)$

$\qquad = 2^3x^3$

Hence,

$(2x)^3 = 2^3x^3$

This example illustrates a second property of exponents.

For any positive integer n,

$(xy)^n = x^n y^n$ *2nd law of exponents*

The second law of exponents can also be extended to three or more factors. For example,

$$(3xy)^5 = 3^5 x^5 y^5$$
$$= 243 x^5 y^5$$

As another example,

$$(2abc)^7 = 2^7 a^7 b^7 c^7 = 128 a^7 b^7 c^7$$

What power of x is $(x^3)^4$? By definition,

$(x^3)^4 = x^3 \cdot x^3 \cdot x^3 \cdot x^3$ *For any number A, $A^4 = A \cdot A \cdot A \cdot A$*
$= x^{3+3+3+3}$ *1st law of exponents*

Thus,

$(x^3)^4 = x^{12}$ $(x^3)^4$ *is the $3 \cdot 4$th, or 12th, power of x*

This illustrates a third property of exponents.

For any positive integers m and n,

$(x^m)^n = x^{m \cdot n}$ *3rd law of exponents*

Problem 2 Find each of the following.

(a) $(x^3 y^4)^5$ (b) $\left(\dfrac{5}{2} a^5\right)^3$ (c) $(-0.2 a^3 b^7)^4$

SOLUTION:

(a) $(x^3 y^4)^5 = (x^3)^5 (y^4)^5$ *2nd law*
$= x^{3 \cdot 5} y^{4 \cdot 5}$ *3rd law*
$= x^{15} y^{20}$

Thus,

$(x^3 y^4)^5 = x^{15} y^{20}$

(b) $\left(\dfrac{5}{2} a^5\right)^3 = \left(\dfrac{5}{2}\right)^3 (a^5)^3$ *2nd law*

$= \dfrac{5^3}{2^3} a^{5 \cdot 3}$ *3rd law:* $\left(\dfrac{5}{2}\right)^3 = \dfrac{5}{2} \cdot \dfrac{5}{2} \cdot \dfrac{5}{2} = \dfrac{5^3}{2^3}$

$= \dfrac{125}{8} a^{15}$

Hence,

$$\left(\frac{5}{2}a^5\right)^3 = \frac{125}{8}a^{15}$$

(c) $(-0.2a^3b^7)^4 = (-0.2)^4(a^3)^4(b^7)^4$ *2nd law*
$= 0.0016a^{3\cdot 4}b^{7\cdot 4}$ *3rd law*
$= 0.0016a^{12}b^{28}$

Thus,

$$(-0.2a^3b^7)^4 = 0.0016a^{12}b^{28}$$

The monomials

$$-3x^4, \quad -(3x)^4, \quad (-3x)^4$$

are all different. The first is equal to a monomial with the coefficient -3,

$$-3x^4 = (-3)x^4$$

The second has equivalent forms

$$-(3x)^4 = -(3^4x^4)$$
$$= -81x^4 \quad \text{or} \; (-81)x^4$$

The third has equivalent forms

$$(-3x)^4 = (-3)^4x^4$$
$$= 81x^4$$

Quick Reinforcement

Perform the indicated operations.

(a) $(-5a^2b^3)\cdot(10ab^2)$ (b) $(4xy^2)^2\cdot(-3x^2y)^3$

Answers   (a) $-50a^3b^5$ (b) $-432x^8y^7$

EXERCISES 8-3

In each of the following exercises, multiply the monomials. Simplify your answer as much as possible.

💡 **1.** (a) $3x^2 \cdot 15x^3$ (b) $4x \cdot 3x^4$
2. (a) $-2x^2 \cdot 7x$ (b) $6x^3 \cdot (-9x^4)$
3. (a) $(-\frac{7}{2}x^2)\cdot(-\frac{1}{4}x^3)$ (b) $(5.2x^3)\cdot(1.3x^2y)$
4. (a) $(-0.3xy)\cdot(0.4xy)$ (b) $(-\frac{2}{3}x^2y)\cdot(\frac{5}{6}x^2y^2)$

5. (a) $(3a^2b^3) \cdot (5ab^7)$ (b) $(12.5p^3r^3) \cdot (-2.1p^2r^4)$
6. (a) $(45x^6) \cdot (7x^2y^5z^9)$ (b) $(25a^2b^3) \cdot (-4ab^4c^2)$
7. (a) $(-\frac{3}{5}x^2y) \cdot (-\frac{2}{3}xy^3)$ (b) $(-\frac{9}{7}x^7y^8) \cdot (-21x^3y^4z^5)$
8. (a) $(5.8a^3b) \cdot (0.5ab^2c) \cdot (1.1ab^5)$ (b) $(9.2x^3y^9z) \cdot (1.5xy^5z^2) \cdot (4.0xyz)$
9. (a) $(-\frac{3}{4}x^2y^5) \cdot (-\frac{2}{5}xz^4) \cdot (-\frac{5}{3}y^3z)$ (b) $(-\frac{5}{6}ac^2) \cdot (\frac{3}{10}b^2c^6) \cdot (-\frac{1}{4}a^5b^3)$
10. (a) $(13x^3y^8z^{10}) \cdot (9w^2z^6) \cdot (10y^5)$ (b) $(20xyz) \cdot (5wxy^2) \cdot (10wyz)$

In each exercise below, use the laws of exponents to simplify.

11. (a) $(x^2y)^4$ (b) $(ab^3)^2$
12. (a) $(\frac{1}{2}p^2r)^3$ (b) $(0.2x^3y)^4$
13. (a) $(1.1abc^2)^3$ (b) $(-\frac{2}{3}m^3n)^2$
14. (a) $(-a^3b^2)^2$ (b) $-(a^3b^2)^2$
15. (a) $(3x^2)^0$ (b) $(-4x^2y^4)^0$
16. (a) $(-0.3p^2q^4r)^3$ (b) $(2x^3y^4z)^4$
17. (a) $(-\frac{1}{4}a^2bc^5)^4$ (b) $(-1.5xy^2z^3)^3$
18. (a) $(2.1x^3y^6z^2)^2$ (b) $(\frac{3}{4}ab^4c^3)^2$
19. (a) $(a^2b)^3 \cdot (ab^2)^2$ (b) $(xy^2)^2 \cdot (x^2y^3)^4$
20. (a) $(-mn^4)^3 \cdot (m^2n^2)^4$ (b) $(p^2qr^3)^2 \cdot (-pqr^5)^3$
21. (a) $(-4xy^2z^3)^3 \cdot (-\frac{1}{2}x^3z^3)^3$ (b) $(\frac{1}{3}a^2b)^3 \cdot (-3a^3b^2)^4$
22. (a) $-(3x^2y^5)^3 \cdot (-x^6y^3)^4$ (b) $(0.5x^2y)^2 \cdot (-0.2xy^4)^3 \cdot (-0.4x^3)^2$
23. (a) $(8a^2b^5) \cdot (2ab^4)^3 \cdot (a^6b)^4$ (b) $(-a^3b^6)^2 \cdot (5a^4b)^3 \cdot (a^{10}b^9)^3$
24. (a) $(x^4y^2z)^2 \cdot (-3xy^4z)^3 \cdot (2x^5)^5$ (b) $(3x^6y)^4 \cdot (-4x^2z^4)^2 \cdot (y^7z^2)^4$

🖩 Calculator Activities

Perform the indicated operations and simplify as completely as possible.

1. (a) $(44x^2y) \cdot (55xy)^3$ (b) $(93a^3b^2) \cdot (35ab^4)$
2. (a) $(-3.25m^3n^2) \cdot (4.11mn^5)$ (b) $(0.226x^3y^5z) \cdot (-1.338xy^3z^2)$
3. (a) $(0.27xy^2z) \cdot (15x^2y) \cdot (-3.2z^3)$ (b) $(-5.27a^2bc) \cdot (94ab^3) \cdot (0.22c^5)$
4. (a) $(31x^2y)^3$ (b) $(0.36a^4b^2)^4$
5. (a) $(-2.1ay^2)^2 \cdot (-3.2axy)$ (b) $(13abc^2)^3 \cdot (22a^2b^5)^4 \cdot (-\frac{1}{2}c^3)^3$

8-4 PRODUCTS OF POLYNOMIALS

You can multiply a monomial by a binomial in the following way.

$$3x \cdot (4x^2 + 5x) = 3x \cdot 4x^2 + 3x \cdot 5x \quad \text{Distributive property}$$
$$= 12x^3 + 15x^2$$

8-4: PRODUCTS OF POLYNOMIALS

$$5ab^2 \cdot (3a + 2b) = 5ab^2 \cdot 3a + 5ab^2 \cdot 2b$$
$$= 15a^2b^2 + 10ab^3$$

It is no more difficult to multiply a monomial by any polynomial. Simply multiply the monomial by every term of the polynomial. Here are some examples.

$$7x^2 \cdot (3x^2 - 4x + 5) = 7x^2 \cdot 3x^2 + 7x^2 \cdot (-4x) + 7x^2 \cdot 5$$
$$= 21x^4 - 28x^3 + 35x^2$$
$$3a^2b \cdot (4a^2 + 7ab - 10b^2) = 3a^2b \cdot 4a^2 + 3a^2b \cdot 7ab + 3a^2b \cdot (-10b^2)$$
$$= 12a^4b + 21a^3b^2 - 30a^2b^3$$

How do you multiply two polynomials? First try multiplying two binomials:

$$(a + b)(x + y) = (a + b)x + (a + b)y \qquad \textit{Distributive property}$$
$$= (ax + bx) + (ay + by) \qquad \textit{Distributive property}$$

Thus,

$$(a + b)(x + y) = ax + bx + ay + by \qquad \begin{array}{l}\textit{Each term of } a + b \textit{ is multiplied}\\ \textit{by each term of } x + y\end{array}$$

This result can be extended to polynomials.

To find the product of two polynomials, multiply each term of one polynomial by each term of the other and add the resulting terms.

Problem 1 Find each product and simplify. (a) $(3x - 2y)(4x + y)$
(b) $(a + 2)(5a + 7)$ (c) $(x^2 + 2xy)(xy - 2y^2)$

SOLUTION:

(a) $(3x - 2y)(4x + y) = 3x \cdot 4x + 3x \cdot y + (-2y)(4x) + (-2y) \cdot y$
$$= 12x^2 + 3xy - 8xy - 2y^2 \qquad \blacksquare \; 3xy - 8xy = (3 - 8)xy$$
$$= 12x^2 - 5xy - 2y^2 \qquad \qquad \qquad = -5xy$$

Thus,

$$(3x - 2y)(4x + y) = 12x^2 - 5xy - 2y^2 \qquad \textit{Simplified product}$$

(b) $(a + 2)(5a + 7) = a \cdot 5a + a \cdot 7 + 2 \cdot 5a + 2 \cdot 7$
$$= 5a^2 + 7a + 10a + 14 \qquad \blacksquare \; 7a + 10a = (7 + 10)a$$
$$= 5a^2 + 17a + 14 \qquad \qquad \qquad = 17a$$

Thus,

$$(a + 2)(5a + 7) = 5a^2 + 17a + 14 \qquad \textit{Standard form}$$

(c) $(x^2 + 2xy)(xy - 2y^2) = x^2 \cdot xy + x^2 \cdot (-2y^2) + 2xy \cdot xy + 2xy \cdot (-2y^2)$
$$= x^3y - 2x^2y^2 + 2x^2y^2 - 4xy^3$$
$$= x^3y - 4xy^3 \qquad \blacksquare \; -2x^2y^2 + 2x^2y^2 = 0$$

Thus,

$$(x^2 + 2xy)(xy - 2y^2) = x^3y - 4xy^3$$

Problem 2 Find the product $(x - 4)(2x^2 - 5x + 3)$.

SOLUTION:

Multiply each term of the binomial by each term of the trinomial.

$(x - 4)(2x^2 - 5x + 3) = x \cdot 2x^2 + x \cdot (-5x) + x \cdot 3 + (-4) \cdot 2x^2$
$$+ (-4) \cdot (-5x) + (-4) \cdot 3$$
$$= 2x^3 - 5x^2 + 3x - 8x^2 + 20x - 12$$
$$\blacksquare \; -5x^2 - 8x^2 = -13x^2; \; 3x + 20x = 23x$$
$$= 2x^3 - 13x^2 + 23x - 12$$

Thus,

$$(x - 4)(2x^2 - 5x + 3) = 2x^3 - 13x^2 + 23x - 12$$

The method above can be called the *horizontal method* for multiplying two polynomials. You can also use a *vertical method*, similar to that used for multiplying numbers in arithmetic. For example, you find $21 \cdot 32$ as follows:

$$\begin{array}{r} 21 \\ \underline{32} \\ 42 \\ \underline{63} \\ 672 \end{array} \begin{array}{l} \\ \\ = 21 \cdot 2 \\ = 21 \cdot 3 \\ = 21 \cdot 32 \end{array}$$

The vertical method for polynomials is illustrated below.

Problem 3 Find each product using the vertical method.

(a) $(3x + 5)(2x - 7)$ \qquad (b) $(x + 4y)(x^2 - 2xy + 3y^2)$

SOLUTION:

(a)
$$\begin{array}{r} 2x - 7 \\ 3x + 5 \hline \end{array}$$

$$\begin{array}{ll} 10x - 35 & = (2x - 7) \cdot 5 \\ 6x^2 - 21x & = (2x - 7) \cdot 3x \\ \hline 6x^2 - 11x - 35 & \end{array}$$ *Line up like terms vertically; add*

Thus,

$$(3x + 5)(2x - 7) = 6x^2 - 11x - 35$$

(b)
$$\begin{array}{r} x^2 - 2xy + 3y^2 \\ x + 4y \hline \end{array}$$

$$\begin{array}{ll} 4x^2y - 8xy^2 + 12y^3 & = (x^2 - 2xy + 3y^2) \cdot 4y \\ x^3 - 2x^2y + 3xy^2 & = (x^2 - 2xy + 3y^2) \cdot x \\ \hline x^3 + 2x^2y - 5xy^2 + 12y^3 & \end{array}$$ *Line up like terms vertically; add*

Thus,

$$(x + 4y)(x^2 - 2xy + 3y^2) = x^3 + 2x^2y - 5xy^2 + 12y^3$$

Quick Reinforcement

Find the products.

(a) $(4x - a)(x + 2a)$ \hspace{2em} (b) $(x^2 - 3x + 2)(4x - 1)$

Answers (a) $4x^2 + 7ax - 2a^2$ (b) $4x^3 - 13x^2 + 11x - 2$

Sometimes you encounter sums, differences, and products of polynomials all in the same problem!

Problem 4 Carry out the indicated operations.

(a) $5(2x - 1) + x(4 - x)$

(b) $6a\left(-\dfrac{7}{2}a + \dfrac{1}{2}b\right) + \dfrac{1}{2}b(2a - 10b)$

(c) $(5xy - 2) - 2(4x - 1)(y + 1)$

(d) $-1.2(t^2 - 2t + 3) + 0.4(t^2 - t) - 1.6(2t - 3)$

266 CHAPTER 8: POLYNOMIALS

SOLUTION:

(a) $5(2x - 1) + x(4 - x) = (5 \cdot 2x - 5) + (4x - x^2)$
$= 10x - 5 + 4x - x^2$
$= -x^2 + 14x - 5$

(b) $6a\left(-\dfrac{7}{2}a + \dfrac{1}{2}b\right) + \dfrac{1}{2}b(2a - 10b) = 6a \cdot \left(-\dfrac{7}{2}a\right) + 6a \cdot \dfrac{1}{2}b + \dfrac{1}{2}b \cdot 2a$
$+ \dfrac{1}{2}b \cdot (-10b)$
$= -21a^2 + 3ab + ab - 5b^2$
$= -21a^2 + 4ab - 5b^2$

(c) $(5xy - 2) - 2(4x - 1)(y + 1) = (5xy - 2) - 2(4xy + 4x - y - 1)$
$= 5xy - 2 - 8xy - 8x + 2y + 2$
$= -3xy - 8x + 2y$

(d) $-1.2(t^2 - 2t + 3) + 0.4(t^2 - t) - 1.6(2t - 3)$
$= (-1.2t^2 + 2.4t - 3.6) + (0.4t^2 - 0.4t) + (-3.2t + 4.8)$
$= (-1.2t^2 + 0.4t^2) + (2.4t - 0.4t - 3.2t) + (-3.6 + 4.8)$
$= -0.8t^2 - 1.2t + 1.2$

Quick Reinforcement

Perform the indicated operations.

(a) $4x(2x - 1) + x(x + 1) - 2$ (b) $(3x - 2)(x - 1) - (x + 1)(3x + 2)$

Answers (a) $9x^2 - 3x - 2$ (b) $-10x$

EXERCISES 8-4

In each of the following exercises, find the indicated product using the horizontal method.

💡 **1.** (a) $13y(y + 2)$ (b) $7x(3x + 4)$
     2. (a) $\dfrac{1}{2}z(6z - 8)$ (b) $\dfrac{2}{3}a(-9a + 12)$
     3. (a) $4t^2(t^2 + 5t + 2)$ (b) $-3a^2(2a^2 + 3a - 6)$

4. (a) $10ab^2(3.2a^2 + 4.1ab - 2.3b^2)$ (b) $80x^2y(0.4x^2 - 0.2x + 0.1)$
5. (a) $-5x^3y^2(6x^2 - 4)$ (b) $7a^3b(-a^6b^2 + b)$
6. (a) $21a^5(2a^3b + 5ab^4)$ (b) $9x^3y^2(x^2 - 3x + 5)$
7. (a) $(2x + 3)(x - 5)$ (b) $(7y - 4)(6y + 3)$
8. (a) $(x + 9)(4x - 1)$ (b) $(2y + 5)(4y - 3)$
9. (a) $(4a + 6b)(2a - b)$ (b) $(5x + 3)(2x - 4)$
10. (a) $(x + z)(x - z)$ (b) $(y + 3x)(y - 3x)$
11. (a) $(2y + 7)(y - 2)$ (b) $(4m + 1)(2m - 1)$
12. (a) $(2x + 5)(7x^2 + 2x - 3)$ (b) $(3t + w)(t^2 - 5t + 2)$
13. (a) $(5y - 4)(3y^2 - 4y + 8)$ (b) $(x - 3)(2x^2 + 3x - 1)$
14. (a) $(-3m + 1)(4m^2 - 7)$ (b) $(m^2 - 4)(2m^2 + 5)$
15. (a) $(x^2y + 3x)(xy + 1)$ (b) $(ab^2 - 2b)(ab + 3)$
16. (a) $(p^2 + 3p - 7)(6p^2 - 2p - 1)$
    (b) $(9m^2 - mn + 2n^2)(m^2 + 3mn - 4n^2)$

In each of the following exercises, find the indicated product using the vertical method.

17. (a) $(3x + 4)(-4x + 7)$ (b) $(5x - 2)(3x + 9)$
18. (a) $(6y - 1)(y + 8)$ (b) $(2y + 9)(4y - 3)$
19. (a) $(-m + n)(5m - 3n)$ (b) $(10t + 3)(5t - 6)$
20. (a) $(2x - 3)(2x + 3)$ (b) $(5y + a)(5y - a)$
21. (a) $(2a + 1)(a^2 + 3a - 5)$ (b) $(-x + 5)(3x^2 + 2x - 6)$
22. (a) $(5y + z)(4y^2 + 8y + 1)$ (b) $(-4m - 3)(-3m^2 + 8m - 4)$
23. (a) $(-3x + 4)(x^3 + 2x^2 - 1)$ (b) $(4x^2 - 2y)(x^3 - 5x + 3)$
24. (a) $(2x - 3)(x + 1)(3x + 1)$ (b) $(4y - 3)(y + 7)(5y - 2)$
25. (a) $(-a + 2b)(8a - 7b)(4a + b)$ (b) $(2p + 8)(p - 3)(3p + 5)$
26. (a) $(t^2 - 5t + 7)(-2t^2 + 4t - 3)$ (b) $(3s^2 + 4st - 5t^2)(s^2 - st + 2t^2)$

Perform the indicated operations and simplify as completely as possible.

27. (a) $4(2x - 3) - 3(5 - 7x)$ (b) $\frac{1}{2}(4x + 9y) - \frac{1}{4}(12x - 5y)$
28. (a) $3x(4x + 5) + 2x(x - 5)$ (b) $9a(a + 2) - 4a(3a - 7)$
29. (a) $(2m + 3) + (m - 2)(m + 5)$ (b) $-(3z + 1) + (5z + 2)(z - 7)$
30. (a) $8p(\frac{1}{2}p - \frac{1}{4}) - 6p(\frac{1}{4}p + \frac{1}{2})$ (b) $12y(\frac{1}{3}y - \frac{5}{6}) - (y + 3)(2y - 1)$
31. (a) $3.6t(2t + 3) - 4.1t(5t - 9) + (-9t + 5.1)$
    (b) $(3x + 4)(6x + 1) - (7t - 1)(4t + 6)$
32. (a) $(20m - 1)(3m + 30) + 4(15m + 2)(10m - 6)$
    (b) $(4x + 1)(3x - 7)(x + 1) + (x - 5)(7x + 1)(7x - 1)$
33. (a) $(3a^2 + 2a - 4)(5a^2 - a + 9) - (2a^2 - 7a - 1)(-a^2 - 2a + 11)$
    (b) $(5x^2 - 3xy + 6y^2)(4x^2 + xy - y^2) + (12x^4 - 6y^4)$

Calculator Activities

Determine the product in each of the following.

1. (a) $(0.8x - 15)(14.2x + 28)$ (b) $(29ab - 12x)(14ab + 19x)$
2. (a) $(3.28a^2 - 0.5a + 2b^2)(4.1a^2 - 5.8b^2)$
   (b) $(74x - 200)(18x + 15)(0.219x - 7.22)$
3. (a) $(45y^2 + 1)(60y - 25)$
   (b) $(77m + 32)(81m - 19)(65m + 21)$

Perform the indicated operations and simplify as completely as possible.

4. (a) $(1.22x + 5.23)(0.95x - 1.10) - (3.34x - 5.53)(2.25x + 7.14)$
   (b) $(902a + 43)(65a - 83) - 24(32a^2 - 79a + 615)$

8-5 SOME SPECIAL PRODUCTS

If you multiply two binomials, you get a polynomial consisting of four terms. Of course, some of the terms might be alike, in which case they can be combined. Then the product is perhaps a trinomial, or even a binomial. The examples that follow show the various possibilities.

First, consider the product of the binomials $a + b$ and $c + d$:

$$(a + b)(c + d) = ac + ad + bc + bd \quad \text{Four different terms}$$

The method just displayed for finding the product of two binomials is called the **foil method:**

ac is the product of the **f**irst terms

ad is the product of the **o**uter terms

bc is the product of the **i**nner terms

bd is the product of the **l**ast terms

Using this method, you can find the product of any two binomials by inspection.

Problem 1 Find each product.

(a) $(x - 2)(x + 3)$ (b) $(2y - 5)(3y + 1)$

(c) $(ab - x)(ab + x)$ (d) $(5x - y)(2x + 3)$

8-5: SOME SPECIAL PRODUCTS

SOLUTION:

(a) $(x-2)(x+3) = \overbrace{x \cdot x}^{f} + \overbrace{x \cdot 3}^{o} + \overbrace{(-2) \cdot x}^{i} + \overbrace{(-2) \cdot 3}^{l}$
$= x^2 + 3x - 2x - 6$
$= x^2 + x - 6$ *Trinomial*

(b) $(2y-5)(3y+1) = \overbrace{2y \cdot 3y}^{f} + \overbrace{2y \cdot 1}^{o} + \overbrace{(-5) \cdot 3y}^{i} + \overbrace{(-5) \cdot 1}^{l}$
$= 6y^2 + 2y - 15y - 5$
$= 6y^2 - 13y - 5$ *Trinomial*

(c) $(ab-x)(ab+x) = \overbrace{ab \cdot ab}^{f} + \overbrace{ab \cdot x}^{o} + \overbrace{(-x) \cdot ab}^{i} + \overbrace{(-x) \cdot x}^{l}$
$= a^2b^2 + abx - abx - x^2$
$= a^2b^2 - x^2$ *Binomial*

(d) $(5x-y)(2x+3) = \overbrace{5x \cdot 2x}^{f} + \overbrace{5x \cdot 3}^{o} + \overbrace{(-y) \cdot 2x}^{i} + \overbrace{(-y) \cdot 3}^{l}$
$= 10x^2 + 15x - 2xy - 3y$ *Four different terms*

If two binomials are the same or similar, then their product can be found quite easily. First, consider the product of a binomial times itself, such as $(x + y)(x + y)$:

$(x+y)(x+y) = x^2 + xy + yx + y^2$ ■ $xy + yx = 2xy$
$= x^2 + 2xy + y^2$

Thus, the square of a binomial is a trinomial;

$(x+y)^2 = x^2 + 2xy + y^2$

The pattern for the square of any binomial expression $A + B$ is as follows:

$\boxed{(A+B)^2 = A^2 + 2AB + B^2}$ *A and B represent any mathematical expressions*

For example,

$(2x+5y)^2 = (2x)^2 + 2(2x)(5y) + (5y)^2$ *Let $A = 2x$ and $B = 5y$*
$= 4x^2 + 20xy + 25y^2$

Thus,

$$(2x + 5y)^2 = 4x^2 + 20xy + 25y^2$$

The pattern for the square of $A - B$ is similar:

$$\boxed{(A - B)^2 = A^2 - 2AB + B^2}$$

■ $(A - B)(A - B) = A^2 + A \cdot (-B) + (-B) \cdot A + (-B)^2 = A^2 - AB - AB + B^2$
$$= A^2 - 2AB + B^2$$

For example,

$(3a - 7b)^2 = (3a)^2 - 2(3a)(7b) + (7b)^2 \quad$ Let $A = 3a$ and $B = 7b$
$ = 9a^2 - 42ab + 49b^2$

Thus,

$$(3a - 7b)^2 = 9a^2 - 42ab + 49b^2$$

Problem 2 Find each of the following.

(a) $(4a + 7)^2$ (b) $(\frac{5}{3}x - 3y)^2$
(c) $(3y + 8b)^2$ (d) $(0.4x^2 - 0.6y^3)^2$

SOLUTION:

(a) $(4a + 7)^2 = (4a)^2 + 2 \cdot 4a \cdot 7 + 7^2$
$ = 16a^2 + 56a + 49$

(b) $\left(\frac{5}{3}x - 3y\right)^2 = \left(\frac{5}{3}x\right)^2 - 2 \cdot \frac{5}{3}x \cdot 3y + (3y)^2$
$\phantom{\left(\frac{5}{3}x - 3y\right)^2} = \frac{25}{9}x^2 - 10xy + 9y^2$

(c) $(3y + 8b)^2 = (3y)^2 + 2 \cdot 3y \cdot 8b + (8b)^2$
$ = 9y^2 + 48by + 64b^2$

(d) $(0.4x^2 - 0.6y^3)^2 = (0.4x^2)^2 - 2 \cdot (0.4x^2) \cdot (0.6y^3) + (0.6y^3)^2$
$ = 0.16x^4 - 0.48x^2y^3 + 0.36y^6$

Now let us find the product of the binomials $x + y$ and $x - y$.

$(x + y)(x - y) = x^2 + x \cdot (-y) + y \cdot x - y^2$
$ = x^2 - xy + xy - y^2$

8-5: SOME SPECIAL PRODUCTS

Thus,

$$(x+y)(x-y) = x^2 - y^2 \quad \blacksquare \; -xy + xy = 0$$

This is a case where the product of two binomials is a binomial! The pattern suggested by the example above is as follows:

$$\boxed{(A+B)(A-B) = A^2 - B^2}$$

In words, the pattern of a product of $A+B$ and $A-B$ is a **difference of two squares**, $A^2 - B^2$.

Problem 3 Find the products.

(a) $(6y+11)(6y-11)$
(b) $(\frac{3}{5}a + \frac{8}{3}b)(\frac{3}{5}a - \frac{8}{3}b)$
(c) $(5xy^2 - 3yz^4)(5xy^2 + 3yz^4)$
(d) $(a+b+c)(a+b-c)$

SOLUTION:

(a) $(6y+11)(6y-11) = (6y)^2 - (11)^2$
$= 36y^2 - 121$

(b) $\left(\frac{3}{5}a + \frac{8}{3}b\right)\left(\frac{3}{5}a - \frac{8}{3}b\right) = \left(\frac{3}{5}a\right)^2 - \left(\frac{8}{3}b\right)^2$
$= \frac{9}{25}a^2 - \frac{64}{9}b^2$

(c) $(5xy^2 - 3yz^4)(5xy^2 + 3yz^4) = (5xy^2)^2 - (3yz^4)^2$
$= 25x^2y^4 - 9y^2z^8$

(d) $(a+b+c)(a+b-c) = [(a+b)+c][(a+b)-c]$
$= (a+b)^2 - c^2$
$= a^2 + 2ab + b^2 - c^2$

Quick Reinforcement

Find each of the following products.

(a) $(3t-4)(5t+2)$
(b) $(4x-y)^2$
(c) $(5ab-y)(5ab+y)$

Answers (a) $15t^2 - 14t - 8$ (b) $16x^2 - 8xy + y^2$ (c) $25a^2b^2 - y^2$

EXERCISES 8-5

Find each of the following products.

1. (a) $(x-3)(2x+1)$ (b) $(4x-5)(2x+3)$
2. (a) $(y+1)(y-1)$ (b) $(3x+2)(3x+2)$
3. (a) $(3m-7)(m+2)$ (b) $(4y-1)(5y+2)$
4. (a) $(2x+5)(2x+5)$ (b) $(3p+1)^2$
5. (a) $(3x-7)(3x+7)$ (b) $(ay+b)(ay-b)$
6. (a) $(4a+2b)(4a+2b)$ (b) $(m-1)(m+1)$
7. (a) $(9x+2)^2$ (b) $(-3y+2z)^2$
8. (a) $(m-1)(m+2)$ (b) $(5x-3)(5x+3)$
9. (a) $(\frac{1}{2}y - \frac{1}{4})(\frac{1}{2}y + \frac{1}{4})$ (b) $(1.2x+0.5)(1.2x-0.5)$
10. (a) $(5-8x)^2$ (b) $(6x+y)(6x+y)$
11. (a) $(6x+y)(6x-y)$ (b) $(\frac{2}{3}a - \frac{3}{4}b)(\frac{2}{3}a + \frac{3}{4}b)$
12. (a) $(-4x-7y)(-4x+7y)$ (b) $(1.1m-1.5n)^2$
13. (a) $(2c-3d)(2c+3d)$ (b) $(6x-3y)(2x+7y)$
14. (a) $(13ab+2y)(13ab-2y)$ (b) $(25p-1)^2$
15. (a) $(0.6x-0.9y)^2$ (b) $(2.5a-4.2b)(0.4a-4.2b)$
16. (a) $(\frac{1}{5}x - \frac{1}{10})(\frac{1}{5}x - \frac{1}{10})$ (b) $(3p-8q)^2$
17. (a) $(0.9m-1.2n)(0.9m+1.2n)$ (b) $(\frac{9}{2}x + \frac{2}{3}y)(\frac{9}{2}x - \frac{2}{3}y)$
18. (a) $(14y+3)(2y-9)$ (b) $(1.6xy-7.1z)(1.6xy+7.1z)$
19. (a) $(4mn^2 - 5np^2)^2$ (b) $(x+y-3)(x+y+3)$
20. (a) $(2m+3n+p)(2m+3n-p)$ (b) $(6x^2y^3 - y^2z^2)(6x^2y^3 + y^2z^2)$

Calculator Activities

1. (a) $(45x-15)(19x+16)$ (b) $(3.28x+1.5)(4.1x+0.67)$
2. (a) $(22t-0.21y)^2$ (b) $(0.039z+44.2)^2$
3. (a) $(981x+462y)(981x-462y)$ (b) $(4.1x-0.15y)(4.1x+0.15y)$
4. (a) $(0.3a^2b + 0.4c)(0.29a^2b + 4.4c)$ (b) $(59x-y+0.2z)(59x-y-0.2z)$

KEY TERMS

Polynomials
Constant term
Coefficient
Monomial
Binomial
Trinomial
Standard form

Degree
Like monomials
First law of exponents
Second law of exponents
Third law of exponents
Foil method
Difference of two squares

REVIEW EXERCISES

Tell whether or not each expression is a polynomial. If it is a polynomial, tell whether it is a monomial, binomial, or trinomial.

1. (a) $3x^2 + 2x - 4$ (b) $5x^3 - \dfrac{3}{x^2}$
2. (a) $-\tfrac{9}{2}x^4$ (b) $0.3y^5 - 1.1$
3. (a) $9t^3 + 4t^2 - \dfrac{8}{t}$ (b) $\dfrac{5}{x^3}$
4. (a) $1.1y^3 - 3.9$ (b) $4m^3$

List the coefficients and constant term of each polynomial.

5. (a) $1.5x^2 + 2.8x - 0.6$ (b) $-22x^3 + x^2$
6. (a) $8y^5 - 2y^3 + \tfrac{1}{2}y + \tfrac{2}{3}$ (b) $9y^3 - 4y^2 - y + 2$

Express each polynomial in standard form.

7. (a) $5x - 3x^3 + 2$ (b) $4 - y + 2y^2 - y^3$
8. (a) $2.3m^2 + 5.1m + 8.9m^3 - 1.1$ (b) $\tfrac{17}{4} - x^4$

Simplify each polynomial by combining like terms. Express the polynomial in standard form if it involves only one variable.

9. (a) $6x + 3y - 2x + y + 1$ (b) $4x^2 + 2x - 3x^2 + 5$
10. (a) $\tfrac{1}{4}xy + \tfrac{5}{2}x - \tfrac{3}{4}xy - 2y$ (b) $1.3x^2 - 2.5x^2 + 1.9 - 0.1x^2$

Evaluate each of the following polynomials at the given values.

11. (a) $4m^3 - 2m + 5$, for $m = \tfrac{1}{2}$ (b) $y^5 - 2$, for $y = -1$
12. (a) $6.2x^2 - 3.5x - 1.2$, for $x = 2$ (b) $17 - 32x + 21x^2$, for $x = 5$
13. The rate of change of price is called inflation. The price in dollars after t years is given by the polynomial $3t^2 - 2t + 1$.
    (a) Find the price of an item (which originally cost $1, when $t = 0$) after 3 years.
    (b) Find the price of an item after 4 years.
14. The energy consumption of a nation has been estimated as a function of the gross national product P according to the formula $18P^2 + 2P$, where P is measured in trillions of dollars and where the evaluation of the polynomial gives millions of barrels per day of oil.
    (a) Find the energy consumption (in millions of barrels of oil equivalent) when the gross national product P is $1 trillion ($P = 1$).
    (b) Find the energy consumption when $P = 2.5$.

In each of the following exercises, find the indicated sum or difference.

15. (a) $(4x^2 + 3x - 7) + (2x^2 - 5x + 6)$
    (b) $(-2x^2 + 3x + 8) + (5x^2 + x - 11)$
16. (a) $(0.5y^2 + 2.1y - 1.3) - (4.1y^2 + 3.2y - 5.4)$
    (b) $(\frac{4}{5}y^2 - \frac{3}{4}y - 3) - (\frac{3}{5}y^2 + \frac{1}{2}y - 20)$
17. (a) $(-\frac{3}{8}x^2 + \frac{2}{3}x + \frac{5}{7}) + (\frac{1}{6}x - \frac{3}{14})$
    (b) $(6.2x^2 - 1.2x - 3.5) + (8.1x^3 + 4.3x^2 + 1.9x - 4.7)$
18. (a) $(5y^2 + 7y - 6) - (8y^2 + 2y + 1)$
    (b) $(y^4 + 13y^3 - 2y + 12) - (10y^3 + 4y^2 - 9y + 3)$

In each of the following exercises, perform the operations as indicated.

19. (a) $[(4x^2 + 3x - 7) - (x^2 - 7x + 2)] + (8x^2 - 9x - 1)$
    (b) $(9.3y^2 - 1.5) - [(3.2y^2 + 5.1y - 3.8) + (4.1y^2 - 7.8)]$
20. (a) Subtract $8p^2 - 4p + 5$ from the sum of $p^2 - 3p + 2$ and $9p - 11$.
    (b) Find the sum of $3.1a^2b + 5.8ab - 4.1ab^2$, $9.3ab + 6.2ab^2$, and $11.2a^2b - 8.6ab^2$.

In each of the following exercises, multiply the monomials. Simplify your answers as much as possible.

21. (a) $6x^2 \cdot (-5x^5)$ (b) $9x^8 \cdot 2x^6$
22. (a) $(-3.1x^4y^2z) \cdot (2.3xy^5z^2)$ (b) $(\frac{3}{2}a^4b^4c) \cdot (\frac{5}{3}a^2b^2c) \cdot (\frac{2}{5}abc)$

In each exercise, use the laws of exponents to simplify.

23. (a) $(-a^3b^2c)^0$ (b) $(2x^5yz^2)^4$
24. (a) $-(1.3x^4y^2)^2 \cdot (0.2xy^7)^3$ (b) $(\frac{1}{2}a^4b^2c)^2 \cdot (ab^6c^2)^4 \cdot (\frac{1}{3}abc)^3$

In each of the following exercises, find the indicated product.

25. (a) $3x^2(3x - 5)$ (b) $7.1t^3(5.0t^2 - 4.2t + 1.3)$
26. (a) $(4x - 2)(6x + 1)$ (b) $(7y + 3)(2y - 8)$
27. (a) $(2m + 9)(-3m + 4)$ (b) $(\frac{2}{3}r - \frac{1}{2})(\frac{3}{4}r + \frac{1}{4})$
28. (a) $(-5x + 3)(2x^2 - 4x - 1)$ (b) $(7y + 5)(-3y^2 + 11y - 9)$
29. (a) $(2a - 3)(a + 4)(3a - 1)$ (b) $(5s + 2)(5s - 2)(s + 3)$
30. (a) $(7x + 3)(7x - 3)$ (b) $(2.5m - 4.1)(2.5m + 4.1)$
31. (a) $(\frac{5}{8}x + 2)^2$ (b) $(1.1y - 0.6)^2$
32. (a) $(-4x + 11)(-4x - 11)$ (b) $(2x - 3y)(2x + 3y)$

Perform the indicated operations and simplify as completely as possible.

33. (a) $(4x - 2)(4x + 2) + 2x(x^2 - 3x + 5)$
    (b) $(1.1y - 0.4)^2 - (y - 2.1)(3.2y + 5.0)$
34. (a) $(2x + 1)(x - 1)(x + 1) - (3x - 5)(x^2 + 7)$
    (b) $9y(y - 3) + (4y - 3)^2$

35. (a) $(\frac{3}{5}x - \frac{5}{6})(\frac{3}{5}x + \frac{5}{6}) + (\frac{3}{5}x + \frac{5}{6})(\frac{3}{5}x + \frac{5}{6})$
    (b) $(7m - 4)(2m^2 + 3m - 1) - (5m + 2)(3m - 1)(5m - 2)$

Calculator Activities

Evaluate each of the following polynomials for the indicated values of the variables.

1. (a) $m^3 - 4m + 29$, for $m = 12$ (b) $-15x^2 + 13x - 27$, for $x = 27$
2. (a) $1.7a^2b - 3.2ab + 7.5ab^2$, for $a = 1.3$ and $b = 2.6$
   (b) $0.31t^3 - 1.27t^2 - 9.3t + 1.2$, for $t = 0.04$
3. Suppose that the number of bacteria in a petri dish after t hours is given by the polynomial

$$\frac{t^3}{6} + \frac{t^2}{2} + t + 1$$

   (a) Find the number of bacteria after 2.4 hours.
   (b) Find the number of bacteria after 3.7 hours.

Combine the following polynomials.

4. (a) $(753x^3 - 589x^2 + 302) + (603x^2 - 456x + 208)$
   (b) $(1027t^2v - 904tv - 732tv^2) - (820tv^2 + 442tv - 573t^2v)$
5. (a) From the sum of $2.45x^2 + 4.21x - 5.59$ and $6.92x^2 - 3.30x + 6.19$, subtract $1.13x^2 - 6.22x + 5.02$.
   (b) Add $93y^3 + 45y^2 - 20y + 89$ to the sum of $102y^3 - 55y + 38$ and $73y^2 - 60y - 59$.

Perform the indicated operations and simplify as completely as possible.

6. (a) $(93xy^3) \cdot (102xy)^2$ (b) $(35x^2yz^3) \cdot (87xyz^5)$
7. (a) $(-5.3a^3b^2)^3$ (b) $(0.13xy^3z)^4$
8. (a) $(12ab^2)^3 \cdot (23ab)^2$ (b) $(2.4x^2y)(3.2xy^2)^3 \cdot (5.1x^2y^2)^2$
9. (a) $(238x - 75)(143x + 90)$ (b) $(17ab + 83x)(56ab - 14x)$
10. (a) $(1.2x - 3.7)(2.3x^2 - 5.1x - 9.4)$
    (b) $(4.3a - 6.9)(3.2a + 8.0)(3.1a + 3.7)$
11. (a) $(78b + 25)(54b - 30) - (28b + 12)(28b - 12)$
    (b) $(34x + 25y)^2 - (17x - 56y)(17x + 56y)$
12. (a) $(1.3p + 4.2q)(1.3p - 4.2q) + (7.3p - 2.4q)(1.2p + 5.8q)$
    (b) $(0.3x - 5.7y + 8.1)(0.3x - 5.7y - 8.1)$

9 Dividing and Factoring Polynomials

Randy Palmer has hit a baseball and will next calculate the height of the ball after it has gone x meters: $-0.009x^2 + 1.25x$. How far will the ball have traveled when it hits the ground (has a height of 0)?

9-1 DIVISION BY A MONOMIAL

The three laws of exponents introduced in Chapter 8 tell how to multiply powers of numbers or variables. But how do you divide powers? For example, what is $2^8 \div 2^5$?

$$\frac{2^8}{2^5} = \frac{2 \cdot 2 \cdot 2 \cdot 2 \cdot 2 \cdot 2 \cdot 2 \cdot 2}{2 \cdot 2 \cdot 2 \cdot 2 \cdot 2}$$

$$= \frac{2}{2} \cdot \frac{2}{2} \cdot \frac{2}{2} \cdot \frac{2}{2} \cdot \frac{2}{2} \cdot 2 \cdot 2 \cdot 2 \qquad \text{Each } \frac{2}{2} = 1$$

$$= 2 \cdot 2 \cdot 2$$

$$= 2^3$$

Thus,

$$\frac{2^8}{2^5} = 2^3 \qquad \blacksquare \ 8 - 5 = 3$$

277

Next,

$$\frac{x^{16}}{x^{11}} = \frac{\overbrace{x \cdot x \cdot \ldots \cdot x}^{16x\text{'s}}}{\underbrace{x \cdot x \cdot \ldots \cdot x}_{11x\text{'s}}}$$

$$= \frac{\overbrace{x \cdot x \cdot \ldots \cdot x}^{11x\text{'s}}}{\underbrace{x \cdot x \cdot \ldots \cdot x}_{11x\text{'s}}} \cdot x \cdot x \cdot x \cdot x \cdot x \qquad \text{Each } \frac{x}{x} = 1$$

$$= x \cdot x \cdot x \cdot x \cdot x$$

$$= x^5$$

Thus,

$$\frac{x^{16}}{x^{11}} = x^{16-11}, \text{ or } x^5$$

These examples illustrate another property of exponents. For any positive integers m and n,

$$\frac{x^m}{x^n} = x^{m-n} \qquad \text{if } m > n$$

This property holds for zero and negative integer exponents as well as for positive exponents. Recall that

$$x^0 = 1 \qquad \text{if } x \neq 0$$

and

$$x^{-n} = \frac{1}{x^n} \qquad \text{if } x \neq 0 \text{ and } n \text{ is any positive integer}$$

For example,

$$3^{-2} = \frac{1}{3^2}, \text{ or } \frac{1}{9}$$

$$7^0 = 1$$

$$2^{-5} = \frac{1}{2^5}, \text{ or } \frac{1}{32}$$

The definitions above of x^0 and x^{-n} are motivated by the first law of exponents. Consider, for example, $2^0 \cdot 2^3$. If the first law is to hold, then

$$2^0 \cdot 2^3 = 2^{0+3} = 2^3$$

Thus,
$$2^0 \cdot 8 = 8$$
$$2^0 \cdot 8 \cdot \frac{1}{8} = 8 \cdot \frac{1}{8} \qquad \textit{Multiply each side by } \tfrac{1}{8}$$
$$2^0 \cdot 1 = 1$$
$$2^0 = 1$$

Similarly, you can show that

$x^0 = 1 \qquad$ for every $x \neq 0$

It is also clear that

$$2^{-3} \cdot 2^3 = 2^{-3+3} = 2^0 = 1$$

Thus, 2^{-3} is the multiplicative inverse of 2^3; that is,

$$2^{-3} = \frac{1}{2^3}$$

Similarly, you can show that

$x^{-n} = \dfrac{1}{x^n} \qquad$ if $x \neq 0$ and n is any positive integer

Using negative exponents,

$$\frac{x^7}{x^{13}} = \frac{\overbrace{x \cdot x \cdot \ldots \cdot x}^{7 \ x\text{'s}}}{\underbrace{(x \cdot x \cdot \ldots \cdot x)}_{7 \ x\text{'s}} \cdot \underbrace{(x \cdot x \cdot \ldots \cdot x)}_{6 \ x\text{'s}}} \qquad \frac{x^7}{x^7} = 1$$

$$= \frac{1}{\underbrace{x \cdot x \cdot \ldots \cdot x}_{6 \ x\text{'s}}}$$

$$= \frac{1}{x^6}, \quad \text{or} \quad x^{-6}$$

Thus,

$$\frac{x^7}{x^{13}} = x^{7-13}, \quad \text{or} \quad x^{-6}$$

These examples illustrate the following property of exponents.

For all integers m and n,

$\dfrac{x^m}{x^n} = x^{m-n} \qquad$ 4th law of exponents

280 CHAPTER 9: DIVIDING AND FACTORING POLYNOMIALS

It can be shown that the four laws of exponents hold for all integer exponents, positive, negative, and zero.

Problem 1 Divide as indicated.

(a) $4x^3y^2 \div 2x^2y^2$ 	(b) $36a^7 \div 27a^3$
(c) $12x^{10}y^5 \div 9x^4y^9$ 	(d) $6a^4b^7 \div 2a^4b^8$
(e) $14x^2y^8 \div 2x^4y^2$

SOLUTION:

(a) $\dfrac{4x^3y^2}{2x^2y^2} = \dfrac{4}{2} \cdot \dfrac{x^3}{x^2} \cdot \dfrac{y^2}{y^2}$ ■ $4x^3y^2 \div 2x^2y^2 = \dfrac{4x^3y^2}{2x^2y^2}$

$= 2x^{3-2}y^{2-2}$

$= 2x^1 \cdot 1 \qquad y^0 = 1$

$= 2x$

Thus,

$$4x^3y^2 \div 2x^2y^2 = 2x$$

(b) $\dfrac{36a^7}{27a^3} = \dfrac{36}{27} \cdot \dfrac{a^7}{a^3}$

$= \dfrac{4}{3}a^{7-3}$ ■ $\dfrac{36}{27} = \dfrac{4 \cdot 9}{3 \cdot 9} = \dfrac{4}{3} \cdot 1$

$= \dfrac{4}{3}a^4$

Thus,

$$36a^7 \div 27a^3 = \dfrac{4}{3}a^4$$

(c) $\dfrac{12x^{10}y^5}{9x^4y^9} = \dfrac{12}{9} \cdot \dfrac{x^{10}}{x^4} \cdot \dfrac{y^5}{y^9}$

$= \dfrac{4}{3}x^{10-4}y^{5-9}$

$= \dfrac{4}{3}x^6y^{-4}$

Thus,

$$12x^{10}y^5 \div 9x^4y^9 = \dfrac{4}{3}x^6y^{-4}, \quad \text{or} \quad \dfrac{4}{3}\dfrac{x^6}{y^4} \quad ■ \ y^{-4} = \dfrac{1}{y^4}$$

(d) $\dfrac{6a^4b^7}{2a^4b^8} = \dfrac{6}{2} \cdot \dfrac{a^4}{a^4} \cdot \dfrac{b^7}{b^8}$

$= 3a^{4-4}b^{7-8}$

$= 3a^0b^{-1}$

$= 3 \cdot 1 \cdot \dfrac{1}{b}$

$= 3 \cdot \dfrac{1}{b}$

$= \dfrac{3}{b}$

Thus,

$6a^4b^7 \div 2a^4b^8 = \dfrac{3}{b}$

(e) $\dfrac{14x^2y^8}{2x^4y^2} = \dfrac{14}{2} \cdot \dfrac{x^2}{x^4} \cdot \dfrac{y^8}{y^2}$

$= 7x^{2-4}y^{8-2}$

$= 7x^{-2}y^6$

$= 7 \cdot \dfrac{1}{x^2} \cdot y^6$

$= \dfrac{7y^6}{x^2}$

Thus,

$14x^2y^8 \div 2x^4y^2 = \dfrac{7y^6}{x^2}$

You divide a polynomial by a monomial in the following manner.

$(6x^2 - 5x + 8) \div 4x = (6x^2 - 5x + 8) \cdot \dfrac{1}{4x}$

$= 6x^2 \cdot \dfrac{1}{4x} - 5x \cdot \dfrac{1}{4x} + 8 \cdot \dfrac{1}{4x}$

$= \dfrac{6x^2}{4x} - \dfrac{5x}{4x} + \dfrac{8}{4x}$

$= \dfrac{3}{2}x - \dfrac{5}{4} + \dfrac{2}{x}$

Thus,

$\dfrac{6x^2 - 5x + 8}{4x} = \dfrac{3}{2}x - \dfrac{5}{4} + \dfrac{2}{x}$

A quotient of two polynomials is not necessarily a polynomial, as this example shows. A polynomial cannot contain a term such as $2/x$. The expression above is called a **rational expression.** You will study such expressions in the next chapter.

Problem 2 Find the quotient $(2x^3y^4 - 10x^5y^3) \div 20x^2y^4$.

SOLUTION:

$$(2x^3y^4 - 10x^5y^3) \div 20x^2y^4 = (2x^3y^4 - 10x^5y^3) \cdot \frac{1}{20x^2y^4}$$

$$= \frac{2x^3y^4}{20x^2y^4} - \frac{10x^5y^3}{20x^2y^4}$$

$$= \frac{x^{3-2}y^{4-4}}{10} - \frac{x^{5-2}y^{3-4}}{2}$$

$$= \frac{xy^0}{10} - \frac{x^3y^{-1}}{2}$$

$$= \frac{x}{10} - \frac{x^3}{2y} \qquad \blacksquare \ y^0 = 1, \ y^{-1} = \frac{1}{y}$$

Thus,

$$(2x^3y^4 - 10x^5y^3) \div 20x^2y^4 = \frac{x}{10} - \frac{x^3}{2y}$$

Quick Reinforcement

Perform the indicated operations and simplify, using the law of exponents.

(a) $\dfrac{3a^3b^2}{4a^2b^4}$ (b) $(5x^2y - 6xy) \div xy$

Answers (a) $\dfrac{3a}{4b^2}$ (b) $5x - 6$

EXERCISES 9-1

Perform each indicated operation and simplify, using the laws of exponents.

1. (a) $\dfrac{2x^2}{4x}$ (b) $\dfrac{25y^3}{5y^2}$
2. (a) $6x^4y \div 3xy$ (b) $14x^3y^4 \div 2x^2y$

3. (a) $3.2x^3y \div 1.6x^3y$ (b) $450ab^4 \div 25ab^4$
4. (a) $\dfrac{28a^3b^4}{49ab^2}$ (b) $\dfrac{-18x^5y^8}{5xy^3}$
5. (a) $\dfrac{15m^{-3}n^2}{35mn^{-5}}$ (b) $\dfrac{9tx^{-2}}{3t^{-1}x}$
6. (a) $15r^9t^5 \div 45r^{12}t^6$ (b) $81a^7b^3 \div (-9a^3b^8)$
7. (a) $-22x^{-2}y \div 121xy^2$ (b) $-4.5x^3y^{-3} \div (-0.5x^{-1}y)$
8. (a) $7x^4y^2z \div 3xy^2z^3$ (b) $144r^8s^4t \div 12r^2s^2t^2$
9. (a) $\dfrac{72p^4qr^{-1}}{-4p^{-1}r^3}$ (b) $\dfrac{-80m^3n^2}{-16m^{-5}n}$
10. (a) $-5.4x^3y^{-2} \div (-0.9xy^3z^2)$ (b) $1.69x^3z^2 \div 1.30x^5y$
11. (a) $(36x^5 + 12x^3) \div 6x^2$ (b) $(49y^5 - 28y^2) \div 7y^2$
12. (a) $\dfrac{6.3m^4 + 4.9m^2}{0.7m^2}$ (b) $\dfrac{125a^3 - 75a^2}{-25a}$
13. (a) $(56r^4 - 32r^3) \div (-8r^2)$ (b) $(-42x^2y^4 + 18xy^3) \div 6xy^2$
14. (a) $(55x^5y^7 + 33x^3y^6) \div (22x^2y^2)$ (b) $(-25x^5z + 30x^3z^2) \div (-10xz^2)$
15. (a) $\dfrac{56x^4y^5 + 21x^3y^2 - 28x^2y^2}{7x^2y^2}$ (b) $\dfrac{64r^3t^5 - 20r^2t^4 + 12rt}{4r^2t^2}$
16. (a) $(-48a^2b^2c^3 + 54ab^2c^2 + b) \div 6ab$
    (b) $(m^4n^2p^5 - m^3n^3p^4 + m^2n^3p^2) \div 2mn^2$
17. (a) $\dfrac{4.8x^3y^2z - 1.2x^2y - 1.8x}{0.6x^2y}$ (b) $\dfrac{8.5m^3n^2p^2 + 5.1m^2n^2p - 11.9m^2n}{1.7m^2n}$
18. (a) $(4m^8n^6p^4 + 10m^6n^4p^3 - 12m^2np) \div (-6m^2np^2)$
    (b) $(9x^2y^4z^9 - 8xy^3z^4 + 11xy^2z^2) \div (3xy^2z^3)$

Calculator Activities

Perform the indicated operations and simplify, using the laws of exponents.

1. (a) $\dfrac{418x^5y^3}{11x^{-2}y^5}$ (b) $\dfrac{2.31ab^2c^{-1}}{1.5ab^4c^{10}}$
2. (a) $\dfrac{756x^2y^2 - 108xy^4}{0.48x^2y^3}$
   (b) $(0.33ab^2c^3 + 26.4a^2bc - 88ab^2) \div 1.1a^2b^2$
3. (a) $\dfrac{1764a^6b^3}{7a^2b^{-1}}$ (b) $\dfrac{5.29x^2y^2}{0.4x^2}$
4. (a) $\dfrac{0.7x^3y^4 - 4.2x^2y^2 - 4.9xy}{-1.4x^2y^2}$
   (b) $\dfrac{-16.15a^7b^3 + 8.721a^5b^2 - 20.672a^3}{4.845a^2b^2}$

5. Evaluate the expression

$$\frac{118x^5y^2}{-4xy^3}$$

(a) for $x = 14$ and $y = 0.012$ (b) for $x = 1.32$ and $y = 20.71$

9-2 DIVISION OF POLYNOMIALS

Most of us learned to divide numbers by long division. For example, divide 528 by 44.

```
     12
44)528
    44
    ‾‾
    88
    88
    ‾‾
     0
```

dividend → $\dfrac{528}{44} = 12$ ← quotient

divisor ↗

✓ CHECK:

$44 \cdot 12 = 528 \qquad \dfrac{a}{b} = c \text{ if } a = b \cdot c$

A somewhat analogous division process exists for polynomials in one variable. This process depends on your ability to multiply polynomials. For example, divide $x^2 - x - 6$ by $x - 3$.

```
            x + 2
    x - 3)x² -  x - 6
          x² - 3x
          ‾‾‾‾‾‾‾
               2x - 6
               2x - 6
               ‾‾‾‾‾‾
                    0
```

x goes into x^2 x times; x times $x - 3$ is $x^2 - 3x$; enter x and $x^2 - 3x$ as shown

Subtract $x^2 - 3x$ from dividend, getting $2x$, and bring down the -6

x goes into $2x$ two times; 2 times $x - 3$ is $2x - 6$; enter 2 and $2x - 6$ as shown

Subtract $2x - 6$ from $2x - 6$

Thus,

dividend → $\dfrac{x^2 - x - 6}{x - 3} = x + 2$ ← quotient ✓

divisor ↑

✓ CHECK:

$x^2 - x - 6 = (x - 3)(x + 2)$

With long division there is sometimes a remainder. Consider the following example.

9-2: DIVISION OF POLYNOMIALS

$$\begin{array}{r}76\\23{\overline{\smash{\big)}\,1759}}\\\underline{161}\\149\\\underline{138}\\11\end{array}$$

dividend ↓ remainder ↓

$$\frac{1759}{23} = 76 + \frac{11}{23}$$

↑ divisor   ↑ divisor

quotient

Remainder must be less than divisor

Now divide $2x^2 + 7x + 5$ by $x + 2$.

divisor → $x + 2 \overline{\smash{\big)}\, 2x^2 + 7x + 5}$ ← dividend $\quad$ *x goes into $2x^2$ $2x$ times; that is, $2x^2 \div x = 2x$; put $2x$ on top*

with $2x$ on top.

Continuing,

$$\begin{array}{r}2x + 3\\x + 2{\overline{\smash{\big)}\,2x^2 + 7x + 5}}\\\underline{-\ 2x^2 + 4x}\\3x + 5\\\underline{-\ 3x + 6}\\-1\end{array}$$

Multiply $x + 2$ by $2x$, getting $2x^2 + 4x$; subtract it from the dividend, and bring down the 5

x goes into $3x$ 3 times; that is, $3x \div x = 3$; add 3 to top

Multiply $x + 2$ by 3, getting $3x + 6$; subtract it from $3x + 5$

x doesn't go into -1, so the division ends

What you have done above can be expressed as follows:

dividend ↓   remainder ↓

$$\frac{2x^2 + 7x + 5}{x + 2} = (2x + 3) + \frac{-1}{x + 2} \quad \checkmark$$

↑ divisor   ↑ quotient   ↑ divisor

√ CHECK:

Multiply each side of the equation above by $x + 2$, getting

$$2x^2 + 7x + 5 = (2x + 3)(x + 2) + (-1)$$
$$= 2x^2 + 4x + 3x + 6 - 1$$
$$= 2x^2 + 7x + 5 \quad \checkmark \qquad \textit{Both sides equal the dividend}$$

Problem 1 Divide $11y + 6y^2 + 2$ by $1 + 2y$.

Before carrying out the division, you must express each polynomial in standard form, that is, rearrange the terms so that the term containing the highest power of y comes first, the term containing the next to the highest power comes second, and so on. You can then divide $6y^2 + 11y + 2$ by $2y + 1$.

SOLUTION:

$$\begin{array}{r} 3y + 4 \\ 2y + 1 \overline{\smash{)}6y^2 + 11y + 2} \\ \underline{6y^2 + 3y} \\ 8y + 2 \\ \underline{8y + 4} \\ -2 \end{array}$$

  $2y$ goes into $6y^2$ $3y$ times; multiply $2y + 1$ by $3y$; getting $6y^2 + 3y$; put $3y$ on top
Subtract $6y^2 + 3y$ from dividend, getting $8y$ and bring down the $+2$; $2y$ goes into $8y$ 4 times; multiply $2y + 1$ by 4, getting $8y + 4$; add 4 to top; subtract $8y + 4$ from $8y + 2$, getting -2; $2y$ doesn't go into -2, so the division ends

What you have done above can be expressed as follows:

$$\underset{\underset{\text{divisor}}{\uparrow}}{\underset{\text{dividend}}{\underbrace{\frac{6y^2 + 11y + 2}{2y + 1}}}} = \underset{\underset{\text{quotient}}{\uparrow}}{(3y + 4)} + \underset{\underset{\text{divisor}}{\uparrow}}{\underset{\text{remainder}}{\frac{-2}{2y + 1}}}$$

✓ CHECK:

Multiply each side above by $2y + 1$, obtaining

$$\begin{aligned} 6y^2 + 11y + 2 &= (3y + 4)(2y + 1) + (-2) \\ &= 6y^2 + 3y + 8y + 4 - 2 \\ &= 6y^2 + 11y + 2 \checkmark \end{aligned}$$

Both sides equal the dividend

Quick Reinforcement

Perform the long division $2x - 5 \overline{\smash{)}4x^2 - 2x + 3}$

Answer $2x + 4 + \dfrac{23}{2x - 5}$

EXERCISES 9-2

In each of the exercises below, carry out the indicated division. Check your answer.

1. (a) $(x^2 - 5x + 3) \div (x + 3)$ (b) $(x^2 - 4x + 11) \div (x - 4)$
2. (a) $(m^2 - 2m - 24) \div (m + 4)$ (b) $(x^2 + 2x - 24) \div (x + 6)$
3. (a) $(-5 + 8x + x^2) \div (x - 3)$ (b) $(16m^2 - 49) \div (4m - 7)$

4. (a) $(x^2 - 5x - 6) \div (x + 1)$ (b) $(4 + 3x + 5x^2) \div (2 + x)$
5. (a) $(9y^2 + 6y - 8) \div (3y - 2)$ (b) $(2x^2 + 9x - 35) \div (x + 7)$
6. (a) $(2x^2 + 7xy - 4y^2) \div (x + 4y)$ (b) $(2x^2 - 5xy - 12y^2) \div (2x + 3y)$
7. (a) $(4x^2 - 4x + 2) \div (2x - 1)$ (b) $(10x^3 + 7x - 12) \div (2x + 3)$
8. (a) $(2x^2 + 5x - 3) \div (2x - 1)$ (b) $(-2x^2 + 7x + 4) \div (-x + 4)$
9. (a) $(12 + 8x + x^2) \div (6 + x)$ (b) $(6x^2 + 11x - 10) \div (5 + 2x)$
10. (a) $(x^3 - 2x^2 - 5x - 12) \div (x - 4)$ (b) $(2y^2 - 11y - 21) \div (2y + 3)$
11. (a) $(y^3 - 5y^2 + 4y + 7) \div (y^2 + 2)$ (b) $(x^3 + 1) \div (x + 1)$
12. (a) $(x^3 + 64) \div (x + 4)$ (b) $(6a^4 - 32a^2 - 9) \div (3a^2 - 1)$
13. (a) $(25x^3 + xy^2 + y^3) \div (5x + 2y)$ (b) $(3 - 7x + 6x^2) \div (3x + 1)$
14. (a) $(4p^4 - 10p - 9p^2 - 10) \div (2p + 3)$
    (b) $(12m^2 - 19m^3 - 4m + 12m^5 - 3) \div (4m^2 - 1)$
15. (a) $(-8y^2 + 6y^4 + 16y - 5y^3 - 8) \div (3y^2 - 4 + 2y)$
    (b) $(24t^4 - 7 - 13t + 8t^2) \div (6t^2 + 3t + 5)$
16. (a) $(30a^3 - 61a^2b - 82ab^2 + 77b^3) \div (10a - 7b)$
    (b) $(60m^3 - 5m^2n + 360mn^2 - 30n^3) \div (12m - n)$

Calculator Activities

Carry out the indicated division and check your answer.

1. (a) $(x^2 - 5x + 6) \div (0.1x - 0.2)$
   (b) $(180x^2 - 570x + 59) \div (12x + 19)$
2. (a) $(6.76x^2 - 20.41x - 14.57) \div (1.3x - 4.7)$
   (b) $(2795y^2 - 2908xy - 7663x^2) \div (65y + 79x)$

9-3 MONOMIAL FACTORS

The integer 28 can be expressed as a product of two integers in many ways. For example,

$28 = 4 \cdot 7$ 28 *is called a* **multiple** *of* 4 *and* 7

We call 4 and 7 **factors** of 28. The process of expressing an integer as a product of integers is called **factoring**.

If a polynomial C can be expressed as a product of two polynomials A and B,

$C = A \cdot B$

then C has been factored; A and B are called factors of C, and C is called a multiple of A and B.

You can easily find factors of a monomial. For example, the monomial

$3x^2y^3$

has 3, x^2, and y^3 as factors. Also, xy, y, $3x$, $3x^2y$, and many more monomials, are factors of $3x^2y^3$.

Given two or more monomials, you can ask for their **greatest common factor (GCF).** It is the monomial with the largest coefficient and greatest power of the variables that is a factor of each given monomial. You can assume the coefficients are integers unless told otherwise. For example, the GCF of the monomials

$3x^2y^4$ and $6x^3y^3$

is $3x^2y^3$.

$GCF(3x^2y^4, 6x^3y^3) = 3x^2y^3$ ■ $GCF(3,6) = 3$, $GCF(x^2, x^3) = x^2$, $GCF(y^4, y^3) = y^3$

Now find the GCF of

$30a^5b^9$, $42a^6b^7$, $60a^8b^4$

What is the GCF of 30, 42, and 60? Isn't it 6? The greatest power of a that appears in all three monomials is a^5; of b is b^4. Thus,

$GCF(30a^5b^9, 42a^6b^7, 60a^8b^4) = 6a^5b^4$

The polynomial

$4x^2 - 6x^3$

has a **greatest monomial factor** of $2x^2$, and

$4x^2 - 6x^3 = 2x^2(2 - 3x)$ *Check, using distributive law;*
$2x^2 \cdot 2 - 2x^2 \cdot 3x = 4x^2 - 6x^3$

This means that $2x^2$ is the GCF of the *terms* of the polynomial $4x^2 - 6x^3$.

Problem 1 Find the greatest monomial factor of each polynomial and write the polynomial in factored form.

(a) $2x^2 - 4x + 6x^3$ (b) $6xy^2 - 8x^2y^5 + 10x^3y^2$
(c) $15a^4 + 20a^6 - 35a^3$

SOLUTION:

(a) $GCF(2, 4, 6) = 2$, $GCF(x^2, x, x^3) = x$. Therefore, $2x$ is the greatest monomial factor of $2x^2 - 4x + 6x^3$ and

$$2x^2 - 4x + 6x^3 = 2x(x - 2 + 3x^2)$$

(b) GCF$(6, 8, 10) = 2$, GCF$(x, x^2, x^3) = x$, GCF$(y^2, y^5, y^2) = y^2$. Hence, $2xy^2$ is the greatest monomial factor and

$$6xy^2 - 8x^2y^5 + 10x^3y^2 = 2xy^2(3 - 4xy^3 + 5x^2)$$

(c) GCF$(15, 20, 35) = 5$, GCF$(a^4, a^6, a^3) = a^3$. Therefore, $5a^3$ is the greatest monomial factor and

$$15a^4 + 20a^6 - 35a^3 = 5a^3(3a + 4a^3 - 7)$$

You might ask why we would want to factor a polynomial. The principal reason is to break up the polynomial into a product of simpler polynomials. For example, suppose the polynomial

$$x^2 - 18x + 72 \qquad x > 6$$

represents a company's profits (in hundreds of dollars) when they produce x hundreds of items. If the president of the company knows how to factor, he knows that

$$x^2 - 18x + 72 = (x - 6)(x - 12) \quad \blacksquare \quad (x - 6)(x - 12) = x^2 - 12x - 6x + 72$$

He sees at a glance that the profits are negative if

$$6 < x < 12$$

For example, if $x = 7$,

$$(7 - 6)(7 - 12) = -5$$

and the company has a loss of $500. The profits are zero if $x = 12$

$$(12 - 6)(12 - 12) = 6 \cdot 0 = 0$$

and positive if $x > 12$. For example, if $x = 20$,

$$(20 - 6)(20 - 12) = 14 \cdot 8, \quad \text{or} \quad 112$$

and the company has a profit of $11,200.

Quick Reinforcement

Factor.

(a) $10a^2b - 5ab^3$

(b) $12x^3y - 18x^2y^2 + 36x^3y$

Answers (a) $5ab(2a - b^2)$ (b) $6x^2y(2x - 3y + 6x)$

EXERCISES 9-3

Find the greatest monomial factor of each polynomial, and express the polynomial in factored form.

1. (a) $4x^2 - 6x$ (b) $12x^3 - 4x^4$
2. (a) $21y^5 + 14y^3$ (b) $3.2m^4 + 1.6m^2$
3. (a) $\frac{1}{4}p^3 - \frac{3}{4}p$ (b) $\frac{5}{6}x^2 + \frac{1}{6}x^4$
4. (a) $17a^5 + 19a^4$ (b) $21m^2 - 43m^5$
5. (a) $-32a^6 + 16a^4 - 24a^3$ (b) $36b^5 + 42b^3 - 24b^2$
6. (a) $1.2x^2 - 3.6x^4 + 0.6x^5$ (b) $25y^7 - 125y^5 + 50y^4$
7. (a) $45y^7x^3 - 18y^5x$ (b) $-12x^3y + 30x^2$
8. (a) $4x^2 - 6xy$ (b) $9a^2b^3 - 5ab^2$
9. (a) $17x^3y^5 + 5x^2y^3 - 8x^2y^2$ (b) $12x^4 - 15x^3 - 3x^2y$
10. (a) $11x^5y^5 - 22x^3y^3 + 55x^2y$ (b) $-81x^4z^2 - 54x^3z$
11. (a) $\frac{5}{9}m^6n^2p + \frac{1}{9}m^2n^2p - \frac{7}{9}mn^2p$ (b) $-\frac{9}{11}s^6t^2 - \frac{3}{11}s^3t^2 + \frac{1}{3}s^2t$
12. (a) $5a^2 + 30b^3 - 15a^2b^2$ (b) $7m^5n^3 - 35m^3n^2 + 7m^2n^2$
13. (a) $-9x^7 + 81x^5 - 24x^3$ (b) $40t^5s^3 - 24t^3s^2 + 16t^2s$
14. (a) $6.2a^3b^5c^2 - 21.7a^2c^2 + 15.5ab^2c$ (b) $-8.7mn^3p + 11.6m^2p + 5.8mnp$
15. (a) $24a^3b^9 - 36a^2b^5 + 144ab$ (b) $13x^4y^2 + 17x^3y^3 - 9x^2y^6$

Calculator Activities

Find the greatest monomial factor of each polynomial, and express the polynomial in factored form.

1. (a) $180y^3z^2 - 228y^2z^2$ (b) $120a^2b^3c^4 - 48ab^2c^3 - 576a^2b^3c^2$
2. (a) $6.24mn^2 - 5.2mn + 0.13m^2n$ (b) $0.001x^4y^4z^3 + 0.01x^2y^3z^2 - 0.1x^3y^2z^4$

9-4 SPECIAL BINOMIAL FACTORS

Monomial factors of a polynomial are easy to find, as you saw in the preceding section. You will see in this section that it is easy to find binomial factors for certain polynomials.

In Chapter 8 you learned that the product of $A + B$ and $A - B$ has a simple form,

$$(A + B)(A - B) = A^2 - B^2$$

You called such a product a **difference of two squares.**

Starting with a polynomial expressed as a difference of two squares, you can factor it as a product of two polynomials. Thus, if the given polynomial has the form

$$A^2 - B^2$$

for some polynomials A and B, then it can be *factored* as

$$A^2 - B^2 = (A + B)(A - B)$$

For example, the binomial

$$4x^2 - 81y^2$$

is a difference of two squares:

$$4x^2 - 81y^2 = (2x)^2 - (9y)^2$$

Therefore, it can be factored as

$$4x^2 - 81y^2 = (2x + 9y)(2x - 9y)$$

A binomial can be factored in this way, provided:

1. It is a *difference* of two monomials.
2. The two monomials are *perfect squares*.

Problem 1 Factor, if possible.

(a) $36a^2 - 1$ (b) $4 + 9y^2$
(c) $36x^2 - 49y^2$ (d) $4a^3 - 25b^4$

SOLUTION:

(a) $36a^2 - 1$ is a difference of two monomials, $36a^2$ and 1. Each monomial is a perfect square:

$$36a^2 = (6a)^2, \quad 1 = 1^2$$

Therefore, the binomial can be factored:

$$36a^2 - 1 = (6a + 1)(6a - 1) \quad \blacksquare \; A^2 - B^2 = (A + B)(A - B)$$

(b) $4 + 9y^2$ is a *sum* of two monomials. While it can be expressed as

$$4 + 9y^2 = 4 - (-9y^2)$$

the monomial $-9y^2$ is not a perfect square. The binomial cannot be factored as above.

(c) $36x^2 - 49y^2 = (6x)^2 - (7y)^2$ *Difference of two monomials; each monomial is a perfect square*

Therefore,

$$36x^2 - 49y^2 = (6x + 7y)(6x - 7y) \quad \blacksquare \; A^2 - B^2 = (A + B)(A - B)$$

(d) $4a^3 - 25b^4$

is a difference of two monomials. The monomial $25b^4$ is a perfect square: $25b^4 = (5b^2)^2$. However, the monomial $4a^3$ is not a perfect square: a^3 is *not* the square of a power of a. Thus, the monomial cannot be factored.

A binomial such as

$200x^3 - 98xy^2$ ■ *$200x^3$ and $98xy^2$ are not perfect squares*

can be factored, although it is not a difference of two squares. First of all, it has a monomial factor:

$200x^3 - 98xy^2 = 2x(100x^2 - 49y^2)$ ■ *$100x^2$ and $49y^2$ are perfect squares*

There is now a difference of two squares within the parentheses, and the binomial can be further factored:

$200x^3 - 98xy^2 = 2x(10x + 7y)(10x - 7y)$

Recall that the square of a binomial $A + B$ is a trinomial:

$(A + B)^2 = A^2 + 2AB + B^2$

Now reverse the steps; start with a trinomial and try to factor it as a square of a binomial. For example, consider the trinomial

$9x^2 + 6x + 1$

Does it fit the pattern $A^2 + 2AB + B^2$ above? It does if you select A and B as shown below:

$$\begin{array}{ccc} 9x^2 + & 6x & + 1 \\ \updownarrow & \updownarrow & \updownarrow \\ A^2 + & 2AB & + B^2 \end{array}$$ ■ *$A = 3x, B = 1$; then $2AB = 2 \cdot 3x \cdot 1 = 6x$*

Therefore,

$9x^2 + 6x + 1 = (3x + 1)^2$ *A perfect-square trinomial*

The trinomial

$A^2 - 2AB + B^2$

is also a perfect square:

$A^2 - 2AB + B^2 = (A - B)^2$

For example,

$9x^2 - 6x + 1 = (3x - 1)^2$

A trinomial is a **perfect square** provided:

9-4: SPECIAL BINOMIAL FACTORS

1. Two of the terms are squares of monomials: call them A^2 and B^2.
2. The third term is either $2AB$ or $-2AB$.

If (1) and (2) are satisfied, the given trinomial equals $(A + B)^2$ or $(A - B)^2$.

Problem 2 Factor, if possible, as perfect squares.

(a) $16x^2 + 8x + 1$ (b) $x^2 - 6xy + 9y^2$
(c) $4a^2 + 10ab + 25b^2$ (d) $2x^2 + 4x + 1$
(e) $81y^4 - 144y^2x + 64x^2$

SOLUTION:

(a) $16x^2$ and 1 are squares of monomials:

$16x^2 = (4x)^2$, $1 = 1^2$ ■ $A = 4x, B = 1$

Is the third term either $2AB$ or $-2AB$? Yes, $8x = 2AB = 2 \cdot 4x \cdot 1$. Hence,

$16x^2 + 8x + 1 = (4x + 1)^2$

(b) x^2 and $9y^2$ and squares of monomials:

$x^2 = (x)^2$, $9y^2 = (3y)^2$ ■ $A = x, B = 3y$

Is the third term either $2AB$ or $-2AB$? Yes, $-6xy = -2AB = -2 \cdot x \cdot 3y = -6xy$. Therefore,

$x^2 - 6xy + 9y^2 = (x - 3y)^2$

(c) $4a^2$ and $25b^2$ are monomial squares:

$4a^2 = (2a)^2$, $25b^2 = (5b)^2$ ■ $A = 2a, B = 5b$

Is the third term either $2AB$ or $-2AB$? No. $2AB = 2 \cdot 2a \cdot 5b = 20ab$. This is not $10ab$. The polynomial $4a^2 + 10ab + 25b^2$ is not a perfect square.

(d) Since $2x^2$ and $4x$ are not squares of monomials, the trinomial is not a perfect square.

(e) $81y^4$ and $64x^2$ are monomial squares:

$81y^4 = (9y^2)^2$, $64x^2 = (8x)^2$ ■ $A = 9y^2, B = 8x$

The third term is $-2AB$: $-144y^2x = -2 \cdot 9y^2 \cdot 8x$. Therefore,

$81y^4 - 144y^2x + 64x^2 = (9y^2 - 8x)^2$

A trinomial such as

$-36x^2 + 6x + 54x^3$

is not a perfect square. However, it does have a monomial factor:

$$-36x^2 + 6x + 54x^3 = 6x(-6x + 1 + 9x^2)$$

Perhaps the new trinomial $-6x + 1 + 9x^2$ is a perfect binomial square. Rewriting it in standard form,

$9x^2 - 6x + 1$ $\quad$ $9x^2$ and 1 are monomial squares; $9x^2 = A^2$, $1 = B^2$, where $A = 3x$, $B = 1$, $-2AB = -6x$

Therefore,

$$9x^2 - 6x + 1 = (3x - 1)^2$$

The given trinomial factors as

$-36x^2 + 6x + 54x^3 = 6x(3x - 1)^2$ $\quad$ *When you factor a polynomial, always look first to see if it has a monomial factor!*

If A, B, and C are polynomials, then AC, BC, and $AC + BC$ are also polynomials. The last polynomial, $AC + BC$, can be factored:

$AC + BC = (A + B)C$ $\quad$ *Distributive property*

This is like factoring out a monomial, except, of course, C need not be a monomial. For example,

$$x(x + 2) + y(x + 2) = (x + y)(x + 2)$$
$$A \cdot C \quad + B \cdot C \quad = (A + B) \cdot C$$

$$2x(x^2 - x) + 3(x^2 - x) = (2x + 3)(x^2 - x)$$
$$A \cdot C \quad + B \cdot C \quad = (A + B) \cdot C$$

Starting with a given polynomial, try to group its terms in such a way that it has the form $AC + BC$ for some polynomials A, B, and C. If you can, then you can factor the given polynomial as above. This process is called **factoring by grouping.** For example, try to factor by grouping $3x + 3y + ax + ay$:

$$3x + 3y + ax + ay = (3x + 3y) + (ax + ay)$$
$$= 3(x + y) + a(x + y)$$
$$A \cdot C \quad + B \cdot C$$
$$= (3 + a)(x + y)$$
$$(A + B) \cdot C$$

Thus,

$$3x + 3y + ax + ay = (3 + a)(x + y)$$

Problem 3 Factor, if possible, by grouping.

(a) $7a - 7b - xa + xb$ (b) $x^2 + 3x + 4x + 12$
(c) $x^3 + y^2 + x^2y + xy$ (d) $6 - ab + 2a - 3b$

SOLUTION:

(a) $7a - 7b - xa + xb = (7a - 7b) - (xa - xb)$
$\qquad\qquad\qquad\qquad\;\; = 7(a - b) - x(a - b)$
$\qquad\qquad\qquad\qquad\;\; = (7 - x)(a - b)$ ■ $AC - BC = (A - B)C$

(b) $x^2 + 3x + 4x + 12 = (x^2 + 3x) + (4x + 12)$
$\qquad\qquad\qquad\qquad\; = x(x + 3) + 4(x + 3)$
$\qquad\qquad\qquad\qquad\; = (x + 4)(x + 3)$ ■ $AC + BC = (A + B)C$

(c) $x^3 + y^2 + x^2y + xy = (x^3 + y^2) + (x^2y + xy)$ Not of form $AC + BC$

Regroup the terms!

$\qquad\qquad\qquad\qquad\; = (x^3 + x^2y) + (y^2 + xy)$
$\qquad\qquad\qquad\qquad\; = x^2(x + y) + y(y + x)$ Of form $AC + BC$
$\qquad\qquad\qquad\qquad\; = (x^2 + y)(x + y)$

(d) $6 - ab + 2a - 3b = (6 - ab) + (2a - 3b)$ No good: try again
$\qquad\qquad\qquad\;\;\; = (6 + 2a) - (ab + 3b)$
$\qquad\qquad\qquad\;\;\; = 2(3 + a) - b(a + 3)$ Now of form $AC + BC$
$\qquad\qquad\qquad\;\;\; = (2 - b)(3 + a)$

Quick Reinforcement

Completely factor the following polynomials.

(a) $9a^2 - 4b^2$
(b) $12x^2 - 12x + 3$
(c) $28a^3b^4 - 7ab^2$
(d) $3a + 9b + ab + 3b^2$

Answers (a) $(3a - 2b)(3a + 2b)$ (b) $3(2x - 1)^2$ (c) $7ab^2(2ab - 1)(2ab + 1)$ (d) $(3 + b)(a + 3b)$

EXERCISES 9-4

Factor, if possible, each of the following, using the difference of squares pattern.

1. (a) $81x^2 - 25$ (b) $100a^2 - 49$
2. (a) $16 - 9y^2$ (b) $23 - 25x^2$
3. (a) $64x^2 - 121y^2$ (b) $144b^2 - 25$
4. (a) $36p^2 + 81$ (b) $625m^2 - 1$
5. (a) $32a^2 - 1$ (b) $169t^2 - 100s^2$
6. (a) $100s^2 + 169t^2$ (b) $1 - m^2$
7. (a) $x^4 - y^4$ (b) $4a^4 - 81b^4$
8. (a) $t^3 - 1$ (b) $16m^4 - 1$

Factor, if possible, each of the following as perfect squares.

9. (a) $x^2 + 4x + 4$ (b) $y^2 - 4y + 4$
10. (a) $a^2 - 10y + 25$ (b) $t^2 + 14t + 49$
11. (a) $y^2 + 25y + 25$ (b) $p^2 - 22p + 121$
12. (a) $4x^2 - 20x + 25$ (b) $9x^2 + 24x + 16$
13. (a) $49x^2 - 42x + 9$ (b) $81x^2 + 38x + 4$
14. (a) $36y^2 + 12y + 1$ (b) $144m^2 - 120m + 25$
15. (a) $36t^2 - 132t + 121$ (b) $4m^2 + 20mn + 25n^2$
16. (a) $9p^2 - 40pq + 49q^2$ (b) $81t^2 - 36tr + 4r^2$

Factor, if possible, by grouping.

17. (a) $9x + 9y + ax + ay$ (b) $ka - kb - 11a + 11b$
18. (a) $m^2 + 4m - 3m - 12$ (b) $y^2 - 5y + 8y - 40$
19. (a) $32 + 4r - 8t - tr$ (b) $6 - 9b + 4a - 6ab$
20. (a) $10x^2 - 12x + 5x - 6$ (b) $4x^2 - 28xy - 3xy + 21y^2$

Factor as completely as possible. Look for common monomial factors first, then check for difference of squares, perfect squares, and grouping.

21. (a) $4x^2 - 9y^2$ (b) $49a^2 + 14a + 1$
22. (a) $3x^3 + 12x^2 + 12x$ (b) $11y^3 - 99yz^2$
23. (a) $50m^3 + 20m^2 + 2m$ (b) $10x^2 - 30y^2$
24. (a) $9s^2 + 6st + 4t^2$ (b) $45p^2q^2 - 150pq^2 + 125q^2$
25. (a) $192x^2 - 147z^2$ (b) $99m^2 + 44n^2$
26. (a) $28ax^3 - 70ax^2y + 4bx^3 + 10bx^2y$
    (b) $5m^3t^3 - 15m^3st^2 + 5m^2nt^3 - 15m^2nst^2$
27. (a) $108x^3 - 36x^4 + 3x^5$ (b) $32 - 8x + x^2$
28. (a) $3y^4 + 60y^2 + 300$ (b) $294m^4 - 54m^2n^2$

Calculator Activities

Factor as completely as possible.

1. (a) $361m^2 - 121n^2$ (b) $980x^2y - 1820xy^2 + 845y^3$
2. (a) $1.0201x^2 + 1.8584xy + 0.8464y^2$ (b) $4.1209p^2 - 9.8596q^2$
3. (a) $0.0625x^4 - 0.45x^3y + 0.81x^2y^2$ (b) $1225t^2z^5 - 1470tz^5 + 441z^5$

9-5 FACTORING TRINOMIALS

The trinomial

$$x^2 + 10x + 16$$

is almost, but not quite, a perfect square. (The trinomial $x^2 + 8x + 16$ is a perfect square trinomial.) However, it still might be factorable. In fact,

$$x^2 + 10x + 16 = (x + 8)(x + 2)$$

It can be factored as a product of two different binomials.

There are ways of deciding whether or not a trinomial such as the one above can be factored, as you will see. All polynomials considered below have integers for coefficients of their terms, and only factors having integer coefficients are sought.

Can the trinomial

$$x^2 + 7x + 12$$

be factored as a product of two binomials? It if can, then the product of the first terms of the binomials must equal x^2, the first term of the trinomial, and there must exist integers A and B such that

$$x^2 + 7x + 12 = (x + A)(x + B)$$

By the foil method, $(x + A)(x + B) = x^2 + Bx + Ax + AB = x^2 + (A + B)x + AB$
$\uparrow\uparrow\uparrow\uparrow$
$\text{f}\text{o}\text{i}\text{l}$

That is,

$$x^2 + 7x + 12 = x^2 + (A + B)x + AB$$

The sum of the outer and inner products equals the middle term; the product of the last terms equals the last term of the trinomial

On comparing coefficients of like terms on each side of the equation above, you can see that

$AB = 12, \quad A + B = 7$

Your job is to find integers A and B whose product is 12 and whose sum is 7. Clearly, A and B must be positive integers. Possible choices for A and B such that $AB = 12$ are shown below:

A	B	
1	12	$A + B = 13$
2	6	$A + B = 8$
3	4	$A + B = 7$

For only one of these choices is the sum of A and B equal to 7,

$A = 3, \quad B = 4 \quad$ or $A = 4, B = 3$

Thus,

$x^2 + 7x + 12 = (x + 3)(x + 4) \quad$ *Factored as a product of two binomials*

Problem 1 Factor, if possible.

(a) $t^2 + 9t + 20$ \hspace{2cm} (b) $y^2 + 2y + 6$

SOLUTION:

(a) If

$t^2 + 9t + 20 = (t + A)(t + B) \quad A, B$ integers

then

$t^2 + 9t + 20 = t^2 + (A + B)t + AB$

and

$AB = 20, \quad A + B = 9$

$AB = 20$

A	B	
1	20	
2	10	
4	5	$A + B = 9$

Thus, $A = 4$ and $B = 5$ and

$t^2 + 9t + 20 = (t + 4)(t + 5) \quad$ *Factored as a product of two binomials*

(b) If
$$y^2 + 2y + 6 = (y + A)(y + B)$$
then

$$y^2 + 2y + 6 = y^2 + (A + B)y + AB \quad \blacksquare \quad AB = 6, A + B = 2$$

$AB = 6$

A	B	
1	6	$A + B = 7$
2	3	$A + B = 5$
3	2	$A + B = 5$
6	1	$A + B = 7$

You tried all possible factors of 6. In no case did $A + B = 2$. Therefore, the trinomial $y^2 + 2y + 6$ cannot be factored as a product of two binomials.

The following problem differs from the preceding one in that some of the coefficients are negative numbers.

Problem 2 Factor, if possible.

(a) $x^2 - 9x + 18$ 　　　　　　(b) $x^2 - xy - 20y^2$

SOLUTION:

(a) If
$$x^2 - 9x + 18 = (x + A)(x + B) \quad \blacksquare \quad x^2 - 9x + 18 = x^2 + (A + B)x + AB$$
then

$AB = 18, \quad A + B = -9$ 　　Since $A + B = -9$, A or B or both must be negative; since $AB = 18$, both A and B must be negative

$AB = 18$

A	B	
-1	-18	
-2	-9	
-3	-6	$A + B = -9$

Hence,
$$x^2 - 9x + 18 = (x-3)(x-6)$$

(b) If $x^2 - xy - 20y^2$ can be factored as a product of two binomials, it will have the form

$$x^2 - xy - 20y^2 = (x + Ay)(x + By)$$

■ $(x + Ay)(x + By)$
  $= x^2 + Axy + Bxy + ABy^2$

or

$$x^2 - xy - 20y^2 = x^2 + (A + B)xy + ABy^2$$

The last terms must involve y, since their product contains y^2

■ $-xy = (-1)xy$

Thus,

$AB = -20, \quad A + B = -1$

$AB = -20$

Select A negative, B positive (or vice versa)

A	B	
-20	1	
-10	2	
-5	4	$A + B = -1$

Hence,

$$x^2 - xy - 20y^2 = (x - 5y)(x + 4y) \quad \text{Factored form}$$

In the previous examples, the coefficient of the first term of each polynomial was always 1. Now consider how you would factor a polynomial with a first coefficient different from 1.

Problem 3 Factor, if possible.

(a) $3x^2 - 4xy - 4y^2$ \qquad (b) $10x^2 + 23x - 5$

SOLUTION:

(a) The possible factors of $3x^2$ are x and $3x$. Thus, possible binomial factors of $3x^2 - 4xy - 4y^2$ are $x + Ay$ and $3x + By$.

$$3x^2 - 4xy - 4y^2 = (x + Ay)(3x + By) \qquad A, B \text{ integers}$$

Then

$$3x^2 - 4xy - 4y^2 = 3x^2 + (3A + B)xy + ABy^2$$

■ $(x + Ay)(3x + By)$
  $= 3x^2 + 3Axy + Bxy + ABy^2$

Thus, select integers A and B such that

$AB = -4$ and $3A + B = -4$

The product of the last terms must equal the last term of the trinomial; the sum of the outer and inner products must equal the middle term of the trinomial

Try out possible values of A and B in your head: $A = 1, B = -4$; $A = -1, B = 4$; $A = 2, B = -2$; and so on. $A = -2, B = 2$ does the job because $3A + B = -6 + 2 = -4$. Therefore,

$3x^2 - 4xy - 4y^2 = (x - 2y)(3x + 2y) \qquad A = -2, B = 2$

(b) The possible factors of $10x^2$ are $2x$ and $5x$, or x and $10x$. In the first case,

$10x^2 + 23x - 5 = (2x + A)(5x + B)$

or

$10x^2 + 23x - 5 = 10x^2 + (5A + 2B)x + AB$ ∎ $(2x + A)(5x + B)$
$= 10x^2 + (5A + 2B)x + AB$

Now try to find integers A and B such that

$AB = -5, \qquad 5A + 2B = 23$

Possible values are: $A = -1, B = 5$; $A = 1, B = -5$; $A = 5, B = -1$; and so on. Doesn't $A = 5, B = -1$ work? It does because $5A + 2B = 25 - 2 = 23$. You have shown that $10x^2 + 23x - 5$ can be factored:

$10x^2 + 23x - 5 = (2x + 5)(5x - 1)$

The following problem illustrates the different kinds of trinomials that can be factored by the methods we have studied so far.

Problem 4 Factor completely.

(a) $4p^2q + pq^2 + 4p^3$ \qquad (b) $x^4 - 7x^2 - 18$
(c) $11x^2y^2 - 4xy^3 + 3x^3y$

SOLUTION:

(a) $4p^2q + pq^2 + 4p^3 = p(4pq + q^2 + 4p^2)$ \qquad *Monomial factor*
$= p(4p^2 + 4pq + q^2)$ \qquad *Rearrange terms*
$= p(2p + q)^2$ \qquad *Perfect-square trinomial*

(b) $x^4 = x^2 \cdot x^2$ and the middle term involves x^2, so it seems natural to factor $x^4 - 7x^2 - 18$ as follows:

$$x^4 - 7x^2 - 18 = (x^2 + A)(x^2 + B)$$
$$= x^4 + (A + B)x^2 + AB \quad \text{Use foil}$$

Integers A and B exist if

$AB = -18, \quad A + B = -7$

$A = -9, B = 2$ works. Thus,

$x^4 - 7x^2 - 18 = (x^2 - 9)(x^2 + 2)$

One of the factors, $x^2 - 9$, can be further factored:

$x^4 - 7x^2 - 18 = (x - 3)(x + 3)(x^2 + 2)$

This is the final factored form.

(c) $11x^2y^2 - 4xy^3 + 3x^3y = xy(11xy - 4y^2 + 3x^2)$ *Monomial factor*
$= xy(3x^2 + 11xy - 4y^2)$ *Rearrange terms*
$= xy(x + Ay)(3x + By)$ *$3x^2$ factors as $x \cdot 3x$, $-4y^2$ as $Ay \cdot By$*
$= xy(3x^2 + (3A + B)xy + ABy^2)$

Comparing like terms,

$AB = -4, \quad 3A + B = 11$

One of A, B is positive, the other negative. Doesn't $A = 4, B = -1$ work? Thus, the given trinomial has factored form

$xy(x + 4y)(3x - y)$

Quick Reinforcement

Factor completely.

(a) $x^2 - 2x - 15$ (b) $y - 6 + 12y^2$

(c) $4m^2 - n^2 + 3mn$ (d) $6xy^2 + 4xy - 2x$

Answers (a) $(x-5)(x+3)$ (b) $(3y-2)(4y+3)$ (c) $(4m-n)(m+n)$ (d) $2x(3y-1)(y+1)$

EXERCISES 9-5

Factor each of the following trinomials.

1. (a) $x^2 - x - 20$ (b) $x^2 - 10x + 21$
2. (a) $x^2 + 4x - 45$ (b) $x^2 + 9x + 20$

3. (a) $p^2 - 4p - 12$ (b) $x^2 - 3x - 40$
4. (a) $m^2 + 2m - 35$ (b) $-y^2 - 5y + 66$
5. (a) $-t^2 - 11t + 12$ (b) $z^2 - 11z + 24$
6. (a) $x^2 + 9x - 22$ (b) $-x^2 + 7x + 18$
7. (a) $2x^2 - 8x - 10$ (b) $12y^2 - 7y - 12$
8. (a) $2z^2 + 19z - 10$ (b) $6s^2 + s - 12$
9. (a) $2x^2 + 5x + 2$ (b) $3x^2 + xy - 2y^2$
10. (a) $8x^2 + 47x - 6$ (b) $8y^2 - 10y - 3$
11. (a) $4p^2 - 27pq + 18q^2$ (b) $3y^2 + 10y - 88$
12. (a) $6x^2 + x - 40$ (b) $8p^2 + 2p - 21$
13. (a) $12x^2 - 23x + 5$ (b) $6 - 11x + 4x^2$
14. (a) $2m^2 + 5mn - 3n^2$ (b) $16y^2 - 2y - 5$
15. (a) $3x + 2x^2 + 1$ (b) $5 - 16x + 12x^2$

Factor each of the following polynomials as completely as possible.

16. (a) $4y^2 - 12y + 9$ (b) $-y^2 + 7y - 10$
17. (a) $4x^2 + 3x - 1$ (b) $7x^2y - 7xy - 35y$
 18. (a) $18x^3 - 21x^2 - 30x$ (b) $24y^2 - y - 44$
19. (a) $25x^2 + 90x + 81$ (b) $4t^2 - 900$
20. (a) $64m^4 - 81n^4$ (b) $-4t^2 + 23t + 35$
21. (a) $3x^2y^2 + 10xy^2 - 2y^2$ (b) $15x^2 - 2xy - y^2$
22. (a) $20 + 19y - 6y^2$ (b) $33 - 34n + 8n^2$
23. (a) $12x^3y + 34x^2y - 28xy^2$ (b) $28x^4 + 36x^3 - 12x^2$
24. (a) $6a^2x + 12ax - 21x$ (b) $2x^4y^2 + 6x^3y^2 - 20x^2y^2$
25. (a) $12 - 3m^2$ (b) $4x - x^2 - 3$
26. (a) $x^4 - 1$ (b) $5y^5 - 80y$
27. (a) $27x^3 - 3$ (b) $16x^2 - 40xy + 25y^2$
28. (a) $9a^2 + 24ab + 16b^2$ (b) $2xy + 2x^2y - 60x^3y$
29. (a) $x^4 + 4x^2$ (b) $81a^4 - 256$
30. (a) $36x^3 - 69x^2 + 15x$ (b) $2m^2 - 20m^3 + 50m^4$
31. (a) $2x^4z^4 - 32$ (b) $t^4 - 35t^2 - 36$
32. (a) $y^6 + 5y^4 - 36y^2$ (b) $4c^4 - 16d^4$
33. (a) $a + 2b - ax - 2bx$ (b) $100y^2 - (y + z)^2$
34. (a) $(m + n)^2 - 9m^2$ (b) $3tz - 4bz + 6at - 8ab$

Calculator Activities

Factor as completely as possible.

1. (a) $406x^2 - 81x - 36$ (b) $150x^2 + 2243xy - 105y^2$
2. (a) $810m^2 - 5490m + 720$ (b) $156ab^2 + 1704abc - 132ac^2$

9-6 QUADRATIC EQUATIONS

Are there any squares whose area and perimeter are the same number? Stated algebraically, does the equation

$x^2 - 4x$ See Figure 9-1

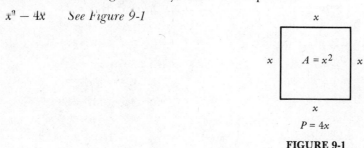

FIGURE 9-1

have a positive solution? You can answer this question as follows.

	$x^2 = 4x$	*Given equation*
	$x^2 - 4x = 0$	*Add $-4x$ to each side*
B		
	$x(x - 4) = 0$	*Factor*
A		
	$x = 0$ or $x - 4 = 0$	*If $A \cdot B = 0$, then $A = 0$ or $B = 0$*
	$x = 0$ or $x = 4$	*$x = 0$ doesn't give a square!*

Thus, the only square whose area and perimeter are the same number is shown in Figure 9-2. Whatever units you select (feet or meters, for example), the length of each side must be 4 units.

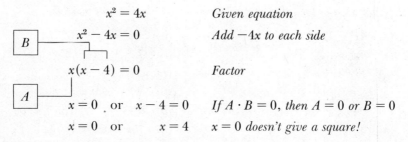

FIGURE 9-2

An expression of the form

$ax^2 + bx + c$ *a, b, c numbers with $a \neq 0$*

is called a **quadratic polynomial** in the variable x. Observe that the greatest exponent of x in a quadratic polynomial is 2. An equation that is equivalent to one of the form

$ax^2 + bx + c = 0$

9-6: QUADRATIC EQUATIONS

is called a **quadratic equation**. For example, our earlier equation,

$$x^2 = 4x$$

is quadratic because it is equivalent to the equation

$x^2 - 4x = 0$ $\quad$ $x^2 - 4x = ax^2 + bx + c$
$\quad\quad\quad\quad\quad$ *if you let $a = 1$, $b = -4$, and $c = 0$*

Many quadratic equations can be solved by factoring, as was done in the earlier example of the square.

Problem 1 Solve the equation $x^2 + 2x - 8 = 0$.

SOLUTION:

Can the quadratic polynomial $x^2 + 2x - 8$ be factored?

$x^2 + 2x - 8 = 0$ $\quad$ $x^2 + 2x - 8 = (x + A)(x + B)$,
$\quad\quad\quad\quad\quad\quad$ *where $AB = -8$, $A + B = 2$;*
$\quad\quad\quad\quad\quad\quad$ *doesn't $A = 4$, $B = -2$ work?*

$(x + 4)(x - 2) = 0$

$x + 4 = 0$ or $x - 2 = 0$ $\quad$ *If a product of two numbers is zero,*
$\quad\quad\quad\quad\quad\quad\quad\quad\quad$ *then one of the numbers is zero*

$x = -4$ or $x = 2$

Thus, the solution set is $\{-4, 2\}$.

✓ CHECK:

Let $x = -4$ $\quad\quad\quad\quad\quad\quad$ Let $x = 2$
$\quad x^2 + 2x - 8 = 0$ $\quad\quad\quad\quad$ $x^2 + 2x - 8 = 0$
$(-4)^2 + 2 \cdot (-4) - 8 = 0$ $\quad\quad$ $2^2 + 2 \cdot 2 - 8 = 0$
$\quad\quad 16 - 8 - 8 = 0$ ✓ $\quad\quad\quad\quad$ $4 + 4 - 8 = 0$ ✓

Problem 2 Solve the equation $x^2 = 4x + 21$.

SOLUTION:

$\boxed{equivalent}$ → $x^2 = 4x + 21$ $\quad$ Given equation

$\quad\quad\quad\quad$ → $x^2 - 4x - 21 = 0$ $\quad$ Add $-4x - 21$ to each side
$\quad\quad\quad\quad\quad\quad\quad\quad\quad\quad\quad\quad$ to get the polynomial equal to zero

Can you factor $x^2 - 4x - 21$? If

$x^2 - 4x - 21 = (x + A)(x + B)$

then

$$AB = -21, \quad A + B = -4 \quad \begin{array}{l} A = -3, B = 7 \text{ doesn't work;} \\ A = 3, B = -7 \text{ does!} \end{array}$$

Thus

$$(x + 3)(x - 7) = 0$$
$$x + 3 = 0 \quad \text{or} \quad x - 7 = 0$$
$$x = -3 \quad \text{or} \quad x = 7$$

The solution set is $\{-3, 7\}$.

√ CHECK:

Let $x = -3$

$$x^2 = 4x + 21$$
$$(-3)^2 = 4 \cdot (-3) + 21$$
$$9 = -12 + 21$$
$$9 = 9 \quad \checkmark$$

Let $x = 7$

$$x^2 = 4x + 21$$
$$7^2 = 4 \cdot 7 + 21$$
$$49 = 28 + 21$$
$$49 = 49 \quad \checkmark$$

Problem 3 Solve the equation $5y(y + 1) = y + 12$.

SOLUTION:

Although the left side of the equation is factored, it is of no help in solving the equation. You must first find an equivalent equation of the form $ay^2 + by + c = 0$ before solving by factoring.

$5y(y + 1) = y + 12$ *Given equation*

$5y^2 + 5y = y + 12$

$5y^2 + 5y - y - 12 = 0$ *Add $-y - 12$ to each side*

$5y^2 + 4y - 12 = 0$ *Now try to factor the left side; it must have the form $(5y + A)(y + B)$, where $A + 5B = 4$, $AB = -12$; doesn't $A = -6, B = 2$ work?*

$(5y - 6)(y + 2) = 0$

$5y - 6 = 0 \quad \text{or} \quad y + 2 = 0$

$y = \dfrac{6}{5} \quad \text{or} \quad y = -2 \qquad \dfrac{6}{5} = 1.2$

The solution set is $\{1.2, -2\}$.

9-6: QUADRATIC EQUATIONS

✓ CHECK:

Let $y = 1.2$

$5y(y + 1) = y + 12$
$6(1.2 + 1) = 1.2 + 12$
$6(2.2) = 13.2$ ✓

Let $y = -2$

$5y(y + 1) = y + 12$
$-10(-2 + 1) = -2 + 12$
$-10(-1) = 10$ ✓

Problem 4 Solve the equation $6x^2 + 3x = 0$.

SOLUTION:

The binomial $6x^2 + 3x$ is not a product of two binomials. However, it does have a monomial factor of $3x$:

$6x^2 + 3x = 3x(2x + 1)$

Thus

$3x(2x + 1) = 0$

$3x = 0$ or $2x + 1 = 0$

$x = 0$ or $x = -\frac{1}{2}$ If $3x = 0$ then $x = 0$

The solution set is $\{0, -0.5\}$.

✓ CHECK:

Let $x = 0$

$6x^2 + 3x = 0$
$6 \cdot 0^2 + 3 \cdot 0 = 0$
$6 \cdot 0 + 0 = 0$
$0 + 0 = 0$ ✓

Let $x = -0.5$

$6x^2 + 3x = 0$
$6 \cdot (-0.5)^2 + 3 \cdot (-0.5) = 0$
$6 \cdot (0.25) + (-1.5) = 0$
$1.5 - 1.5 = 0$ ✓

Problem 5 Mr. Jones has a rectangular front yard having an area of 154 square meters. If its length is 3 meters greater than its width, what are the dimensions of the yard?

SOLUTION:

If you let x designate the width of the yard, then its dimensions are as shown in Figure 9-3. By what is given,

$(x + 3)x = 154$ Length × width = area

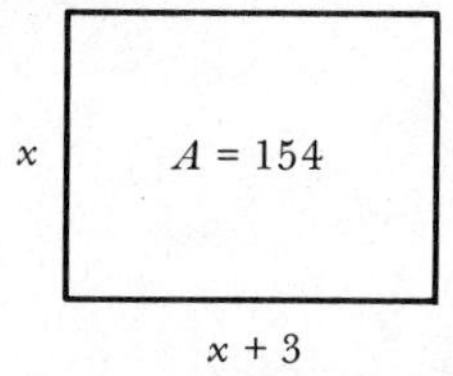

FIGURE 9-3

Thus,

$$x^2 + 3x - 154 = 0$$
$$(x - 11)(x + 14) = 0 \quad -154 = (-11) \cdot 14, \ 14 + (-11) = 3$$
$$x - 11 = 0 \quad \text{or} \quad x + 14 = 0$$
$$x = 11 \quad \text{or} \quad x = -14$$

The answer must be a positive number, so $x = 11$ is the solution.

√ CHECK:

Width = 11 meters, length = 14 meters, area = 11 · 14, or 154, square meters. √

Problem 6 The product of two positive numbers is 12, and their difference is $6\frac{1}{2}$. Find the two numbers.

SOLUTION:

Let the two numbers be x and y. Then

$$\begin{cases} xy = 12 & \text{\textit{Their product is} 12} \\ x - y = \dfrac{13}{2} & \text{\textit{Their difference is} } 6\frac{1}{2} = \frac{13}{2} \end{cases}$$

Thus,

$$x = \frac{13}{2} + y$$

$$\left(\frac{13}{2} + y\right)y = 12 \quad \textit{Replace x in 1st equation by } \tfrac{13}{2} + y \textit{ from 2nd equation}$$

$$\frac{13}{2}y + y^2 = 12$$

$$13y + 2y^2 = 24 \quad \textit{Multiply each side by 2}$$

$$2y^2 + 13y - 24 = 0 \quad \textit{Is it factorable?}$$

$$(2y - 3)(y + 8) = 0 \quad \blacksquare \ 2y^2 + 13y - 24 = (2y + A)(y + B) \textit{ if } AB = -24, \ A + 2B = 13; \ A = -3, \ B = 8 \textit{ works}$$

$2y - 3 = 0$ or $y + 8 = 0$

$y = \dfrac{3}{2}$ or $y = -8$

Since the number must be positive, $y = \dfrac{3}{2}$. Then

$x = \dfrac{13}{2} + \dfrac{3}{2}$

$= \dfrac{16}{2}$, or 8

The two numbers are $\dfrac{3}{2}$ and 8.

✓ CHECK:

$8 \cdot \dfrac{3}{2} = 12$ ✓   Product $= 12$

$8 - \dfrac{3}{2} = 6\dfrac{1}{2}$ ✓   Difference $= 6\dfrac{1}{2}$

Quick Reinforcement

Solve the equations.

(a) $x^2 - x - 20 = 0$

(b) $12x^2 - 5x = 28$

Answers: (a) 5, −4   (b) $\dfrac{7}{4}, -\dfrac{4}{3}$

EXERCISES 9-6

Solve each of the following quadratic equations, and check.

1. (a) $x^2 - 7x + 12 = 0$ (b) $x^2 - 3x + 2 = 0$
2. (a) $x^2 + 6x + 5 = 0$ (b) $x^2 - 5x - 14 = 0$
3. (a) $12y^2 + 5y - 2 = 0$ (b) $y^2 - 16 = 0$
4. (a) $m^2 + 4m = 0$ (b) $t^2 - 7t = 0$
5. (a) $a^2 + 8a + 16 = 0$ (b) $m^2 - 10m + 25 = 0$
6. (a) $4x^2 + 4x + 1 = 0$ (b) $y^2 = y + 6$
7. (a) $m^2 = 5m + 6$ (b) $x^2 = 3 + 2x$
8. (a) $3y + 4 = y^2$ (b) $2y^2 + 3y = 20$
9. (a) $z^2 - 5z = 0$ (b) $x^2 + 6x = 27$
10. (a) $25x^2 + 20x + 4 = 0$ (b) $t(t + 3) = 0$

11. (a) $m(m-11) = 0$ (b) $8p - p^2 = 0$
12. (a) $(x-5)^2 + 2x = 10$ (b) $(x+2)^2 + 4x = 13$
13. (a) $(b+2)^2 = 3b^2 - 2b + 4$ (b) $t^2 + 4t + 12 = 8t + t^2$
14. (a) $5x^2 = 6 - 13x$ (b) $7x^2 = 34x + 5$
15. (a) $(y-2)(y+1) = 4$ (b) $(x+3)(x+4) = 6$
16. (a) $(x+5)^2 = (x-5)^2$ (b) $(x+3)(x-3) = 8x$

Solve each of the following applied problems by using a quadratic equation.

17. (a) Jim lives on a large ranch in Montana. An aerial view of the ranch shows that it is rectangular in shape. The ranch hands know from their years of riding the ranges that the east-west boundary of the ranch is four kilometers longer than the north-south boundary. If the area of the ranch is 252 square kilometers, what are the dimensions of the ranch?

    (b) Find two consecutive even integers such that their product is 26 more than thirteen times their sum.

18. (a) The original plans for a small neighborhood play yard called for it to be rectangular in shape with a length seven meters more than twice the width. However, due to the presence of a large oak tree that no one wanted destroyed, the plans were changed. Three meters were added to the width, and eleven meters were subtracted from the length. The result was a play yard with an area of forty-eight square meters. What were the dimensions on the original plans?

    (b) During a riddle game in geometry class, Melissa posed this problem: "I have a square and a rectangle, and their areas are the same. The width of the rectangle is one centimeter less than the side of the square, and the length of the rectangle is six centimeters less than five times a side of the square. What are the dimensions of the square and rectangle?" Pablo answered, "That's not a geometry problem, it's an algebra problem! But I can solve it." Can you?

19. (a) Given four consecutive odd numbers such that the sum of the first three numbers equals the product of the second and fourth numbers, what are the four numbers?

    (b) Malia wanted to build a large chest to hold her extensive shell and mineral collection. The chest was to be two meters high, and the width was to be one less than the length. The volume of the chest was to be 60 cubic meters. Malia couldn't figure out what the dimensions should be, so she asked her older sister Lia, who computed them for her. What did Lia find?

20. (a) A stone thrown directly upward from the surface of a certain small planetoid with an initial velocity of 80 feet per second reaches a cer-

tain height in t seconds, where the height is given by $80t - t^2$. If we let the height be zero, which would be right before we threw the stone and also when the stone has returned to the planetoid, we can find out how long it will take the stone to come back down. Solve $80t - t^2 = 0$ and find that time.

(b) On a different celestial body, a thrown stone reaches a given height in t seconds, where the height is now shown by $265t - 5t^2$. How long will it take a thrown stone to return to this celestial body?

21. (a) A skateboard manufacturing company found that their revenue was calculated by multiplying the selling price, $5, by the number of items sold, x, giving $R = 5x$. They also computed that their costs were found by the polynomial $x^2 + 2x - 4$. Since the break-even point is found when the revenue equals the costs, they were able to calculate the number of skateboards they needed to sell to break even. How many is that?

(b) The roller skate division of the same company found that their revenue equation was $R = 8x$, while their cost equation was $C = x^2 + 5x - 28$. How many roller skates must they sell to break even?

22. (a) A company wanted to manufacture a serving plate by cutting a square from each corner of a rectangular sheet of metal 10 centimeters by 20 centimeters, and then bending up the sides. If the base area of the plate was to be 144 square centimeters, what were the dimensions of the corners to be cut?

(b) Flat cookie boxes were to be made for Mrs. Yummy's Old-Fashioned Cookie Factory. Mrs. Yummy specified that the area of the base of the box must be 54 square inches (she has not yet learned to think in metric). The company manufacturing the boxes said they would be made from cardboard pieces measuring 9 inches by 12 inches, from which square corners would be removed before the sides were folded up and taped. Mrs. Yummy wanted to know how deep the boxes would be. Can you tell her?

𝔑ostalgia 𝔓roblems

1. (a) Two men were talking of their ages. One said, "I am 94 years old." "Then," replied the younger, "the sum of your age and mine, multiplied by the difference between our ages, will produce 8512." What was the age of the younger?

(b) There is a rectangular field whose breadth is $\frac{5}{8}$ of the length. After laying out $\frac{1}{6}$ of the whole ground for a garden, it was found that there were left 625 square rods for mowing. What are the length and breadth of the field?

Calculator Activities

Solve the following quadratic equations.

1. (a) $198x^2 + 743x - 105 = 0$ (b) $52x^2 - 228 - 199x$
2. (a) $112 = 135y^2 - 246y$ (b) $169a^2 + 91a = 294$
3. The average number of workers absent from the cotton mill for a given day's high temperature of T degrees Fahrenheit is $0.2(T - 60)^2 + 148$.
   (a) How many workers are absent if the temperature is 60°F? 95°F?
   (b) For what temperature would an absentee total of 235 be expected?
4. (a) The Blue Rock State Park is rectangular in shape and has an area of 1300 square kilometers. The longer dimension of the park, which borders on the ocean, is 45 kilometers longer than the inland dimension. What is the length of the shoreline of Blue Rock State Park?
   (b) The product of two consecutive even numbers is one hundred more than fifty times their sum. What are the numbers?

KEY TERMS

Fourth law of exponents
Rational expression
Multiple
Factors
Factoring
Greatest common factor (GCF)

Greatest monomial factor
Difference of two squares
Perfect square
Factoring by grouping
Quadratic polynomial
Quadratic equation

REVIEW EXERCISES

Perform the indicated operations and simplify, using the laws of exponents.

1. (a) $\dfrac{32x^2y}{8xy}$ (b) $78ab^3 \div 13ab^2$
2. (a) $15x^4y^3 \div 5xy^2$ (b) $-105a^4b^9 \div 15a^3b^5$
3. (a) $\dfrac{27m^{-2}n^3}{81m^{-1}n^{-3}}$ (b) $\dfrac{99x^3yz^{-3}}{22x^2y^{-2}z}$
4. (a) $(42r^3s^2 - 28r^2s) \div (14rs)$
   (b) $(-2.4x^6y^5 + 1.8x^4y^3 - 1.2x^2y^2) \div (0.6x^2y)$
5. (a) $\dfrac{96m^3n^4p^2 - 88m^2n^2p^2 + 40m^2p}{-16m^2p}$ (b) $\dfrac{3.5r^5s^3 + 2.1rs^3t - 4.9rst^2}{0.7r^2st}$

In each exercise below, carry out the indicated division.

6. (a) $(5x^2 - 16x + 3) \div (5x - 1)$ (b) $(19y - 10 + 2y^2) \div (2y - 1)$
7. (a) $(6m^2 + m - 12) \div (3 + 2m)$ (b) $(b^4 - 16) \div (b^2 + 4)$

8. (a) $(125y^6 - 64) \div (5y^2 - 4)$
   (b) $(x^4 - 2x^3 - 1 - 4x^5) \div (2x^3 - 2x^2 + 1)$

Find the greatest monomial factor of each polynomial below, and express the polynomial in factored form.

9. (a) $6x^5 + 12x^4 - 18x^3 + 42x^2$
   (b) $3.6y^5z^2 - 2.4y^4z^3 + 1.6y^3z^4 - 6.4y^2z^5$
10. (a) $25a^3b^2c - 15a^3b^3c^2 + 20ab^2c$ (b) $\frac{2}{9}m^5n^2p - \frac{1}{9}m^3np^2 + \frac{7}{9}m^2n^2p^3$

Where possible, factor each of the following polynomials completely.

11. (a) $4x^2 - 20x + 25$ (b) $121y^2 - 144$
12. (a) $72x^2y - 32y^3$ (b) $9x^2 + 14x + 25$
13. (a) $15a^2 + 7a - 4$ (b) $64x^2yz^3 - 48xy^2z^3 + 9y^3z^3$
14. (a) $5x^2 + 18x + 9$ (b) $6x^2 + 7x + 2$
15. (a) $5y^2 - 7y - 6$ (b) $6a^2 - 216b^2$
16. (a) $49x^2 + 144$ (b) $5x^2 + 55x + 150$
17. (a) $-8x^2 - 47x + 6$ (b) $15a^2b^2 - 7ab^2 - 4b^2$
18. (a) $x^5 - x$ (b) $48y^3 + 147y$
19. (a) $y^4 - 26y^3 + 160y^2$ (b) $7x^2 - 7$
20. (a) $121 - 196a^2$ (b) $28 + 3y - 18y^2$

Solve each of the following quadratic equations, and check your answers.

21. (a) $x^2 + 3x + 2 = 0$ (b) $y^2 = 3y + 4$
22. (a) $4z^2 = 5z + 6$ (b) $6x^2 + 2 = 7x$
23. (a) $y(y - 3) = 4(y + 2)$ (b) $x - 2x^2 = (2x)^2 - 1$
24. (a) $x^2 + 3 = 2x + 6$ (b) $3x^2 - 2x - 4 = x^2 - 3x + 2$

Solve by using a quadratic equation.

25. (a) Frieda, addicted to number puzzles, presented her long-suffering twin sister, Gerta, with yet another mind boggler. It was: "Suppose I said I was thinking of three consecutive even integers such that double the sum of all three of them is the same as half the product of the smaller two. Could you tell me the integers?" Her sister, used to these questions by now, had perfected her algebra so that she could answer almost immediately. What was her answer?

    (b) Tired of always being on the receiving end of the number riddles, Gerta gave this one back to Frieda: "Can you find two consecutive positive integers such that 3 times their sum is 6 less than their product?" Frieda couldn't do it! Can you?

26. (a) An object's vertical position at time t, in seconds, is given by $-2t^2 + 9t - 7$. Determine when the object leaves the ground and when it returns.

(b) The height of an object t seconds after being propelled by an initial velocity v_0 is $v_0 t - 16t^2$. If a ball is thrown upward with an initial velocity of 96 feet per second, how long does it take the ball to reach a height of 80 feet on the way up and on the way down?

Calculator Activities

Perform the indicated operations and simplify, using the laws of exponents.

1. (a) $\dfrac{1517x^{-2}y^{15}}{37x^{-12}y^{-3}}$ (b) $\dfrac{6.92a^5b^4c^{-3}}{36.676a^{-1}bc^2}$

2. (a) $\dfrac{804m^4n^2p - 335m^2np^2}{67m^3np}$

   (b) $\dfrac{-9.702xyz^5 + 11.781xy^2z^4 - 19.635xy^3z^2}{-2.31xyz^2}$

3. Carry out the division as indicated. Check your answer.

   (a) $(243x^2 + 513x - 360) \div (27x - 15)$
   (b) $(1.3x^3 - 0.38 + 2.0x - 0.76x^2) \div (x - 0.2)$

Factor as completely as possible.

4. (a) $387x^2 - 1075y^2$ (b) $289x^2y^2 - 782x^2y + 529x^2$
5. (a) $1.89m^2 + 2.04m - 0.33$ (b) $2.4x^2 - 3.7x - 0.5$
6. Solve the following quadratic equations.
   (a) $288t - 96t^2 = 0$ (b) $147x^2 = 60 - 217x$

Solve by using a quadratic equation.

7. (a) Sofia and Leo want to build a patio area around their new rectangular swimming pool. The pool has an area of 108 square meters and the length is three times the width. If the surrounding patio area is to have a constant width and an area of 112 square meters, find the width of the patio.

   (b) A ball thrown up with an initial velocity of 88 feet per second is observed at a height of 96 feet twice. If the height of the ball with this initial velocity is shown by $88t - 16t^2$, at what two times t (on the way up and on the way down) is the ball at a height of 96 feet?

10 Rational Expressions

Adrian Perenon, after figuring how much paint he will need to paint his house (Chapter 1), realized that it would take him approximately 16 hours to do the job. If his nephew, Dave McClendon, helps him, and they are able to finish in only 9 hours, how long would it have taken Dave working alone?

10-1 SIMPLIFYING RATIONAL EXPRESSIONS

A quotient of two integers, such as $\frac{3}{8}$ or $-\frac{4}{7}$, is called a rational number. A quotient of two polynomials is called a **rational expression.** For example,

$$\frac{2y}{x}, \quad \frac{x^2 - 3x + 4}{x + 2}, \quad \frac{a^2 + b^2}{2a + 3b}$$

are rational expressions.

You work with rational expressions in much the same way you work with rational numbers. Recall that two nonzero fractions represent the same rational number if they are related as follows:

$$\frac{a}{b} = \frac{c}{d} \quad \text{if} \quad a \cdot d = b \cdot c$$

For example,

$$\frac{14}{21} = \frac{2}{3} \quad \text{because} \quad 14 \cdot 3 = 21 \cdot 2 \quad \blacksquare = 42$$

You can also tell if two nonzero rational expressions are equal:

$$\frac{A}{B} = \frac{C}{D} \quad \text{if} \quad A \cdot D = B \cdot C \quad A, B, C, D \text{ polynomials}$$

For example,

$$\frac{x^2 y}{xy^2} = \frac{x}{y} \quad \text{because} \quad (x^2 y) \cdot y = (xy^2) \cdot x \quad \blacksquare = x^2 y^2$$

As another example,

$$\frac{x^2 + 2x}{x^2 + 3x} = \frac{x+2}{x+3}$$

because

$$(x^2 + 2x)(x+3) = (x^2 + 3x)(x+2) \quad \blacksquare = x(x+2)(x+3)$$

Every nonzero rational number can be expressed in **simplest form,** that is, in the form

$$\frac{a}{b} \quad \text{where GCF } (a, b) = 1 \quad \begin{array}{l} GCF(a,b) = 1 \text{ means that the greatest} \\ \text{common factor of } a \text{ and } b \text{ is } 1 \end{array}$$

You can find the simplest form of a rational number c/d by first finding the greatest common factor (GCF) of integers c and d. If the GCF of c and d is k, then

$$c = ak, \quad d = bk$$

for some integers a and b. Since

$$\frac{ak}{bk} = \frac{a}{b} \quad \text{Because } (ak)b = (bk)a = abk$$

and the GCF of a and b is 1, a/b is the simplest form of c/d.

For example, you find the simplest form of

$$\frac{20}{35}$$

by first finding the GCF of the integers 20 and 35. Isn't it 5? Thus

$$\frac{20}{35} = \frac{4 \cdot 5}{7 \cdot 5} = \frac{4}{7} \quad \text{The GCF of 4 and 7 equals 1}$$

Thus, $\frac{4}{7}$ is the simplest form of the rational number $\frac{20}{35}$.

Similar statements can be made for rational expressions. Thus, every nonzero rational expression C/D has a simplest form A/B with

10-1: SIMPLIFYING RATIONAL EXPRESSIONS

a GCF of A and B equal to 1. To find the simplest form of C/D, first find the GCF of the two polynomials C and D. If the GCF is P, then

$$C = AP, \quad D = BP$$

for two polynomials A and B. Now

$$\frac{AP}{BP} = \frac{A}{B} \quad \text{Because } (AP)B = (BP)A = ABP$$

with the GCF of A and B equal to 1. Thus, A/B is the simplest form of C/D.

In practice, an easy way to find the simplest form of a rational expression is to factor the numerator and denominator as completely as possible. Then the GCF of the numerator and denominator will be the product of the common factors. For example,

$$\frac{x^2 y}{xy^2} = \frac{x \cdot x \cdot y}{x \cdot y \cdot y} \quad \text{Numerator and denominator factored completely}$$

$$= \frac{x \cdot (xy)}{y \cdot (xy)} \quad \text{GCF of numerator and denominator is } xy$$

$$= \frac{x}{y} \quad \text{Simplest form}$$

Thus,

$$\frac{x^2 y}{xy^2} = \frac{x}{y} \quad \text{Simplest form}$$

Problem Find the simplest form of each rational expression.

(a) $\dfrac{12a^3 b^4}{20a^4 b}$ (b) $\dfrac{x^3 - 5x^2}{3x^2 + 2x^4}$ (c) $\dfrac{x^2 + x - 6}{x^2 + 4x + 3}$

SOLUTION:

(a) $\dfrac{12a^3 b^4}{20a^4 b} = \dfrac{2^2 \cdot 3 \cdot a^3 \cdot b^4}{2^2 \cdot 5 \cdot a^4 \cdot b} = \dfrac{3 \cdot b^3 \cdot \overbrace{2^2 \cdot a^3 \cdot b}^{\text{GCF}}}{5 \cdot a \cdot 2^2 \cdot a^3 \cdot b} = \dfrac{3b^3}{5a}$ ← simplest form

(b) $\dfrac{x^3 - 5x^2}{3x^2 + 2x^4} = \dfrac{(x-5) \cdot \overbrace{x^2}^{\text{GCF}}}{(3 + 2x^2) \cdot x^2} = \dfrac{x-5}{3 + 2x^2}$ ←

(c) $\dfrac{x^2 + x - 6}{x^2 + 4x + 3} = \dfrac{(x-2) \cdot \overbrace{(x+3)}^{\text{GCF}}}{(x+1) \cdot (x+3)} = \dfrac{x-2}{x+1}$ ←

CHAPTER 10: RATIONAL EXPRESSIONS

Quick Reinforcement

Find the simplest form of each rational expression.

(a) $\dfrac{30x^5w^3}{42x^2w^6}$ (b) $\dfrac{5m-10m^2}{5m^2}$ (c) $\dfrac{2y^2-y-3}{y^2-1}$

Answers (a) $\dfrac{5x^3}{7w^3}$ (b) $\dfrac{1-2m}{m}$ (c) $\dfrac{2y-3}{y-1}$

EXERCISES 10-1

Put each rational expression in simplest form.

1. (a) $\dfrac{75x^5}{25x^3}$ (b) $\dfrac{3ab}{-b}$

2. (a) $\dfrac{-400a^3b^4}{-80ab}$ (b) $\dfrac{24x^3y^2}{-6x^2y^4z}$

3. (a) $\dfrac{4m-16}{4m}$ (b) $\dfrac{25xy+30x^2y}{5xy}$

4. (a) $\dfrac{8y^2-12y}{4y}$ (b) $\dfrac{3x^2+18x}{6xy}$

5. (a) $\dfrac{x^2-x}{x^2+2x}$ (b) $\dfrac{3x+2x^2}{2x-x^2}$

6. (a) $\dfrac{x^2+xy}{xy+y^2}$ (b) $\dfrac{3m^2-2mn}{3mn-2n^2}$

7. (a) $\dfrac{x^2-1}{xy+y}$ (b) $\dfrac{4a+4b}{2a^2+2b^2}$

8. (a) $\dfrac{6x^2+9x}{3x^2-12x}$ (b) $\dfrac{n^3-2n^2}{n^2-4n+4}$

9. (a) $\dfrac{y^2-4}{y^2-4y+4}$ (b) $\dfrac{m^2-4mn+3n^2}{m^2-7mn+12n^2}$

10. (a) $\dfrac{x^2-16y^2}{4xy-16y^2}$ (b) $\dfrac{9y^2-4}{3y^2+7y-6}$

11. (a) $\dfrac{a^2b-8ab+15b}{ab-3b}$ (b) $\dfrac{6x^2-7xy+2y^2}{3x^2+xy-2y^2}$

12. (a) $\dfrac{a^2x^4-a^2y^4}{x^4+x^2y^2}$ (b) $\dfrac{6a^2-11a+4}{2a^2+3a-2}$

13. (a) $\dfrac{1-2x}{18xy-36x^2y}$ (b) $\dfrac{x^2+7xy+12y^2}{x^2-xy-20y^2}$

14. (a) $\dfrac{a^2-ab-2b^2}{a^2-3ab+2b^2}$ (b) $\dfrac{6y-3}{2y^2-15y+7}$

15. (a) $\dfrac{7-x}{x^2-49}$ (b) $\dfrac{c^2-2d^2}{2d^2-c^2}$

 Calculator Activities

Put each rational expression in simplest form.

1. (a) $\dfrac{44a-242}{33b-121}$ (b) $\dfrac{45x^2-43x-156}{50x-120}$

2. (a) $\dfrac{208y^2-844y+748}{169y^2-289}$ (b) $\dfrac{69xa-95yb-15ay+437xb}{9a^2+114a+361}$

10-2 MULTIPLYING AND DIVIDING RATIONAL EXPRESSIONS

You recall how to multiply and divide rational numbers:

$$\dfrac{2}{3}\cdot\dfrac{4}{7}=\dfrac{2\cdot 4}{3\cdot 7}=\dfrac{8}{21} \qquad \blacksquare\ \dfrac{a}{b}\cdot\dfrac{c}{d}=\dfrac{a\cdot c}{b\cdot d}$$

$$\dfrac{2}{3}\div\dfrac{4}{7}=\dfrac{2}{3}\cdot\dfrac{7}{4} \qquad \dfrac{a}{b}\div\dfrac{c}{d}=\dfrac{a}{b}\cdot\dfrac{d}{c}=\dfrac{a\cdot d}{b\cdot c},\ \dfrac{c}{d}\ne 0;$$

$$=\dfrac{14}{12},\ \text{or}\ \dfrac{7}{6} \qquad \dfrac{d}{c}\ \text{is called the reciprocal of}\ \dfrac{c}{d}$$

Rational expressions are multiplied and divided in the same way. For example,

$$\dfrac{x^2}{x+2}\cdot\dfrac{x-2}{x}=\dfrac{x^2\cdot(x-2)}{(x+2)\cdot x} \qquad GCF(\text{numerator, denominator})=x$$

You can express this product in the form

$$=\dfrac{x}{x}\cdot\dfrac{x(x-2)}{x+2}=\dfrac{x(x-2)}{x+2} \qquad \blacksquare\ \dfrac{x}{x}=1$$

Thus,

$$\dfrac{x^2}{x+2}\cdot\dfrac{x-2}{x}=\dfrac{x(x-2)}{x+2} \qquad \textit{Simplest form}$$

In words, the product of two rational expressions is the product of their numerators divided by the product of their denominators. It is a good idea to factor each numerator and denominator before you multiply. If the product is in factored form, it can easily be reduced to simplest form.

CHAPTER 10: RATIONAL EXPRESSIONS

Problem 1 Find each product and simplify.

(a) $\dfrac{x^3}{3y^2} \cdot \dfrac{18y}{x^4}$
(b) $\dfrac{a^3 b^2}{a-1} \cdot \dfrac{a^2-1}{a^2 b^3}$

SOLUTION:

(a) $\dfrac{x^3}{3y^2} \cdot \dfrac{18y}{x^4} = \dfrac{x^3 \cdot 18y}{3y^2 \cdot x^4} = \dfrac{2 \cdot 3^2 \cdot x^3 \cdot y}{3 \cdot x^4 \cdot y^2}$,

$$= \dfrac{3 \cdot 2 \cdot 3 \cdot x^3 \cdot y}{x \cdot y \cdot 3 \cdot x^3 \cdot y} = \dfrac{3 \cdot 2}{x \cdot y} \cdot \dfrac{3 \cdot x^3 \cdot y}{3 \cdot x^3 \cdot y}$$

$$= \dfrac{3 \cdot 2}{x \cdot y} = \dfrac{6}{xy}$$

Thus,

$$\dfrac{x^3}{3y^2} \cdot \dfrac{18y}{x^4} = \dfrac{6}{xy} \qquad \text{Simplest form}$$

(b) $\dfrac{a^3 b^2}{a-1} \cdot \dfrac{a^2-1}{a^2 b^3} = \dfrac{a^3 b^2 \cdot (a+1)(a-1)}{(a-1) \cdot a^2 b^3}$

$$= \dfrac{a \cdot (a+1) \cdot a^2 \cdot b^2 \cdot (a-1)}{b \cdot a^2 \cdot b^2 \cdot (a-1)}$$

$$= \dfrac{a \cdot (a+1)}{b} \cdot \dfrac{a^2 \cdot b^2 \cdot (a-1)}{a^2 \cdot b^2 \cdot (a-1)}$$

$$= \dfrac{a(a+1)}{b}$$

Thus,

$$\dfrac{a^3 b^2}{a-1} \cdot \dfrac{a^2-1}{a^2 b^3} = \dfrac{a(a+1)}{b} \qquad \text{Simplest form}$$

Each nonzero rational expression has a reciprocal:

$\dfrac{2x-3y}{x+y}$ has reciprocal $\dfrac{x+y}{2x-3y}$

$\dfrac{a^2-3a+1}{a+2}$ has reciprocal $\dfrac{a+2}{a^2-3a+1}$

The product of a rational expression and its reciprocal is 1:

$$\dfrac{2x-3y}{x+y} \cdot \dfrac{x+y}{2x-3y} = 1 \quad \blacksquare = \dfrac{x+y}{x+y} \cdot \dfrac{2x-3y}{2x-3y} = 1 \cdot 1 = 1$$

$$\dfrac{a^2-3a+1}{a+2} \cdot \dfrac{a+2}{a^2-3a+1} = 1$$

You divide one rational expression by a second by multiplying the first one by the reciprocal of the second one. Before you actually divide, factor each numerator and denominator. Then the quotient, being in factored form, can easily be reduced to simplest form. For example,

$$\frac{x(x+y)}{x-1} \div \frac{(x+y)^2}{x^2-1} = \frac{x(x+y)}{x-1} \cdot \frac{x^2-1}{(x+y)^2}$$

$$= \frac{x+y}{x+y} \cdot \frac{x-1}{x-1} \cdot \frac{x(x+1)}{x+y} \qquad \blacksquare \ x^2 - 1 = (x+1)(x-1)$$

$$= \frac{x(x+1)}{x+y}$$

Thus,

$$\frac{x(x+y)}{x-1} \div \frac{(x+y)^2}{x^2-1} = \frac{x(x+1)}{x+y} \qquad \textit{Simplest form}$$

Problem 2 Find each quotient and simplify.

(a) $\dfrac{12y^6}{a^5} \div \dfrac{18y^3}{5a^2}$

(b) $\dfrac{x^2+x-6}{x^2-x-6} \div \dfrac{x^2-9}{x^2-4}$

(c) $\dfrac{\dfrac{x^2-x}{x+1}}{\dfrac{x^2}{x^2-1}}$

SOLUTION:

(a) $\dfrac{12y^6}{a^5} \div \dfrac{18y^3}{5a^2} = \dfrac{12y^6}{a^5} \cdot \dfrac{5a^2}{18y^3} = \dfrac{2^2 \cdot 3 \cdot 5 \cdot a^2 \cdot y^6}{2 \cdot 3^2 \cdot a^5 \cdot y^3}$

$$= \frac{2 \cdot 5 \cdot y^3 \cdot 2 \cdot 3 \cdot a^2 \cdot y^3}{3 \cdot a^3 \cdot 2 \cdot 3 \cdot a^2 \cdot y^3}$$

$$= \frac{2 \cdot 5 \cdot y^3}{3 \cdot a^3} \cdot \frac{2 \cdot 3 \cdot a^2 \cdot y^3}{2 \cdot 3 \cdot a^2 \cdot y^3}$$

$$= \frac{2 \cdot 5 \cdot y^3}{3 \cdot a^3} = \frac{10y^3}{3a^3}$$

Thus,

$$\frac{12y^6}{a^5} \div \frac{18y^3}{5a^2} = \frac{10y^3}{3a^3} \qquad \textit{Simplest form}$$

(b) Each of the polynomials can be factored:

$$x^2 + x - 6 = (x-2)(x+3), \qquad x^2 - x - 6 = (x+2)(x-3)$$
$$x^2 - 9 = (x+3)(x-3), \qquad x^2 - 4 = (x+2)(x-2)$$

Then

$$\frac{x^2+x-6}{x^2-x-6} \div \frac{x^2-9}{x^2-4} = \frac{(x-2)(x+3)}{(x+2)(x-3)} \cdot \frac{(x+2)(x-2)}{(x+3)(x-3)}$$

$$= \frac{(x+3) \cdot (x+2)}{(x+3) \cdot (x+2)} \cdot \frac{(x-2)^2}{(x-3)^2}$$

$$= \frac{(x-2)^2}{(x-3)^2}$$

Thus,

$$\frac{x^2+x-6}{x^2-x-6} \div \frac{x^2-9}{x^2-4} = \frac{(x-2)^2}{(x-3)^2} \qquad \textit{Simplest form}$$

(c) $\dfrac{\dfrac{x^2-x}{x+1}}{\dfrac{x^2}{x^2-1}} = \dfrac{x^2-x}{x+1} \div \dfrac{x^2}{x^2-1}$ *By definition*

$$= \frac{x(x-1)}{x+1} \cdot \frac{(x-1)(x+1)}{x^2} \qquad \blacksquare \; x^2 - x = x(x-1);$$
$$x^2 - 1 = (x-1)(x+1)$$

$$= \frac{x(x-1)(x-1)(x+1)}{(x+1)x^2} \qquad GCF(num., denom.) = x(x+1)$$

$$= \frac{(x-1)^2}{x} \qquad \textit{Simplest form}$$

Quick Reinforcement

Perform the indicated operation and simplify.

(a) $\dfrac{5x^2y}{x-2} \cdot \dfrac{x^2-4}{xy}$

(b) $\dfrac{2x^2+xy-y^2}{x^3y^2} \div \dfrac{4x+4y}{2xy}$

Answers (a) $5x(x+2)$ (b) $\dfrac{2x-y}{x^2y}$

EXERCISES 10-2

Carry out the indicated operation for each exercise. Simplify your answer.

💡 1. (a) $\dfrac{5}{3a^2b} \cdot \dfrac{21b^2}{7a}$ \qquad (b) $\dfrac{2}{x} \cdot \dfrac{x^3}{8}$

2. (a) $\dfrac{11}{5mn^2} \cdot \dfrac{15mn}{44}$ \qquad (b) $\dfrac{32a^3b}{15c} \cdot \dfrac{45c^2}{8a^2b}$

3. (a) $\dfrac{14x^2}{9y^2} \div \dfrac{25x^2}{36y^2}$ (b) $\dfrac{-10ab^2}{4x^2yz} \div \dfrac{35a^3b^3}{-12xy}$

4. (a) $\dfrac{39x^2y^5}{-12ab} \div \dfrac{-13x^3y^4}{4a^2b}$ (b) $\dfrac{75m^3}{14n^2} \div \dfrac{25mn}{49}$

5. (a) $\dfrac{x^2y}{3xy^2} \cdot \dfrac{3x^2}{y^2}$ (b) $\dfrac{4x^3y^4}{5xy} \cdot \dfrac{10x^2}{2y^3}$

6. (a) $\dfrac{2x+2y}{5} \cdot \dfrac{10}{x+y}$ (b) $\dfrac{-5x^3y}{35xy^3} \cdot \dfrac{63x^2y}{45xy^2}$

7. (a) $\dfrac{2m-3}{m+3} \div \dfrac{5m-2}{m+4}$ (b) $\dfrac{y^2-2y}{y+2} \cdot \dfrac{3y+6}{y-2}$

8. (a) $\dfrac{2a-1}{6a-3} \cdot \dfrac{6a+18}{2a-6}$ (b) $\dfrac{6x-6y}{4x^2-y^2} \cdot \dfrac{6x+y}{3}$

9. (a) $\dfrac{t-2}{4s} \div \dfrac{t^2-t-6}{12s^2}$ (b) $\dfrac{x+y}{x^2-y^2} \div \dfrac{x^2-xy}{x^2-2xy+y^2}$

10. (a) $\dfrac{3x^2+6x-24}{x^2-4x+4} \cdot \dfrac{x^2+5x-14}{x^2+11x+28}$ (b) $\dfrac{x^2-6x+9}{x^2-4x+4} \div \dfrac{x^2-5x+6}{x^2-x-2}$

11. (a) $\dfrac{21x^2-20xy-y^2}{x^2+xy-2y^2} \div \dfrac{21x^2+22xy+y^2}{x^2+xy-2y^2}$

   (b) $\dfrac{y^2-4}{y^2-4y} \div \dfrac{y+2}{3}$

12. (a) $\dfrac{t^2-t-6}{t^2+t-12} \div \dfrac{t^2+2t-3}{t^2+3t-4}$

   (b) $\dfrac{a^2+a-2}{a^2+3a-4} \div \dfrac{a+2}{a+3}$

13. (a) $\dfrac{a^2-6a+9}{a^2-a-6} \div \dfrac{a^2+2a-15}{a^2+2a}$

   (b) $\dfrac{3p-3r}{p^2+pr} \cdot \dfrac{pr+r^2}{3p+3r}$

14. (a) $\dfrac{2a^2-13a+15}{2a^2-a-3} \cdot \dfrac{4a^2-a-5}{3a^2-17a+10}$

   (b) $\dfrac{16-x^2}{x^2+2x-8} \div \dfrac{x^2-2x-8}{4-x^2}$

15. (a) $\dfrac{\dfrac{x^2-3x+2}{x^2-1}}{\dfrac{2-x}{1-x}}$

   (b) $\dfrac{a^2+5a+6}{(a+3)(3a^2+8a+4)} \cdot \dfrac{6a^2-11a-10}{2a^2+a-15}$

16. (a) $\dfrac{6r^2}{4r^2t - 12rt} \cdot \dfrac{r^2 + r - 12}{3r^2 + 12r}$

(b) $\dfrac{\dfrac{2x^2 + 11x + 12}{2x - 7}}{\dfrac{x^2 + 7x + 12}{x - 5}}$

17. (a) $\dfrac{3x^2 - 3x}{5x - 40} \cdot \dfrac{x^2}{3} \div \dfrac{4 - 4x}{x^2 - 7x - 8}$

(b) $\left[\dfrac{x - y}{z - x} \div \dfrac{z - y}{x - z}\right] \cdot \dfrac{y - z}{y - x}$

Calculator Activities

Perform the indicated operations and simplify.

1. (a) $\dfrac{84x^2 - 335x + 75}{3x^2y} \cdot \dfrac{144x^2y^2}{52x^2 - 151x - 165}$

(b) $\dfrac{196u^2z - 169v^2z}{126y^2} \div \dfrac{196u^2 + 364uv + 169v^2}{9z^2y^2}$

2. (a) $\dfrac{208x^3 - 2060x^2 - 200x}{256x^4y^4} \cdot \dfrac{256x - 24}{1664x^2 + 4x - 15}$

(b) $\left[(120xy - 88zy) \cdot \dfrac{8x + 13z}{75x^2 - 190xz + 99z^2}\right] \div \dfrac{8y^2x + 13y^2z}{5x - 9z}$

10-3 ADDING AND SUBTRACTING RATIONAL EXPRESSIONS

You add and subtract rational expressions having the same denominator just as you do rational numbers.

$\dfrac{4}{8} + \dfrac{3}{8} = \dfrac{7}{8}$ Add numerators when denominators are equal

$\dfrac{4}{8} - \dfrac{3}{8} = \dfrac{1}{8}$ Subtract numerators when denominators are equal

$\dfrac{3x}{x + y} + \dfrac{4y}{x + y} = \dfrac{3x + 4y}{x + y}$ Add numerators when denominators are equal

$\dfrac{a + 5b}{3a} - \dfrac{2b}{3a} = \dfrac{a + 5b - 2b}{3a}$ Subtract numerators when denominators are equal

$= \dfrac{a + 3b}{3a}$

To add or subtract two rational numbers having different denominators, first find two equivalent fractions having the same denominators. One way of doing this is illustrated below.

$$\frac{3}{4}+\frac{5}{6}=\frac{3\cdot 6}{4\cdot 6}+\frac{5\cdot 4}{6\cdot 4} \quad \textit{Use the product } 4\cdot 6 \textit{ as the same denominator}$$

$$=\frac{18}{24}+\frac{20}{24}$$

$$=\frac{38}{24}, \text{ or } \frac{19}{12} \quad \textit{Simplifying } \frac{38}{24}$$

Thus

$$\frac{3}{4}+\frac{5}{6}=\frac{19}{12}$$

As another example,

$$\frac{2}{3}-\frac{3}{10}=\frac{2\cdot 10}{3\cdot 10}-\frac{3\cdot 3}{10\cdot 3} \quad \textit{Use the product } 3\cdot 10 \textit{ as the same denominator}$$

$$=\frac{20}{30}-\frac{9}{30}$$

$$=\frac{11}{30}$$

That is,

$$\frac{2}{3}-\frac{3}{10}=\frac{11}{30}$$

There are common denominators other than the products used above. For example, another common denominator of $\frac{3}{4}$ and $\frac{5}{6}$ is 12;

$$\frac{3}{4}+\frac{5}{6}=\frac{3\cdot 3}{4\cdot 3}+\frac{5\cdot 2}{6\cdot 2} \quad 12 \textit{ is a multiple of both } 4 \textit{ and } 6$$

$$=\frac{9}{12}+\frac{10}{12}=\frac{19}{12}$$

Evidently, 12 is the **least common denominator (LCD)** of the two fractions above; that is, there is no positive integer less than 12 that is a multiple of both 4 and 6.

One way to find the LCD of two fractions is to factor each denominator into primes. Then the maximum power of each prime appearing in either denominator is the power of the prime appearing in the LCD. For example,

$$\frac{5}{18}-\frac{1}{24}=\frac{5\cdot 4}{18\cdot 4}-\frac{1\cdot 3}{24\cdot 3} \quad \blacksquare\ 18=2\cdot 3^2,\ 24=2^3\cdot 3\ (2 \textit{ and } 3 \textit{ are primes});$$
$$LCD = 2^3\cdot 3^2,\textit{ or } 72$$

$$=\frac{20}{72}-\frac{3}{72}=\frac{17}{72}$$

Thus,

$$\frac{5}{18} - \frac{1}{24} = \frac{17}{72}$$

Problem 1 Perform each operation.

(a) $\dfrac{7}{12} - \dfrac{5}{14}$ \hspace{2em} (b) $\dfrac{4}{45} + \dfrac{3}{175}$

SOLUTION:

(a) $\dfrac{7}{12} - \dfrac{5}{14} = \dfrac{7}{12} \cdot \dfrac{7}{7} - \dfrac{5}{14} \cdot \dfrac{6}{6}$ ■ $12 = 2^2 \cdot 3$, $14 = 2 \cdot 7$, so $LCD = 2^2 \cdot 3 \cdot 7$, or 84; 2, 3, and 7 are primes

$= \dfrac{49}{84} - \dfrac{30}{84}$

$= \dfrac{19}{84}$

Thus,

$$\frac{7}{12} - \frac{5}{14} = \frac{19}{84}$$

(b) $\dfrac{4}{45} + \dfrac{3}{175} = \dfrac{4}{45} \cdot \dfrac{35}{35} + \dfrac{3}{175} \cdot \dfrac{9}{9}$ ■ $45 = 3^2 \cdot 5$, $175 = 5^2 \cdot 7$, so $LCD = 3^2 \cdot 5^2 \cdot 7$, or 1575; 3, 5, and 7 are primes

$= \dfrac{140}{1575} + \dfrac{27}{1575}$

$= \dfrac{167}{1575}$

Thus,

$$\frac{4}{45} + \frac{3}{175} = \frac{167}{1575}$$

Using the same approach, you add or subtract two rational expressions having different denominators by finding two equivalent rational expressions having a common denominator. For example, add $3/x$ and $2/y$:

$\dfrac{3}{x} + \dfrac{2}{y} = \dfrac{3}{x} \cdot \dfrac{y}{y} + \dfrac{2}{y} \cdot \dfrac{x}{x}$ *Use the product xy as a common denominator*

$= \dfrac{3y}{xy} + \dfrac{2x}{xy} = \dfrac{3y + 2x}{xy}$

Thus,

$$\frac{3}{x} + \frac{2}{y} = \frac{3y + 2x}{xy}$$

As a second example, subtract $1/2x$ from $5/x$:

$$\frac{5}{x} - \frac{1}{2x} = \frac{2 \cdot 5}{2 \cdot x} - \frac{1}{2x} \qquad \text{$2x$ is a common denominator; } \frac{5}{x} = \frac{2 \cdot 5}{2 \cdot x}$$

$$= \frac{10 - 1}{2x} = \frac{9}{2x}$$

Thus,

$$\frac{5}{x} - \frac{1}{2x} = \frac{9}{2x}$$

As a third example, subtract $\dfrac{x}{y + y^2}$ from $\dfrac{y}{x + xy}$.

$$\frac{y}{x + xy} - \frac{x}{y + y^2} = \frac{y}{x(1+y)} - \frac{x}{y(1+y)} \qquad \text{Factor denominators}$$

$$= \frac{y}{x(1+y)} \cdot \frac{y(1+y)}{y(1+y)} - \frac{x}{y(1+y)} \cdot \frac{x(1+y)}{x(1+y)}$$

Use the product $x(1+y) \cdot y(1+y)$ as a common denominator

$$= \frac{y^2(1+y) - x^2(1+y)}{xy(1+y)^2}$$

$$= \frac{(y^2 - x^2)(1+y)}{xy(1+y)^2} \qquad \text{Factor numerator; $(1+y)$ is a factor of each term}$$

$$= \frac{y^2 - x^2}{xy(1+y)} \cdot \frac{1+y}{1+y}$$

Thus,

$$\frac{y}{x + xy} - \frac{x}{y + y^2} = \frac{y^2 - x^2}{xy(1+y)} \qquad \text{Simplest form}$$

You could have used a common denominator that was "smaller" than the product, namely $xy(1+y)$, since it is a multiple of both denominators. Then

$$\frac{y}{x(1+y)} - \frac{x}{y(1+y)} = \frac{y}{x(1+y)} \cdot \frac{y}{y} - \frac{x}{y(1+y)} \cdot \frac{x}{x}$$

$$= \frac{y^2}{xy(1+y)} - \frac{x^2}{xy(1+y)}$$

$$= \frac{y^2 - x^2}{xy(1+y)} \qquad \text{Same answer as above}$$

In the example above, the common denominator $xy(1+y)$ is the **least common denominator** (LCD) in the sense that no common denominator can have fewer than three factors. For example, $y(1+y)$ is not a common denominator: it is not a multiple of $x(1+y)$. It may be shown that any two rational expressions have an LCD. You can find it by factoring each denominator into primes (that is, polynomials that cannot be further factored) and proceeding as you did with integers.

Problem 2 Find each sum.

(a) $\dfrac{x+y}{x-y} + \dfrac{x-y}{x+y}$

(b) $\dfrac{y-2}{y^2+2y} + \dfrac{y-3}{y^2+3y}$

SOLUTION:

(a)
$$\dfrac{x+y}{x-y} + \dfrac{x-y}{x+y} = \dfrac{(x+y)\cdot(x+y)}{(x-y)\cdot(x+y)} + \dfrac{(x-y)\cdot(x-y)}{(x+y)\cdot(x-y)} \qquad \text{The LCD is } (x-y)(x+y)$$

$$= \dfrac{x^2+2xy+y^2}{x^2-y^2} + \dfrac{x^2-2xy+y^2}{x^2-y^2} \qquad \text{Over common denominator}$$

$$= \dfrac{(x^2+2xy+y^2)+(x^2-2xy+y^2)}{x^2-y^2} \qquad \text{Add numerators}$$

$$= \dfrac{x^2+2xy+y^2+x^2-2xy+y^2}{x^2-y^2}$$

$$= \dfrac{2x^2+2y^2}{x^2-y^2}$$

Thus,

$$\dfrac{x+y}{x-y} + \dfrac{x-y}{x+y} = \dfrac{2x^2+2y^2}{x^2-y^2}$$

(b)
$$\dfrac{y-2}{y^2+2y} + \dfrac{y-3}{y^2+3y} = \dfrac{y-2}{y(y+2)} + \dfrac{y-3}{y(y+3)} \qquad \text{The LCD is } y(y+2)(y+3)$$

$$= \dfrac{(y-2)(y+3)}{y(y+2)(y+3)} + \dfrac{(y-3)(y+2)}{y(y+2)(y+3)} \qquad \text{Over common denominator}$$

$$= \dfrac{(y^2+y-6)+(y^2-y-6)}{y(y+2)(y+3)} \qquad \text{Add numerators}$$

$$= \dfrac{y^2+y-6+y^2-y-6}{y(y+2)(y+3)}$$

$$= \dfrac{2y^2-12}{y(y+2)(y+3)}$$

10-3: ADDING AND SUBTRACTING RATIONAL EXPRESSIONS

Thus,

$$\frac{y-2}{y^2+2y} + \frac{y-3}{y^2+3y} = \frac{2y^2-12}{y(y+2)(y+3)}$$

Problem 3 Find each difference.

(a) $\dfrac{a+2b}{a} - \dfrac{b-2a}{b}$ (b) $\dfrac{x-3}{x^2+x} - \dfrac{x}{x^2-1}$

(c) $\dfrac{4y}{y^2-1} - \dfrac{2}{y} - \dfrac{2}{y+1}$

SOLUTION:

(a) $\dfrac{a+2b}{a} - \dfrac{b-2a}{b} = \dfrac{(a+2b)\cdot b}{a\cdot b} - \dfrac{(b-2a)\cdot a}{a\cdot b}$ The LCD is $a \cdot b$

$\qquad = \dfrac{ab+2b^2}{ab} - \dfrac{ab-2a^2}{ab}$ Over common denominator

$\qquad = \dfrac{(ab+2b^2)-(ab-2a^2)}{ab}$ Subtract numerators

$\qquad = \dfrac{ab+2b^2-ab+2a^2}{ab}$

$\qquad = \dfrac{2b^2+2a^2}{ab}$

Thus,

$$\frac{a+2b}{a} - \frac{b-2a}{b} = \frac{2b^2+2a^2}{ab}$$

(b) $\dfrac{x-3}{x^2+x} - \dfrac{x}{x^2-1} = \dfrac{x-3}{x(x+1)} - \dfrac{x}{(x+1)(x-1)}$ The LCD is $x(x+1)(x-1)$

$\qquad = \dfrac{(x-3)\cdot(x-1)}{x(x+1)\cdot(x-1)} - \dfrac{x\cdot x}{(x+1)(x-1)\cdot x}$
  Over common denominator

$\qquad = \dfrac{x^2-4x+3}{x(x^2-1)} - \dfrac{x^2}{x(x^2-1)}$

$\qquad = \dfrac{x^2-4x+3-x^2}{x(x^2-1)}$ Subtract numerators

$\qquad = \dfrac{-4x+3}{x(x^2-1)}$

Thus,

$$\frac{x-3}{x^2+x} - \frac{x}{x^2-1} = \frac{-4x+3}{x(x^2-1)}$$

(c) A sum or difference of three or more rational expressions can be found in the same way.

$$\frac{4y}{y^2-1} - \frac{2}{y} - \frac{2}{y+1} = \frac{4y}{(y-1)(y+1)} \cdot \frac{y}{y} - \frac{2}{y} \cdot \frac{(y-1)(y+1)}{(y-1)(y+1)}$$

$$- \frac{2}{y+1} \cdot \frac{y(y-1)}{y(y-1)} \qquad \text{The LCD is } y(y-1)(y+1)$$

$$= \frac{4y^2}{y(y-1)(y+1)} - \frac{2(y^2-1)}{y(y-1)(y+1)}$$

$$- \frac{2(y^2-y)}{y(y-1)(y+1)}$$

$$= \frac{4y^2 - (2y^2 - 2) - (2y^2 - 2y)}{y(y-1)(y+1)}$$

$$= \frac{4y^2 - 2y^2 + 2 - 2y^2 + 2y}{y(y-1)(y+1)}$$

$$= \frac{2(y+1)}{y(y-1)(y+1)}$$

$$= \frac{2}{y(y-1)}$$

Thus,

$$\frac{4y}{y^2-1} - \frac{2}{y} - \frac{2}{y+1} = \frac{2}{y(y-1)} \qquad \textit{Simplified form}$$

Quick Reinforcement

Carry out the indicated operation.

(a) $\dfrac{4}{y} - \dfrac{y-5}{y}$ 　　(b) $\dfrac{2x}{x+3y} + \dfrac{6xy}{x^2-9y^2}$ 　　(c) $\dfrac{3x}{8x^2-2x-3} - \dfrac{4+x}{2x^2+x}$

Answers (a) $\dfrac{9-y}{y}$ (b) $\dfrac{(x+3y)(x-3y)}{2x^2}$ (c) $\dfrac{-x^2-13x+12}{x(4x-3)(2x+1)}$

EXERCISES 10-3

Carry out the indicated operation for each exercise. Simplify your answer.

1. (a) $\dfrac{2x}{9y} + \dfrac{5+3y}{9y}$ 　　　　　　　(b) $\dfrac{5y}{4x} + \dfrac{1-2x}{4x}$

10-3: ADDING AND SUBTRACTING RATIONAL EXPRESSIONS

2. (a) $\dfrac{3x}{7y} + \dfrac{6y}{x}$ (b) $\dfrac{7x}{3y} + \dfrac{2y}{5x}$

3. (a) $\dfrac{5x+2}{3xy} + \dfrac{2x+1}{2y}$ (b) $\dfrac{2x}{9y} - \dfrac{3-7x}{9y}$

4. (a) $\dfrac{5}{m-n} - \dfrac{4}{n-m}$ (b) $\dfrac{x-1}{x-2} - \dfrac{x-4}{x-3}$

5. (a) $\dfrac{5}{6t-12r} - \dfrac{4}{3t+6r}$ (b) $\dfrac{1}{x+y} + \dfrac{x}{x^2+y^2}$

6. (a) $\dfrac{a}{3b} + \dfrac{a+2}{2b}$ (b) $\dfrac{x}{x+y} - \dfrac{y}{x-y}$

7. (a) $\dfrac{1}{x+y} + \dfrac{y}{x^2-y^2}$ (b) $\dfrac{1}{x-1} - \dfrac{2}{x+1}$

8. (a) $\dfrac{2}{3x} - \dfrac{3}{x+3}$ (b) $\dfrac{3}{2y} - \dfrac{2}{y+2}$

9. (a) $\dfrac{x}{x-1} + \dfrac{2}{x^2-1}$ (b) $\dfrac{x+2}{x} + \dfrac{x}{x+2}$

10. (a) $\dfrac{2}{4x^2-16} + \dfrac{3}{4+2x}$ (b) $\dfrac{10y-18}{y-2} - \dfrac{9y-16}{2-y}$

11. (a) $\dfrac{y-3}{y+2} - \dfrac{y-2}{y+3}$ (b) $\dfrac{3a-1}{2a^2+a-3} - \dfrac{2}{a-1}$

12. (a) $\dfrac{3}{a-b} + \dfrac{4}{2a-b}$ (b) $\dfrac{x^2-4}{x} - \dfrac{x}{5}$

13. (a) $\dfrac{6}{x-y} + \dfrac{4x}{y^2-x^2}$ (b) $\dfrac{a+4}{a-2} + \dfrac{a-7}{a+5}$

14. (a) $\dfrac{1}{m} + \dfrac{2}{m+1} - \dfrac{1}{m^2}$ (b) $\dfrac{3x}{2x+1} + \dfrac{2x}{1-2x} - \dfrac{3}{4x^2-1}$

15. (a) $\dfrac{5}{6x^2} + \dfrac{5}{3x} - \dfrac{1}{9x^3}$ (b) $\dfrac{a+1}{a^2-36} - \dfrac{2a-5}{a^2-8a+12}$

16. (a) $\dfrac{9}{b-5} - \dfrac{7}{b^2-2b-15} + \dfrac{2}{b^2}$ (b) $\dfrac{3}{y^2+y} + \dfrac{2}{y-1} - \dfrac{y+3}{y^2-1}$

17. (a) $\dfrac{c+d}{cd^2} + \dfrac{3c+d}{c^2d}$ (b) $\dfrac{4a+2b}{3ab^2} - \dfrac{5a-3b}{a^2b}$

18. (a) $\dfrac{3m}{m^2-4m+4} + \dfrac{10}{m^2+m-6}$ (b) $\dfrac{5}{3-2y} - \dfrac{y-3}{2y^2-y-3} + \dfrac{3}{2y-3}$

Calculator Activities

1. (a) $\dfrac{13x-12}{5x^2-73x+42} + \dfrac{4x+29}{15x^2-210x}$

   (b) $\dfrac{8y-5x}{484y^2-220xy+25x^2} - \dfrac{22y+5x}{220y^2-28xy-5x^2}$

10-4 COMPLEX FRACTIONS

When a rational number or expression appears as a quotient of two other rational numbers or expressions, it is called a **complex fraction.** For example,

$$\frac{1+\frac{2}{3}}{\frac{4}{5}-\frac{2}{3}} \quad \text{and} \quad \frac{\frac{x}{y}-\frac{2}{x}}{3+\frac{2}{y^2}}$$

are complex fractions. Any such fraction is equivalent to one in the simple form as a quotient of two integers or polynomials. There are two possible ways to find this simple form. One way is to work with the numerator and denominator separately. For example,

$$\frac{1+\frac{2}{3}}{\frac{4}{5}-\frac{2}{3}} = \frac{\frac{3}{3}+\frac{2}{3}}{\frac{4}{5}\cdot\frac{3}{3}-\frac{2}{3}\cdot\frac{5}{5}} = \frac{\frac{5}{3}}{\frac{12}{15}-\frac{10}{15}} = \frac{\frac{5}{3}}{\frac{2}{15}}$$

$$= \frac{5}{3} \cdot \frac{15}{2} = \frac{5 \cdot 3 \cdot 5}{3 \cdot 2} = \frac{25}{2}$$

Thus,

$$\frac{1+\frac{2}{3}}{\frac{4}{5}-\frac{2}{3}} = \frac{25}{2}$$

The other method is to find the LCD of all the simple fractions that make up the complex fraction. Then multiply the numerator and denominator of the complex fraction by this LCD. For example, the complex fraction

$$\frac{\frac{x}{y}-\frac{2}{x}}{3+\frac{2}{y^2}}$$

is made up of the four simple fractions

$$\frac{x}{y}, \frac{2}{x}, 3, \frac{2}{y^2}$$

which have denominators y, x, 1, and y^2. The LCD is xy^2. Now multiply the numerator and denominator of the complex fraction by xy^2:

$$\frac{\frac{x}{y}-\frac{2}{x}}{3+\frac{2}{y^2}} = \frac{\left(\frac{x}{y}-\frac{2}{x}\right) \cdot xy^2}{\left(3+\frac{2}{y^2}\right) \cdot xy^2}$$

$$= \frac{\frac{x}{y} \cdot xy^2 - \frac{2}{x} \cdot xy^2}{3 \cdot xy^2 + \frac{2}{y^2} \cdot xy^2}$$

$$= \frac{x^2y - 2y^2}{3xy^2 + 2x} \qquad \textit{Simplest form}$$

Problem 1 Express each complex fraction in simple form.

(a) $\dfrac{\frac{2}{3} - \frac{1}{2}}{\frac{3}{8} + \frac{5}{6}}$
(b) $\dfrac{\frac{2}{a^2} + \frac{5}{ab}}{\frac{7a}{b^2} - \frac{2b}{a}}$
(c) $\dfrac{\frac{x}{x+1} - \frac{x-1}{x}}{\frac{x+1}{x} + \frac{x}{x-1}}$

SOLUTION:

(a) Use the second method. The simple fractions involved are

$$\frac{2}{3}, \frac{1}{2}, \frac{3}{8}, \frac{5}{6}$$

The denominators are 3, 2, 2^3, and $3 \cdot 2$, so the LCD is $3 \cdot 2^3$, or 24. Then

$$\frac{\frac{2}{3} - \frac{1}{2}}{\frac{3}{8} + \frac{5}{6}} = \frac{(\frac{2}{3} - \frac{1}{2}) \cdot 24}{(\frac{3}{8} + \frac{5}{6}) \cdot 24} = \frac{\frac{2}{3} \cdot 24 - \frac{1}{2} \cdot 24}{\frac{3}{8} \cdot 24 + \frac{5}{6} \cdot 24}$$

$$= \frac{16 - 12}{9 + 20} = \frac{4}{29}$$

Thus,

$$\frac{\frac{2}{3} - \frac{1}{2}}{\frac{3}{8} + \frac{5}{6}} = \frac{4}{29} \qquad \textit{Simplest form}$$

(b) Use the first method this time.

$$\frac{\frac{2}{a^2} + \frac{5}{ab}}{\frac{7a}{b^2} - \frac{2b}{a}} = \frac{\frac{2}{a^2} \cdot \frac{b}{b} + \frac{5}{ab} \cdot \frac{a}{a}}{\frac{7a}{b^2} \cdot \frac{a}{a} - \frac{2b}{a} \cdot \frac{b^2}{b^2}} = \frac{\frac{2b}{a^2b} + \frac{5a}{a^2b}}{\frac{7a^2}{ab^2} - \frac{2b^3}{ab^2}} \qquad \textit{LCD of numerator is } a^2b; \textit{LCD of denominator is } ab^2$$

$$= \frac{\dfrac{2b + 5a}{a^2b}}{\dfrac{7a^2 - 2b^3}{ab^2}} = \frac{2b + 5a}{a^2b} \cdot \frac{ab^2}{7a^2 - 2b^3}$$

$$= \frac{ab^2(2b + 5a)}{a^2b(7a^2 - 2b^3)} = \frac{b(2b + 5a)}{a(7a^2 - 2b^3)}$$

Thus,

$$\frac{\frac{2}{a^2}+\frac{5}{ab}}{\frac{7a}{b^2}-\frac{2b}{a}}=\frac{b(2b+5a)}{a(7a^2-2b^3)} \quad \text{Simplest form}$$

(c) Use the second method again. The simple fractions involved are

$$\frac{x}{x+1}, \quad \frac{x-1}{x}, \quad \frac{x+1}{x}, \quad \frac{x}{x-1}$$

The LCD is $x(x-1)(x+1)$. Then

$$\frac{\frac{x}{x+1}-\frac{x-1}{x}}{\frac{x+1}{x}+\frac{x}{x-1}}=\frac{\left(\frac{x}{x+1}-\frac{x-1}{x}\right)\cdot x(x-1)(x+1)}{\left(\frac{x+1}{x}+\frac{x}{x-1}\right)\cdot x(x-1)(x+1)}$$

$$=\frac{\frac{x}{x+1}\cdot x(x-1)(x+1)-\frac{x-1}{x}\cdot x(x-1)(x+1)}{\frac{x+1}{x}\cdot x(x-1)(x+1)+\frac{x}{x-1}\cdot x(x-1)(x+1)}$$

$$=\frac{x^2(x-1)-(x-1)^2(x+1)}{(x-1)(x+1)^2+x^2(x+1)}=\frac{(x-1)[x^2-(x-1)(x+1)]}{(x+1)[(x-1)(x+1)+x^2]}$$

$$=\frac{(x-1)[x^2-(x^2-1)]}{(x+1)(x^2-1+x^2)}=\frac{(x-1)(x^2-x^2+1)}{(x+1)(x^2-1+x^2)}$$

$$=\frac{x-1}{(x+1)(2x^2-1)}$$

Thus,

$$\frac{\frac{x}{x+1}-\frac{x-1}{x}}{\frac{x+1}{x}+\frac{x}{x-1}}=\frac{x-1}{(x+1)(2x^2-1)} \quad \text{Simplest form}$$

When simplifying complex fractions, use the method you feel most comfortable with.

Quick Reinforcement

Simplify each complex fraction.

(a) $\dfrac{\frac{4}{5}+\frac{1}{4}}{1-\frac{3}{10}}$

(b) $\dfrac{\frac{2}{x}-\frac{3}{x^2}}{5+\frac{1}{x}}$

Answers: (a) $\frac{3}{2}$ (b) $\frac{2x-3}{5x^2+x}$

EXERCISES 10-4

Simplify each complex fraction.

1. (a) $\dfrac{\frac{3}{8} - \frac{1}{4}}{1 + \frac{1}{2}}$ (b) $\dfrac{\frac{4}{5} - \frac{1}{2}}{2 - \frac{1}{3}}$

2. (a) $\dfrac{2 + \frac{3}{7}}{1 - \frac{1}{5}}$ (b) $\dfrac{1\frac{1}{4} + 1\frac{1}{2}}{2\frac{1}{2} - 1\frac{1}{4}}$

3. (a) $\dfrac{x - \frac{y^2}{x}}{1 - \frac{y}{x}}$ (b) $\dfrac{\frac{1}{a} - \frac{1}{b}}{\frac{1}{a} + \frac{1}{b}}$

4. (a) $\dfrac{1 + \frac{3}{a}}{a - \frac{9}{a}}$ (b) $\dfrac{\frac{x^2}{y^2} - 1}{\frac{x}{y} - 1}$

5. (a) $\dfrac{\frac{m}{m+n} - \frac{n}{n-m}}{\frac{n}{m+n} + \frac{m}{n-m}}$ (b) $\dfrac{2 + \frac{4}{x^2}}{\frac{1}{x} + \frac{2}{x^2}}$

6. (a) $\dfrac{\frac{3}{y} + 6}{\frac{2y + 1}{8}}$ (b) $\dfrac{\frac{a + b}{a}}{\frac{1}{b} + \frac{1}{a}}$

7. (a) $\dfrac{\frac{x+1}{x-1} + \frac{x-1}{x+1}}{\frac{x-1}{x+1} - \frac{x+1}{x-1}}$ (b) $\dfrac{\frac{3a}{2a+1} - \frac{4}{a}}{\frac{2}{a} + \frac{3a}{2a-1}}$

8. (a) $\dfrac{\frac{3}{y} + \frac{4}{y^2}}{\frac{4}{y} - 1}$ (b) $\dfrac{x + \frac{1}{y}}{\frac{1}{x} + y}$

Calculator Activities

Simplify each of the following complex fractions.

1. (a) $\dfrac{\frac{15}{44} - \frac{13}{21}}{\frac{4}{9} - 16}$ (b) $\dfrac{\frac{22}{y-1} - \frac{14}{17y^2 - 17}}{\frac{12}{y+1}}$

2. (a) $\dfrac{4 - \dfrac{x}{192x^2 + 148x - 65}}{\dfrac{x-1}{12x+13} - \dfrac{41}{x}}$ (b) $\dfrac{\dfrac{19}{121a^2 - 625} + 9}{2 - \dfrac{3a}{25 - 11a}}$

10-5 PROPERTIES OF EXPONENTS

The laws of exponents are basic rules for working with powers. The first four laws, discussed in Chapters 8 and 9, are listed below for convenience. The letters m and n designate integers, x and y variables.

$x^m \cdot x^n = x^{m+n}$ *1st law of exponents*

$(xy)^n = x^n y^n$ *2nd law of exponents*

$(x^m)^n = x^{mn}$ *3rd law of exponents*

$\dfrac{x^m}{x^n} = x^{m-n}$ *4th law of exponents*

You should also recall the following:

$x^n = \underbrace{x \cdot x \cdot \ldots \cdot x}_{n \text{ x's}}$ n any positive integer

$x^0 = 1$ $x \neq 0$

$x^{-n} = \dfrac{1}{x^n}$ n any positive integer, $x \neq 0$

For example,

$x^5 = x \cdot x \cdot x \cdot x \cdot x$

$x^{-5} = \dfrac{1}{x^5}$

The final law of exponents is concerned with powers of rational expressions. For example, what is the third power of the rational expression x/y?

$\left(\dfrac{x}{y}\right)^3 = \dfrac{x}{y} \cdot \dfrac{x}{y} \cdot \dfrac{x}{y}$ By definition

$\phantom{\left(\dfrac{x}{y}\right)^3} = \dfrac{x \cdot x \cdot x}{y \cdot y \cdot y}$ By the rule for multiplying rational expressions

$\phantom{\left(\dfrac{x}{y}\right)^3} = \dfrac{x^3}{y^3}$

Thus,

$\left(\dfrac{x}{y}\right)^3 = \dfrac{x^3}{y^3}$

This example illustrates the **fifth law of exponents.**

$\left(\dfrac{x}{y}\right)^n = \dfrac{x^n}{y^n}$ *5th law of exponents*

10-5: PROPERTIES OF EXPONENTS

Problem 1 Express each rational expression as a quotient of two polynomials.

(a) $\left(\dfrac{3x}{2y^2}\right)^4$
(b) $\left(\dfrac{5a^3}{4b^4}\right)^5$
(c) $\left(\dfrac{2y^2(x-1)}{3x^3(y+1)}\right)^3$

SOLUTION:

(a) $\left(\dfrac{3x}{2y^2}\right)^4 = \dfrac{(3x)^4}{(2y^2)^4}$ 5th law of exponents

$= \dfrac{3^4 x^4}{2^4 (y^2)^4}$ 2nd law of exponents

$= \dfrac{3^4 x^4}{2^4 y^8} = \dfrac{81 x^4}{16 y^8}$ 3rd law of exponents

Thus,

$$\left(\dfrac{3x}{2y^2}\right)^4 = \dfrac{81 x^4}{16 y^8}$$

(b) $\left(\dfrac{5a^3}{4b^4}\right)^5 = \dfrac{(5a^3)^5}{(4b^4)^5}$ 5th law of exponents

$= \dfrac{5^5 (a^3)^5}{4^5 (b^4)^5}$ 2nd law of exponents

$= \dfrac{5^5 a^{15}}{4^5 b^{20}}$ 3rd law of exponents

$= \dfrac{3125 a^{15}}{1024 b^{20}}$

That is,

$$\left(\dfrac{5a^3}{4b^4}\right)^5 = \dfrac{3125 a^{15}}{1024 b^{20}}$$

(c) $\left(\dfrac{2y^2(x-1)}{3x^3(y+1)}\right)^3 = \dfrac{[2y^2(x-1)]^3}{[3x^3(y+1)]^3}$ 5th law of exponents

$= \dfrac{2^3 (y^2)^3 (x-1)^3}{3^3 (x^3)^3 (y+1)^3}$ 2nd law of exponents

$= \dfrac{2^3 y^6 (x-1)^3}{3^3 x^9 (y+1)^3}$ 3rd law of exponents

$= \dfrac{8 y^6 (x-1)^3}{27 x^9 (y+1)^3}$

Thus,

$$\left(\dfrac{2y^2(x-1)}{3x^3(y+1)}\right)^3 = \dfrac{8 y^6 (x-1)^3}{27 x^9 (y+1)^3}$$

CHAPTER 10: RATIONAL EXPRESSIONS

In Chapter 2 you saw that scientific notation requires the use of negative as well as positive exponents. This example illustrates why it is worthwhile to study rational expressions involving negative exponents. For example, the expression $5x^{-2}y/4xy^{-3}$ equals a quotient of two monomials free of negative exponents.

$$\frac{5x^{-2}y}{4xy^{-3}} = \frac{5}{4} \cdot \frac{x^{-2}}{x^1} \cdot \frac{y^1}{y^{-3}} \qquad \blacksquare \; x = x^1,\, y = y^1$$

$$= \frac{5}{4} x^{-2-1} y^{1-(-3)} \qquad \text{4th law of exponents}$$

$$= \frac{5}{4} x^{-3} y^4$$

$$= \frac{5y^4}{4x^3} \qquad \blacksquare \; x^{-3} = \frac{1}{x^3}$$

As a second example, simplify $\dfrac{(a^2 b^{-1})^3}{a^{-2} b^4}$.

$$\frac{(a^2 b^{-1})^3}{a^{-2} b^4} = \frac{(a^2)^3 (b^{-1})^3}{a^{-2} b^4} \qquad \text{2nd law of exponents}$$

$$= \frac{a^6 b^{-3}}{a^{-2} b^4} \qquad \text{3rd law of exponents}$$

$$= a^{6-(-2)} b^{-3-4} \qquad \text{4th law of exponents}$$

$$= a^8 b^{-7}$$

$$= \frac{a^8}{b^7} \qquad \blacksquare \; b^{-7} = \frac{1}{b^7}$$

Problem 2 Express each rational expression as a quotient of two polynomials.

(a) $(2x^2 y^{-2})^{-3}$ \qquad (b) $\left(\dfrac{5a^{-3} b}{3ab^2}\right)^{-2}$

SOLUTION:

(a) $(2x^2 y^{-2})^{-3} = \left(\dfrac{2x^2}{y^2}\right)^{-3} \qquad \blacksquare \; y^{-2} = \dfrac{1}{y^2}$

$$= \left(\frac{y^2}{2x^2}\right)^3 \qquad \blacksquare \; \left(\frac{A}{B}\right)^{-3} = \left(\frac{B}{A}\right)^3$$

$$= \frac{(y^2)^3}{(2x^2)^3} \qquad \text{5th law of exponents}$$

$$= \frac{y^6}{2^3 x^6} \qquad \text{2nd and 3rd laws of exponents}$$

$$= \frac{y^6}{8x^6}$$

Thus,

$$(2x^2y^{-2})^{-3} = \frac{y^6}{8x^6}$$

(b) $\left(\dfrac{5a^{-3}b}{3ab^2}\right)^{-2} = \left(\dfrac{3ab^2}{5a^{-3}b}\right)^2$

$\phantom{(b) \left(\dfrac{5a^{-3}b}{3ab^2}\right)^{-2}} = \dfrac{(3ab^2)^2}{(5a^{-3}b)^2}$ 5th law of exponents

$\phantom{(b) \left(\dfrac{5a^{-3}b}{3ab^2}\right)^{-2}} = \dfrac{3^2a^2b^4}{5^2a^{-6}b^2}$ 2nd and 3rd laws of exponents

$\phantom{(b) \left(\dfrac{5a^{-3}b}{3ab^2}\right)^{-2}} = \dfrac{9a^8b^2}{25}$ ∎ $\dfrac{a^2}{a^{-6}} = a^8, \dfrac{b^4}{b^2} = b^2$

Thus,

$$\left(\frac{5a^{-3}b}{3ab^2}\right)^{-2} = \frac{9}{25}a^8b^2$$

Quick Reinforcement

Use the laws of exponents to simplify.

(a) $\left(\dfrac{5xy^2}{10x^2y^4}\right)^2$

(b) $\left(\dfrac{4x^{-2}y}{15xy^{-2}}\right)^{-1}$

Answers (a) $\dfrac{1}{4x^2y^4}$ (b) $\dfrac{15x^3}{4y^3}$

EXERCISES 10-5

Use the laws of exponents to simplify each of the following into a quotient of two polynomials.

1. (a) $\left(\dfrac{2x}{3y^2}\right)^2$
   (b) $\left(\dfrac{6x^2}{4y^3}\right)^3$

2. (a) $\left(\dfrac{4a^2b}{5ab^3}\right)^3$
   (b) $\left(-\dfrac{2p^5r^3s}{3p^2rs}\right)^4$

3. (a) $\left(\dfrac{3x^2y^{-2}}{9xy}\right)^2$
   (b) $\left(-\dfrac{25a^{-1}b^3c}{15a^2b^{-1}c}\right)^3$

4. (a) $\left(\dfrac{(x+1)^2}{(x-1)^2}\right)^{-1}$
   (b) $\left(\dfrac{3x(x-2)}{2y(y+1)}\right)^3$

5. (a) $\left(\dfrac{3x^3y}{5y^3}\right)^{-2}$
   (b) $\left(\dfrac{2x^4y^6z}{x^2}\right)^0$

6. (a) $\left(\dfrac{-4x^3y^{-2}z^6}{3x^{-2}yz^9}\right)^3$ (b) $\left(\dfrac{a^6b^3c^{-1}}{-3x^2y^{-3}z}\right)^4$

7. (a) $\left(\dfrac{(9x+1)^2}{(x+3)^3}\right)^0$ (b) $\left(\dfrac{(3a-3b)^3}{(2a+3b)^2}\right)^2$

💡 8. (a) $\left(\dfrac{x^{-3}(-4x+1)^2}{y^{-2}(2y-3)^3}\right)^{-2}$ (b) $\left(\dfrac{(x^2-1)^2}{(x+1)^2}\right)^{-1}$

9. (a) $\left(\dfrac{4x(z+3)^3}{2z^3(x+1)^2}\right)^2$ (b) $\left(\dfrac{(6x-1)^3}{6(x+1)^2}\right)^3$

10. (a) $\left(\dfrac{2x^3(11x-5)^0}{3x^4(2x+7)^0}\right)^{-1}$ (b) $\left(\dfrac{(-3a^5b^2)^2}{(2a^3b^3)^3}\right)^{-2}$

11. (a) $\left(\dfrac{5a^{-3}b^2}{(a+b)^2}\right)^3$ (b) $\left(\dfrac{x^2(5x+y)^2}{y^3(3x-y)^3}\right)^3$

12. (a) $\left(\dfrac{(16x^2+24x+9)^2}{(16x^2-9)^3}\right)^2$ (b) $\left(\dfrac{-3x^2y(a+b)^3}{x^3y^{-1}(a-b)^2}\right)^4$

Calculator Activities

Use the laws of exponents to simplify each of the following.

1. (a) $\left(\dfrac{11x^{12}y^3}{12x^{10}y^5}\right)^3$ (b) $\left(\dfrac{15a^{-3}b^{-2}c}{8ab^2c^3}\right)^4$

2. (a) $\left(\dfrac{210xy^4}{15x^2y^{-1}}\right)^{-2}$ (b) $\left(\dfrac{10u^{-2}(v+1)^2z}{220u(v+1)^4}\right)^{-1}$

10-6 EVALUATING RATIONAL EXPRESSIONS

You evaluate a rational expression by giving values to its variables.

Problem 1 Evaluate the rational expression $\dfrac{x-4}{x^2+1}$ when (a) $x = 3$ (b) $x = -2$ (c) $x = 7$

SOLUTION:

$\dfrac{x-4}{x^2+1}$

(a) $\dfrac{3-4}{3^2+1} = \dfrac{-1}{10}$, or -0.1 ■ $x = 3$

(b) $\dfrac{-2-4}{(-2)^2+1} = \dfrac{-6}{5}$, or -1.2 ■ $x = -2$

(c) $\dfrac{7-4}{7^2+1} = \dfrac{3}{50}$, or 0.06 ■ $x = 7$

Problem 2 Evaluate the rational expression $\dfrac{2x+y}{3x+2y}$ when (a) $x=2$, $y=1$ (b) $x=-2$, $y=4$ (c) $x=3$, $y=-3$

SOLUTION:

$$\dfrac{2x+y}{3x+2y}$$

(a) $\dfrac{2\cdot 2 + 1}{3\cdot 2 + 2\cdot 1} = \dfrac{5}{8}$ ■ $x=2$, $y=1$

(b) $\dfrac{2\cdot(-2)+4}{3\cdot(-2)+2\cdot 4} = \dfrac{-4+4}{-6+8} = \dfrac{0}{2}$, or 0 ■ $x=-2$, $y=4$

(c) $\dfrac{2\cdot 3 + (-3)}{3\cdot 3 + 2\cdot(-3)} = \dfrac{3}{3}$, or 1 ■ $x=3$, $y=-3$

A polynomial has values for all values of the variables it contains. This is not the case for rational expressions. A rational expression can have no value for certain values of its variable(s). For example, the rational expression

$$\dfrac{x^2 - x + 2}{2x - 6}$$

has no value when $x = 3$:

$$\dfrac{3^2 - 3 + 2}{2\cdot 3 - 6} = \dfrac{8}{0}, \quad \text{undefined} \quad \textit{Division by zero is undefined}$$

As another example, the expression

$$\dfrac{4x - 7y}{5x + 3y}$$

has no value when $x = -3$, $y = 5$:

$$\dfrac{4\cdot(-3) - 7\cdot 5}{5\cdot(-3) + 3\cdot 5} = \dfrac{-47}{0}, \quad \text{undefined}$$

Problem 3 Find values of the variables for which each rational expression has no value.

(a) $\dfrac{2x+3}{x-7}$ (b) $\dfrac{x^2+4}{x^2-4}$ (c) $\dfrac{3a+8b}{a-11b}$ (d) $\dfrac{2y-3}{2y^2+3y-2}$

SOLUTION:

(a) For what value of x does $x - 7$ equal zero? For $x = 7$, of course!

$$\dfrac{2\cdot 7 + 3}{7 - 7} = \dfrac{17}{0}, \quad \text{undefined}$$

(b) The expression will have no value when the denominator equals zero. The denominator $x^2 - 4$ is zero for every solution of the quadratic equation

$$x^2 - 4 = 0$$

Thus,

$$(x-2)(x+2) = 0$$

and

$$x = 2 \quad \text{or} \quad x = -2$$

The rational expression

$$\frac{x^2 + 4}{x^2 - 4}$$

has no value when $x = 2$ or $x = -2$.

(c) Find values of a and b for which

$$a - 11b = 0$$

Thus,

$$a = 11b$$

and some solutions are

$$a = 11, \quad b = 1$$
$$a = 22, \quad b = 2$$

For these (and many other) values of a and b, the rational expression $\dfrac{3a + 8b}{a - 11b}$ has no value.

(d) The denominator is zero for every solution of the quadratic equation

$$2y^2 + 3y - 2 = 0$$

Factoring,

$$(2y - 1)(y + 2) = 0$$

and

$$2y - 1 = 0 \quad \text{or} \quad y + 2 = 0$$

Thus, if either $y = \frac{1}{2}$ or $y = -2$, the given rational expression has no value.

Quick Reinforcement

(a) Evaluate the rational expression $\dfrac{5x^2 - 2xy + y^2}{4x - 2y}$ for $x = 0$ and $y = -1$.

(b) Determine the values of x for which the rational expression $\dfrac{4x^2 - 16}{3x^2 - 2x - 8}$ has no value.

Answers (a) $-\dfrac{1}{2}$ (b) $2, -\dfrac{4}{3}$

EXERCISES 10-6

Evaluate each rational expression for the indicated values of the variables.

1. $\dfrac{3x - 2}{x + 4}$
   (a) for $x = 3$ (b) for $x = 0$ (c) for $x = -\dfrac{3}{2}$

2. $\dfrac{a^2 - a - 2}{3a + 2}$
   (a) for $a = 1$ (b) for $a = 2.3$ (c) for $a = -2$

3. $\dfrac{3x + y}{5x - 2y}$
   (a) for $x = 1, y = \dfrac{2}{5}$ (b) for $x = -1, y = -2$ (c) for $x = -3, y = 2$

4. $\dfrac{3x^2 + 2x - 1}{2x^2 - 3xy + y^2}$
   (a) for $x = 1, y = 0$ (b) for $x = 0, y = 3.5$ (c) for $x = -1, y = 2$

5. $\dfrac{(5x + 1)(3x - 2)}{2x - 3}$
   (a) for $x = 0$ (b) for $x = -\dfrac{1}{3}$ (c) for $x = 3$

6. $\dfrac{3p^2 + 2pr - 4r^2}{2p - 5r}$
   (a) for $p = 0, r = -1$ (b) for $p = 2, r = 1.2$ (c) for $p = -3, r = -\dfrac{1}{2}$

7. $\dfrac{-2a^2 + 5ab + 4b^2}{a - b}$
   (a) for $a = 0, b = 2$ (b) for $a = 4, b = 3$ (c) for $a = \dfrac{7}{2}, b = 8$

8. $\dfrac{5x^2 + 3x - 7}{3y^2 - 2y + 6}$
   (a) for $y = 0, x = 1.5$ (b) for $y = -1, x = 0$ (c) for $y = 2, x = \dfrac{2}{3}$

Find values of the variables for which each rational expression has no value.

9. (a) $\dfrac{8-5x}{2x+5}$ (b) $\dfrac{17-3x}{2+9x}$

10. (a) $\dfrac{a^2+a+2}{a^2+a-2}$ (b) $\dfrac{u(u+4)}{a^2-7a+10}$

11. (a) $\dfrac{x^2-x-6}{x^2+x-12}$ (b) $\dfrac{x^2+7x+12}{x^2-9}$

12. (a) $\dfrac{y^2-9}{y^2-y-20}$ (b) $\dfrac{21x^2+22xy+y^2}{x^2+xy-2y^2}$

13. (a) $\dfrac{a^2-3a}{2a^2+11a+5}$ (b) $\dfrac{x^2+7xy+12y^2}{x^2y+3xy^2}$

14. (a) $\dfrac{r^2-3rs+2s^2}{rs-2s^2}$ (b) $\dfrac{a^2b+ab^2}{a^2-16b^2}$

15. (a) $\dfrac{x-4}{x^2-8x+16}$ (b) $\dfrac{p^2+7pq+12q^2}{p^2-pq-20q^2}$

16. (a) $\dfrac{6a-3}{2a^2-15a+7}$ (b) $\dfrac{ab(a-b)}{a^2b-ab^2}$

Calculator Activities

Evaluate each rational expression for the indicated values of the variables.

1. $\dfrac{7x-3}{x+2}$
    (a) For $x = 3.51$ (b) For $x = 0.035$

2. $\dfrac{a^3-2a+1}{2a-1}$
    (a) For $a = 43$ (b) For $a = -125$

3. $\dfrac{4x-3y}{6x-y}$
    (a) For $x = 1.33, y = 9.07$ (b) For $x = 302, y = 77$

4. $\dfrac{17t^2-3t+9}{12t^2+8}$
    (a) For $t = 0.021$ (b) For $t = 97$

5. $\dfrac{15x^2-12y}{145x-33y^2}$
    (a) For $x = 11, y = 12$ (b) For $x = -3.01, y = 2.6$

6. Find values of the variables for which each rational expression has no value.
    (a) $\dfrac{15x-12}{210x^2+107x-65}$ (b) $\dfrac{4x+21y}{231x^2-395xy+156y^2}$

10-7 RATIONAL EQUATIONS AND FORMULAS

An equation involving rational expressions, such as

$$\frac{4}{x+3} = \frac{2}{3x-11}$$

is called a **rational equation.** If each side of the equation is either a simple rational expression or a sum of such expressions, then you can solve the equation by multiplying each side of the equation by the LCD of the simple rational expressions. For the example above, the LCD of the two simple fractions is

$$(x+3)(3x-11)$$

Multiply each side of the equation by this polynomial:

$$\frac{4}{x+3} \cdot (x+3)(3x-11) = \frac{2}{3x-11} \cdot (x+3)(3x-11)$$

$$4(3x-11) = 2(x+3)$$

$$12x - 44 = 2x + 6$$

$$10x = 50$$

$$x = 5$$

The possible solution is $x = 5$.

√ CHECK:

Let $x = 5$ in the given equation.

$$\frac{4}{5+3} = \frac{2}{3 \cdot 5 - 11}$$

$$\frac{4}{8} = \frac{2}{4} \quad \checkmark$$

You should check each possible solution of a rational equation to make sure that the denominators of the simple rational expressions involved have nonzero values. Consider, for example, the following equation.

$$\frac{x}{x-3} = \frac{3}{x-3}$$

Proceeding as above, multiply each side by the LCD, $x - 3$:

$$\frac{x}{x-3} \cdot (x-3) = \frac{3}{x-3} \cdot (x-3)$$

$$x = 3 \quad \text{Possible solution}$$

✓ CHECK:

Let $x = 3$ in the given equation:

$$\frac{3}{3-3} = \frac{3}{3-3}$$

$$\frac{3}{0} = \frac{3}{0} \qquad \frac{3}{0} \text{ is undefined}$$

Thus, $x = 3$ is not a solution of the equation

$$\frac{x}{x-3} = \frac{3}{x-3}$$

This equation has no solution.

Problem 1 Solve each rational equation.

(a) $\dfrac{5}{2x+9} + \dfrac{3}{3-4x} = 0$ \qquad (b) $\dfrac{2x}{x+1} = 2 - \dfrac{5}{2x}$

(c) $\dfrac{5}{z-3} - \dfrac{30}{z^2-9} = 1$

SOLUTION:

(a) The LCD is $(2x+9)(3-4x)$. Multiply each side by this LCD.

$$\frac{5}{2x+9} \cdot (2x+9)(3-4x) + \frac{3}{3-4x} \cdot (2x+9)(3-4x) = 0$$

$$5(3-4x) + 3(2x+9) = 0$$

$$15 - 20x + 6x + 27 = 0$$

$$42 = 14x$$

$$3 = x$$

Thus, the given equation has possible solution $x = 3$.

✓ CHECK:

Replace x by 3 in the given equation.

$$\frac{5}{6+9} + \frac{3}{3-12} = 0$$

$$\frac{5}{15} - \frac{3}{9} = 0$$

$$\frac{1}{3} - \frac{1}{3} = 0 \quad \checkmark$$

Hence, 3 is a solution.

(b) The denominators are $x+1$, 1, and $2x$. The LCD is $2x(x+1)$. To solve, multiply each side of the given equation by $2x(x+1)$:

$$\frac{2x}{x+1} \cdot 2x(x+1) = \left(2 - \frac{5}{2x}\right) \cdot 2x(x+1)$$

$$2x \cdot 2x = 2 \cdot 2x(x+1) - \frac{5}{2x} \cdot 2x(x+1)$$

$$4x^2 = 4x(x+1) - 5(x+1)$$

$$4x^2 = 4x^2 + 4x - 5x - 5$$

$$0 = -x - 5$$

$$x = -5$$

Thus, the possible solution of the given equation is $x = -5$.

✓ CHECK:

Replace x by -5 in the given equation.

$$\frac{2 \cdot (-5)}{-5+1} = 2 - \frac{5}{2 \cdot (-5)}$$

$$\frac{-10}{-4} = 2 - \frac{5}{-10}$$

$$2\frac{1}{2} = 2 + \frac{1}{2} \quad ✓$$

This shows that $x = -5$ is a solution of the rational equation

$$\frac{2x}{x+1} = 2 - \frac{5}{2x}$$

(c) The LCD is $z^2 - 9$. To solve, multiply each side by $z^2 - 9$:

$$\left(\frac{5}{z-3} - \frac{30}{z^2-9}\right)(z^2-9) = z^2 - 9$$

$$\frac{5}{z-3} \cdot (z^2-9) - \frac{30}{z^2-9} \cdot (z^2-9) = z^2 - 9$$

$$5(z+3) - 30 = z^2 - 9 \qquad ■ \; z^2 - 9 = (z+3)(z-3)$$

$$5z + 15 - 30 = z^2 - 9$$

$$z^2 - 5z + 6 = 0$$

$$(z-2)(z-3) = 0$$

$$z - 2 = 0 \quad \text{or} \quad z - 3 = 0$$

$$z = 2 \quad \text{or} \quad z = 3$$

The possible solutions of the given equation are 2 and 3.

✓ CHECK:

Let $z = 2$ in the given equation.

$$\frac{5}{2-3} - \frac{30}{4-9} = 1$$

$$-5 + \frac{30}{5} = 1$$

$$-5 + 6 = 1 \quad \checkmark$$

Thus, 2 is a solution.

Let $z = 3$ in the given equation.

$$\frac{5}{3-3} - \frac{30}{9-9} = 1$$

$$\frac{5}{0} - \frac{30}{0} = 1 \qquad \textit{Division by 0 is undefined}$$

Thus, 3 is not a solution. The given equation has only one solution, 2.

A rational equation in two or more variables can frequently be solved for one variable in terms of the others. This allows you to find solutions of the equation. For example, solve the following equation for b:

$$\frac{4ab}{d} = 15$$

$$\frac{4ab}{d} \cdot d = 15 \cdot d \qquad \textit{LCD is d; multiply each side by d}$$

$$4ab = 15d$$

$$\frac{1}{4a} \cdot 4ab = \frac{1}{4a} \cdot 15d \qquad \textit{Isolate b on the left side}$$

$$b = \frac{15d}{4a}$$

When $a = 5$ and $d = 2$,

$$b = \frac{15 \cdot 2}{4 \cdot 5} = \frac{3}{2}$$

Thus, $a = 5$, $b = \frac{3}{2}$, $d = 2$ is a solution of the given equation.

Many formulas are in the form of rational equations. Two examples are

$$P = \frac{A}{1 + rt} \qquad \text{and} \qquad I = \frac{nE}{R + nr}$$

Sometimes it is convenient to express a formula in a different way, to solve it for another one of its variables. Some examples follow.

Problem 2 Solve the formula for the indicated variable.

(a) $P = \dfrac{A}{1+rt}$ for r (b) $I = \dfrac{nE}{R+nr}$ for n

SOLUTION:

(a)
$$P = \dfrac{A}{1+rt}$$
Given formula; LCD $= 1 + rt$

$$P \cdot (1+rt) = \dfrac{A}{1+rt} \cdot (1+rt)$$
Multiply each side by $1 + rt$

$$P + Prt = A$$

$$Prt = A - P$$
Subtract P from each side

$$r = \dfrac{A-P}{Pt}$$
Multiply each side by $1/Pt$ for desired formula

(b)
$$I = \dfrac{nE}{R+nr}$$
Given formula; LCD $= R + nr$

$$I \cdot (R+nr) = \dfrac{nE}{R+nr} \cdot (R+nr)$$
Multiply each side by $R + nr$

$$IR + Inr = nE$$

$$IR = nE - Inr$$
Collect terms involving n

$$IR = n(E - Ir)$$
Factor out n

$$n = \dfrac{IR}{E - Ir}$$
Multiply each side by $\dfrac{1}{E-Ir}$ for desired formula

Quick Reinforcement

(a) Solve the equation $\dfrac{1}{x} - \dfrac{1}{x^2 - 6x} = \dfrac{2}{x}$

(b) Solve the formula $\dfrac{1}{R} = \dfrac{1}{r_1} + \dfrac{1}{r_2}$ for r_1

Answers: (a) $x = 5$ (b) $r_1 = \dfrac{Rr_2}{r_2 - R}$

EXERCISES 10-7

Solve each of the following rational equations. Check your answers.

1. (a) $\dfrac{3}{x+4} = \dfrac{2}{x+6}$ (b) $\dfrac{7}{8x-9} = \dfrac{2}{2x-3}$

2. (a) $\dfrac{2}{3x} + \dfrac{x+1}{x^2} = \dfrac{1}{x}$ (b) $\dfrac{9}{x-2} = 3$

3. (a) $\dfrac{8a-1}{6a+8} = \dfrac{3}{4}$ (b) $\dfrac{2}{x} = \dfrac{x}{5x-12}$

4. (a) $\dfrac{1}{2y} + \dfrac{1}{8} = \dfrac{2}{y} - \dfrac{1}{4}$ (b) $\dfrac{8}{5-a} - \dfrac{6}{a-1} = 1$

5. (a) $\dfrac{1}{2a} - \dfrac{2}{3a} = \dfrac{1}{24}$ (b) $\dfrac{5}{y} - 1 = \dfrac{4}{y}$

6. (a) $\dfrac{y+3}{-y} = \dfrac{-y-9}{y+4}$ (b) $4y - \dfrac{y-1}{y} = 3$

7. (a) $\dfrac{7}{3y+1} = \dfrac{3}{2y-1}$ (b) $\dfrac{2-x}{x} = 0$

8. (a) $\dfrac{r+3}{r} = \dfrac{r+9}{r+4}$ (b) $\dfrac{4}{3x-2} + \dfrac{7}{3x} - \dfrac{1}{x} = 0$

9. (a) $\dfrac{5}{y-3} = \dfrac{33-y}{y^2-6y+9}$ (b) $\dfrac{x+1}{x+4} = \dfrac{2}{x-2}$

10. (a) $\dfrac{5}{3x+1} = \dfrac{1}{2x+3}$ (b) $\dfrac{2x-3}{2x^2-5x-3} = \dfrac{3}{2x-6} + \dfrac{1}{4x+2}$

11. (a) $\dfrac{2}{y-2} = 3 - \dfrac{5}{2-y}$ (b) $\dfrac{6}{x+2} + \dfrac{2}{x-2} = \dfrac{4x}{x^2-4}$

12. (a) $\dfrac{a+2}{2a+6} + \dfrac{3}{a+3} = \dfrac{3}{2}$ (b) $\dfrac{x+3}{x} - \dfrac{x}{x-3} = \dfrac{9}{x^2-3x}$

13. (a) $\dfrac{x}{2x+2} = \dfrac{-2x}{4x+4} + \dfrac{x}{x+1}$ (b) $3 - \dfrac{2x-3}{3} = \dfrac{5-x}{2}$

14. (a) $\dfrac{1}{a+1} + \dfrac{3}{a-1} = \dfrac{4a+2}{a^2-1}$ (b) $\dfrac{t-8}{t-3} - \dfrac{t+8}{t+3} = \dfrac{t}{9-t^2}$

Solve each of the following formulas for the given variable.

15. $k = \dfrac{1}{2}\left(\dfrac{e_1}{b_1} + \dfrac{e_2}{b_2}\right)$
   (a) for e_1 (b) for b_2

16. $a^2 = \dfrac{2T}{(d_1-d_2)g}$
   (a) for g (b) for d_2

17. $I = \dfrac{E}{R + \dfrac{r}{2}}$

    (a) for R (b) for r

18. $\dfrac{1}{F} = (n-1)\left(\dfrac{1}{r_1} + \dfrac{1}{r_2}\right)$

    (a) for r_1 (b) for n

19. $\dfrac{n_2}{v} + \dfrac{n_1}{u} = \dfrac{n_2 - n_1}{r}$

    (a) for u (b) for n_2

20. $\dfrac{v}{R} = \dfrac{1}{(n+a)^2} - \dfrac{1}{(m+b)^2}$

    (a) for v (b) for R

Calculator Activities

1. Solve the following equations.

    (a) $\dfrac{55}{2x} - \dfrac{19}{x-13} = \dfrac{43}{x}$

    (b) $\dfrac{15}{168x^2 + 86x - 65} - \dfrac{14}{14x^2 + 13x} = \dfrac{22}{12x^2 - 5x}$

10-8 APPLIED PROBLEMS

Sometimes, applied problems can be solved by solving rational equations as the following illustrate.

Problem 1 A swimming pool has two pipes by which to fill the pool with water. The larger pipe alone can fill the pool in 10 hours, the smaller pipe alone in 15 hours. How long does it take to fill the pool when both pipes are open?

SOLUTION:

You may use the flowchart on page 208.

Two pipes to pool
Larger pipe alone fills pool in 10 hours   *Bits of information*
Smaller pipe alone fills pool in 15 hours

$x =$ number of hours it takes the two pipes together to fill pool *Variable*

How are x, 10, and 15 related? *Question*

$\dfrac{1}{10} =$ part of pool filled each hour by larger pipe

$\dfrac{1}{15} =$ part of pool filled each hour by smaller pipe *Algebraic language*

$\dfrac{1}{x} =$ part of pool filled each hour by both pipes

$\dfrac{1}{10} + \dfrac{1}{15} = \dfrac{1}{x}$ *Equation relating bits of information*

Now solve this equation:

$$\dfrac{1}{10} + \dfrac{1}{15} = \dfrac{1}{x} \qquad \text{Given equation; LCD} = 30x$$

$$\dfrac{1}{10} \cdot 30x + \dfrac{1}{15} \cdot 30x = \dfrac{1}{x} \cdot 30x \qquad \text{Multiply each side by } 30x$$

$$3x + 2x = 30$$

$$5x = 30$$

$$x = 6$$

Thus, it takes 6 hours for both pipes to fill the pool.

√ CHECK:

$$\dfrac{1}{10} + \dfrac{1}{15} = \dfrac{1}{6}$$

$$\dfrac{3}{30} + \dfrac{2}{30} = \dfrac{1}{6}$$

$$\dfrac{5}{30} = \dfrac{1}{6} \quad \checkmark$$

Problem 2 Two caravans travel 240 kilometers in a total (collective) time of 5 hours. One of the caravans travels 40 km/hr faster than the other. Determine the rate at which each caravan travels.

SOLUTION:

Not knowing the rate of the slower caravan, you designate it by r. Then the rate of the other caravan is $r + 40$. The following table shows the distance, rate, and time relationships.

	Distance	÷ Rate	= Time
Caravan 1	240	r	$\dfrac{240}{r}$
Caravan 2	240	$r + 40$	$\dfrac{240}{r+40}$

By what is given, the sum of the times of the two caravans is 5 hours. Thus,

$$\frac{240}{r} + \frac{240}{r+40} = 5 \qquad \text{Solve this equation;}\ LCD = r(r+40)$$

$$\frac{240}{r} \cdot r(r+40) + \frac{240}{r+40} \cdot r(r+40) = 5 \cdot r(r+40) \qquad \text{Multiply by LCD}$$

$$240(r+40) + 240r = 5r(r+40)$$

$$48(r+40) + 48r = r(r+40) \qquad \text{Divide each term by 5}$$

$$48r + 1920 + 48r = r^2 + 40r$$

$$r^2 - 56r - 1920 = 0 \qquad \begin{array}{l}1920 = 2^7 \cdot 3 \cdot 5;\ \text{find two}\\ \text{numbers whose product is } -1920\\ \text{and whose sum is } -56;\\ \text{how about } -2^4 \cdot 5 \text{ and } 2^3 \cdot 3?\end{array}$$

$$(r-80)(r+24) = 0$$

$$r = 80 \quad \text{or} \quad r = -24 \qquad \text{Can't have a negative rate}$$

$$r = 80$$

The slower caravan is traveling at 80 km/hr and the faster caravan at 80 + 40, or 120 km/hr.

✓ CHECK:

$$\frac{240}{80} = 3 \text{ hours}, \qquad \frac{240}{120} = 2 \text{ hours}, \qquad 3 + 2 = 5 \quad \checkmark$$

Problem 3 An amount of acreage is divided among three people. The first person receives ten acres more than one-third of the original amount and the second person receives five acres more than half of what remains. If the third person receives 100 acres, what was the original amount of acreage?

SOLUTION:

Let x represent the original amount of acreage. A table can be constructed.

	Amount of acreage
First person	$\frac{1}{3}x + 10$
Second person	$\frac{1}{2}[x - (\frac{1}{3}x + 10)] + 5$
Third person	100

Thus,

$$\frac{1}{3}x + 10 + \frac{1}{2}\left[x - \left(\frac{1}{3}x + 10\right)\right] + 5 + 100 = x \quad \text{Total amount is } x$$

$$6 \cdot \left(\frac{1}{3}x + 10\right) + 6 \cdot \frac{1}{2}\left[x - \left(\frac{1}{3}x + 10\right)\right] + 6 \cdot 5 + 6 \cdot 100 = 6x \quad \text{Multiply by the LCD, 6}$$

$$2x + 60 + 3\left[\frac{2}{3}x - 10\right] + 630 = 6x$$

$$2x + 60 + 2x - 30 + 630 = 6x$$

$$4x + 660 = 6x$$

$$660 = 2x$$

$$330 = x$$

Therefore, the original amount of acreage is 330 acres.

Quick Reinforcement

Two trains completed their trips in a total time of 4 hours. One train traveled 320 kilometers and the other train 460 kilometers. They averaged the same speed. Determine this speed.

Answer 195 km/hr

EXERCISES 10-8

Find a rational equation to solve each of the following applied problems.

1. (a) Jaime found that if he spent seven hours each weekend doing the landscape maintenance on the family home, he could get the job done. If his wife, Julia, helps him, they can finish the job in four hours. How long would it take Julia working alone?

   (b) When the outlet valve is open, it takes thirty-two hours to empty the Olympic-size pool at the college. When the pool is totally empty, it takes the inlet pipe twenty hours to fill it again. The maintenance crew has emptied the pool and set the inlet pipe open to refill it, but they forgot to close the outlet pipe. With this situation, how long will it take to refill the pool?

2. (a) What number when added to both the numerator and denominator of the fraction $\frac{2}{3}$ will give the reciprocal of the given fraction?

   (b) The numerator of a given fraction is $\frac{1}{12}$ of the denominator. But when the denominator has 3 subtracted from it and the numerator has 4 added to it, we end up with the fraction $\frac{3}{4}$. What was the original fraction?

3. (a) An old riverboat, now used as a tourist attraction on the Ohio River, can travel 26 kilometers up the river in the same amount of time it journeys 38 kilometers down the river. If the current of the Ohio River is 3 km/hr, what is the speed of the riverboat in still water?

   (b) Lee Ann Chiang traveled by airplane to visit her grandparents. The airplane averaged 150 km/hr going to her grandparents, but on the trip home, which took 1 hour less, the airplane flew at 200 km/hr. How far away from her home do Lee Ann's grandparents live?

4. (a) On a calm day, Elisa can pedal her ten-speed bicycle at 12 km/hr. When the wind is blowing, however, it takes her as long to go 6 kilometers against the wind as it does to go 30 kilometers with the wind. How fast is the wind blowing?

   (b) Tanya flew her airplane 500 kilometers against the wind in the same time it took her to fly it 600 kilometers with the wind. If the speed of the wind was 10 km/hr, what was the average speed of her airplane?

5. (a) The Black Devil Creek has a 3 km/hr current. Scott takes his motorboat out on the creek, and it takes him as long to go 12 kilometers downstream as it does to go 8 kilometers upstream. What is the speed of Scott's boat in still water?

   (b) The supersonic airplane goes 900 kilometers with the wind in the identical time that it takes to go 700 kilometers flying against the wind. If the wind has a speed of 200 km/hr, what is the speed of the supersonic airplane in still air?

6. (a) The relatively new work crew at the Olympic Pool Company can install a new pool in eight days. For a rush job, the veteran crew was asked to lend a hand, and the job was completed in three days. How long would it take the veteran crew working alone?

   (b) A calculus class was given a take-home exam and were told that they could work in groups of up to three students, if they wished. Louisa estimated that the exam would take her three hours working alone; Richard figured it would take him 4 hours alone; and Marco planned on five hours alone. If these three decided to combine their efforts, how long should it take them working together?

7. (a) An experienced painter can paint the same number of houses in eight days that his new apprentice can paint in twelve days. If they decided to work together on a house, how long would it take them to paint one that would have taken the experienced painter 4 days?

   (b) Three keypunch machines of different models can be used to keypunch 100 cards. Model #201 can do the 100 cards in 12 minutes, model #33 can do the cards in 15 minutes, and the oldest machine, model #4, can do the batch of 100 cards in 20 minutes. If the three machines were working simultaneously, how long would it take them to complete the 100 cards?

8. (a) How many cookies are needed for a party if four of the six persons receive one-third, one-eighth, one-fourth, and one-fifth, respectively, of the total number, while the fifth person receives ten cookies and only one cookie remains for the sixth person?

   (b) While playing with a string of beads, two babies broke the string and one-third of the beads fell on the floor. A fifth of the beads were caught in the cushions of the couch, a sixth were picked up by one of the mothers, a tenth were picked up by the other mother, and six beads remained on the string. How many beads were originally on the string?

Nostalgia Problems

1. (a) Two men commenced trade together; the first put in $40 more than the second, and the stock of the first was to that of the second as 5 to 4; what was the stock of each?

   (b) A farmer wishes to mix 116 bushels of provender, consisting of rye, barley, and oats, so that the mixture may contain $\frac{5}{7}$ as much barley as oats and $\frac{1}{2}$ as much rye as barley. How much of each kind of grain must there be in the mixture?

2. (a) An estate is to be divided among 4 children in the following manner: to the first, $200 more than $\frac{1}{4}$ of the whole; to the second, $340 more than $\frac{1}{5}$ of the whole; to the third, $300 more than $\frac{1}{6}$ of the whole; and to the fourth, $400 more than $\frac{1}{8}$ of the whole. What is the value of the estate?

   (b) Of a detachment of soldiers, $\frac{2}{3}$ are on actual duty, $\frac{1}{8}$ of them are sick, $\frac{1}{5}$ of the remainder absent on leave, and the rest, which is 380, have deserted; what was the number of men in the detachment?

KEY TERMS

Rational expression
Simplest form
Least common denominator (LCD)

Complex fraction
Fifth law of exponents
Rational equation

REVIEW EXERCISES

Put each rational expression in simplest form.

1. (a) $\dfrac{a^2 - ab}{a^2 b - ab^2}$ (b) $\dfrac{2x^2 - 10}{8x - 40}$

2. (a) $\dfrac{m^2 n - 8mn + 15n}{mn - 3n}$ (b) $\dfrac{x^3 + 7x^2 + 10x}{x^3 - 25x}$

Carry out the indicated operation for each exercise. Simplify your answers.

3. (a) $\dfrac{2a^2 + 4a}{12a^2 b} \cdot \dfrac{6a}{a^2 + 6a + 8}$ (b) $\dfrac{x + 3}{x^3 + 3x^2} \cdot \dfrac{x^3}{x - 3}$

4. (a) $\dfrac{m^2 - 6m + 9}{m^2 - m - 6} \div \dfrac{m^2 + 2m - 15}{m^2 + 2m}$ (b) $\dfrac{a - b}{4b} \div \dfrac{a^2 - a - 6}{12b^2}$

5. (a) $\dfrac{2}{x + 1} + \dfrac{3}{x - 2}$ (b) $\dfrac{x}{x - 2} + \dfrac{2x}{x + 3}$

6. (a) $\dfrac{5a + b}{3a - b} - \dfrac{4b^2}{b - 3a}$ (b) $\dfrac{5y - 3}{2y + 1} - \dfrac{3y + 1}{y - 4}$

7. (a) $\dfrac{5t + 2}{t^2 + 6t + 9} - \dfrac{3t - 1}{t^2 + t - 6}$ (b) $\dfrac{x - 1}{3x^2 + 6x} - \dfrac{6}{x + 2}$

8. (a) $\dfrac{3}{y^2 + y} - \dfrac{y + 3}{y^2 - 1} + \dfrac{2}{y - 1}$ (b) $\dfrac{4a}{a^2 - 1} - \dfrac{2}{a} - \dfrac{2}{a + 1}$

9. (a) $\dfrac{2a}{b - 2a} \cdot \left[\dfrac{b^2 - 4a^2}{b + 2a} \div (b + 2a) \right]$ (b) $\left[\dfrac{x + 1}{x^2 - 1} \div \dfrac{x + 1}{x^2 - 2x + 1} \right] \div \dfrac{x + 1}{x - 1}$

10. (a) $\dfrac{5 + \dfrac{2}{x}}{\dfrac{3}{x^2} - 1}$ (b) $\dfrac{\dfrac{2}{a} - \dfrac{3}{a^2}}{\dfrac{3}{a} - \dfrac{2}{a^2}}$

11. (a) $\dfrac{\dfrac{2y}{5} + \dfrac{4}{y^2}}{2 - \dfrac{3}{y}}$ (b) $\dfrac{\dfrac{1}{2x + 1} - \dfrac{4}{4x^2 - 1}}{\dfrac{1}{2x + 1} + \dfrac{3}{2x - 1}}$

12. (a) $\left[\dfrac{2x^3 y^2}{4xy} \right]^2$ (b) $\left[\dfrac{x^4 y}{3x^2 y^3} \right]^5$

13. (a) $\left[\dfrac{25a^2 b^{-1} c}{30a^{-1} b^3 c} \right]^3$ (b) $\left[\dfrac{9x^6 y^7 z^{-1}}{72x^2 y z^{-2}} \right]^0$

14. (a) $\left[\dfrac{2r^2 (s - t)^3}{3r^{-2}(s - t)^{-1}} \right]^2$ (b) $\left[\dfrac{32p^{-3} q (2p - 5q)^{-2}}{8p^2 q^{-1}(2p - 5q)^3} \right]^{-1}$

Evaluate each rational expression for the indicated variables.

15. $\dfrac{5x - 3}{4x + 2}$

(a) for $x = -1$ (b) for $x = 2.5$ (c) for $x = 0$

CHAPTER 10: RATIONAL EXPRESSIONS

16. $\dfrac{-3a^2 + 4a - 1}{4a^2 + 4a + 1}$

    (a) for $a = 0$ (b) for $a = -3$ (c) for $a = \tfrac{5}{2}$

Find the values of the variables for which each rational expression has no value.

17. (a) $\dfrac{4x - 1}{9x + 3}$ (b) $\dfrac{5a + 3}{12a - 7}$

18. (a) $\dfrac{y^2 + 3y - 5}{2y^2 - 7y - 15}$ (b) $\dfrac{3m^2 - 4}{16m^2 - 9}$

Solve each of the following rational equations. Check your answers.

19. (a) $\dfrac{x - 5}{6x - 6} = \dfrac{1}{9} - \dfrac{x - 3}{4x - 4}$ (b) $\dfrac{x + 1}{x + 4} = \dfrac{2}{x - 2}$

20. (a) $\dfrac{6x + 7}{9} + \dfrac{7x - 13}{6x + 3} = \dfrac{2x + 4}{3}$ (b) $\dfrac{3}{x - 4} + \dfrac{2}{x + 2} = \dfrac{8}{x^2 - 2x - 8}$

Solve each of the following formulas for the given variable.

21. $\dfrac{s}{k} - a - b = c$

    (a) for k (b) for s

22. $\dfrac{1}{w} - \dfrac{t}{v} = T$

    (a) for v (b) for w

Find a rational equation to solve each of the following applied problems.

23. (a) A large keypunching job, an important rush order for the firm, would take Marie five hours but would take Arnold eight hours. It was decided to put both Marie and Arnold at work on the job so that it could be finished for their client as soon as possible. How long will it take?

    (b) Bruce knows that normally he can paddle his canoe at a rate of 4 km/hr if he is in still water. But at Bear Creek, it takes him the same time to paddle 8 kilometers upstream as it does to paddle 24 kilometers downstream! What is the speed of the current at Bear Creek?

Calculator Activities

1. Put each rational expression in simplest form.

    (a) $\dfrac{459a^2 - 28ab - 247b^2}{324a^2b + 228ab^2}$ (b) $\dfrac{441x^2z - 315xz + 225z}{441x^2z^2 - 225z^2}$

Perform the indicated operation and simplify.

2. (a) $\dfrac{93x^2 - 99x}{29y + 36} \cdot \dfrac{116y^2 + 144y}{372x^2y - 396xy}$

(b) $\dfrac{484m^2 + 396mn + 81n^2}{36n^2 - 88mn} \div \dfrac{484m^2 - 81n^2}{8n^2}$

3. (a) $\dfrac{3}{102x - 7} + \dfrac{4x}{42x - 612x^2}$

(b) $\dfrac{2x}{99x^2 - 50x - 21} - \dfrac{5}{44x^2y + 133xy + 33y}$

4. (a) $\dfrac{\dfrac{13}{9x} - \dfrac{7}{45x^2}}{\dfrac{34}{9x^2 - 18x} + 6}$
(b) $\left[\dfrac{134x^{21}y^{-33}z^0}{335x^{12}y^5z^{-52}}\right]^{11}$

Evaluate the rational expression for the indicated values.

5. $\dfrac{32x^3 - 5x + 13}{135x^2y + 66xy - 133y}$
   (a) for $x = -2.13$, $y = 4.03$ (b) for $x = 13$, $y = 12$

6. Find the values for the variables in problem 5 above for which the rational expression has no value.

Find a rational equation to solve each applied problem.

7. (a) To get a major overpass repair job finished in the minimum amount of time, the construction company decided to put their two best crews on the job at the same time. The foreman of the better crew estimated that his men alone could finish the job in 5.3 hours, while the less experienced crew's foreman estimated 8.2 hours if left to themselves. How long would it take the two crews working together (assuming it was worked out efficiently!)?

   (b) The school swimming pool has a large intake pipe that can fill the pool with fresh water in six hours and 40 minutes. The pool's maintenance workers know that it takes eight and one-third hours for the outtake pipe to empty the pool. One afternoon, after emptying the pool, a new employee opened the intake pipe to refill the pool, but forgot to close the outtake pipe. If no one noticed the error, how long would it take the pool to refill completely?

11

Radicals

A geological survey determined that the distance between points A and B was 12.5 kilometers, and the distance between points B and C was 9.6 kilometers. What is the distance between points A and C?

11-1 REAL NUMBERS

There are three different kinds of real numbers: integers, rational numbers, and irrational numbers. Integers have been used extensively in previous chapters to describe various aspects of algebra. When integers did not suffice, we used rational numbers. Up to this point, irrational numbers have been used sparingly.

Every real number can be expressed in decimal notation. Rational numbers are represented by either **finite decimals** or **infinite repeating decimals**. For example,

$\frac{4}{5} = 0.8, \qquad -\frac{9}{2} = -2.5 \qquad$ *Finite decimals*

$\frac{5}{6} = 0.83333\ldots \qquad\qquad$ *Infinite repeating decimal; 3 repeats infinitely*

$-\dfrac{24}{11} = -2.18181818\ldots$ *Infinite repeating decimal; 18 repeats infinitely*

Irrational numbers are represented by infinite decimals that do not repeat (**infinite nonrepeating decimals**). For example,

$\sqrt{23} = 4.79583152\ldots$ *Infinite nonrepeating decimal*
$\pi = 3.14159265\ldots$* *Infinite nonrepeating decimal*
$0.101001000100001\ldots$ *Infinite nonrepeating decimal*

You can find the repeating decimal of a rational number by long division. For example, you would find the decimal representation of $\dfrac{5}{22}$ as follows:

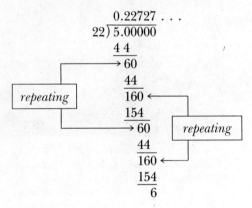

$\dfrac{5}{22} = 0.227272727\ldots$ *27 repeats infinitely*

Check this result with a calculator.

Given an infinite repeating decimal, you can find the rational number it represents by using algebra.

Problem 1 Find the rational number represented by the infinite repeating decimal.

(a) $0.45454545\ldots$ (b) $2.74444\ldots$

SOLUTION:

(a) If x is the rational number represented by $0.45454545\ldots$, then

$x = 0.4545\ldots$
$10x = 4.5454545\ldots$
$100x = 45.454545\ldots$

---

*The first 100,000 digits of the infinite decimal for π have been found by use of a computer.

Observe that x and $100x$ have the same fractional part, $0.454545\ldots$. Hence, $100x - x$ is an integer.

$100x - x = 45$

Thus,

$99x = 45$

$x = \dfrac{45}{99}$, or $\dfrac{5}{11}$ *You can check your answer on a calculator*

Thus, $\dfrac{5}{11}$ is the rational number represented by $0.45454545\ldots$.

(b) Proceed as above. Your task is to multiply x by appropriate powers of 10 (10, 100, 1000, and so on) to find two numbers with the same fractional part. Then subtract one from the other.

$x = 2.74444\ldots$

$10x = 27.4444\ldots$

$100x = 274.444\ldots$ *$10x$ and $100x$ have the same fractional part*

$100x - 10x = 274 - 27$ *Subtract $10x$ from $100x$*

$90x = 247$

$x = \dfrac{247}{90}$ *Check it on a calculator*

Thus, $\dfrac{247}{90}$ is the rational number represented by $2.74444\ldots$.

Square roots, indicated by the **radical sign** $\sqrt{}$, were discussed in Chapter 2. You recall that for any positive numbers x and y,

$\sqrt{y} = x$ if and only if $x^2 = y$

For example,

$\sqrt{100} = 10$ because $10^2 = 100$

$\sqrt{\dfrac{4}{9}} = \dfrac{2}{3}$ because $\left(\dfrac{2}{3}\right)^2 = \dfrac{4}{9}$

Rational numbers such as 100 and $\dfrac{4}{9}$ are perfect squares; thus, the square root of each is again a rational number. Rational numbers such as 23, $\dfrac{33}{9}$ and 0.001 are not perfect squares. Therefore,

$\sqrt{23}$, $\sqrt{\dfrac{33}{9}}$, $\sqrt{0.001}$

are irrational numbers.

There are roots other than square roots, such as third roots, fourth roots, and fifth roots. These are defined as follows.

If y is an nth *power* of x,

$$y = x^n$$

then x is called an nth *root* of y.

For example,

$64 = 4^3$, so 4 is a third root of 64   *3rd root usually called* **cube** *root*

$\dfrac{4}{25} = \left(\dfrac{2}{5}\right)^2$, so $\dfrac{2}{5}$ is a second root of $\dfrac{4}{25}$ *2nd root usually called* **square** *root*

$1{,}000{,}000 = 10^6$, so 10 is a sixth root of 1,000,000

An additional property relating roots and real numbers is as follows. It is true for every integer n greater than 1.

Every positive real number has a unique positive nth root. If n is an odd integer, then every negative real number has a unique negative nth root.

The radical sign $\sqrt[n]{}$ is used to designate nth roots. Thus,

$\sqrt[n]{x}$ *The 2nd root is denoted by $\sqrt{x}$ and is called the square root*

denotes the unique positive nth root of x if $x > 0$, and the unique negative nth root if n is odd and $x < 0$. Of course, $\sqrt[n]{0} = 0$. We call n the **index** and x the **radicand** of $\sqrt[n]{x}$. For example,

$\sqrt{81} = 9$ $9^2 = 81$ *and* $81 > 0$
$\sqrt[3]{-125} = -5$ $(-5)^3 = -125$ *and* $-125 < 0$
$\sqrt[4]{16} = 2$ $2^4 = 16$ *and* $16 > 0$

Problem 2 Find the indicated root.

(a) $\sqrt[3]{\dfrac{27}{64}}$ (b) $\sqrt[5]{-32}$

(c) $\sqrt{\dfrac{169}{121}}$ (d) $\sqrt[3]{0.001}$

SOLUTION:

(a) $\sqrt[3]{\dfrac{27}{64}} = \dfrac{3}{4}$ $\left(\dfrac{3}{4}\right)^3 = \dfrac{27}{64}$

(b) $\sqrt[5]{-32} = -2$ $(-2)^5 = -32$

(c) $\sqrt{\dfrac{169}{121}} = \dfrac{13}{11}$ $\left(\dfrac{13}{11}\right)^2 = \dfrac{169}{121}$

(d) $\sqrt[3]{0.001} = 0.1$ $(0.1)^3 = 0.001$

One thing you cannot do is take the square root, or fourth root, or sixth root, or other even roots of a negative number. For example, there is no real number x that is a square root of -1. (Remember, $x^2 \geq 0$ and thus x^2 cannot equal -1.) However, it is true that positive numbers have negative square roots, fourth roots, sixth roots, and so on. For example, -7 is a square root of 49 because $(-7)^2 = 49$. Since $\sqrt{49}$ designates the *positive* square root of 49, $-\sqrt{49}$ designates the negative square root of 49:

$$\sqrt{49} = 7$$
$$-\sqrt{49} = -7$$

Thus every positive number x has two square roots designated by

$$\sqrt{x} \text{ and } -\sqrt{x}$$

As you saw in Chapter 2, every irrational number can be approximated by a rational number. For example,

$$\sqrt{7} = 2.64575131 \ldots \quad \textit{Infinite nonrepeating decimal}$$

is approximately equal to each of the rational numbers below:

2	A one-digit approximation of $\sqrt{7}$
2.6	A two-digit approximation of $\sqrt{7}$
2.64	A three-digit approximation of $\sqrt{7}$
2.645751	A seven-digit approximation of $\sqrt{7}$

A mini-calculator having a $\sqrt{}$ key gives at least eight-digit approximations of square roots of rational numbers whose numerators and denominators are positive integers made up of no more than eight digits. Roots other than square roots are approximated in Calculator Activities.

Quick Reinforcement

(a) Find the decimal representation of $\frac{1}{7}$ by long division. Check with a calculator.

(b) Express the repeating decimal $1.4212121\ldots$ as a quotient of two integers.

(c) Evaluate (i) $\sqrt[4]{16}$ (ii) $\sqrt[3]{-27}$ (iii) $\sqrt{\frac{9}{4}}$

Answers (a) $0.142857142857142857\ldots$ (b) $\frac{469}{330}$ (c) (i) 2 (ii) -3 (iii) $\frac{3}{2}$

EXERCISES 11-1

In each of the following exercises, find the decimal representation of the rational number. Check with a calculator.

1. (a) $\frac{12}{25}$ (b) $\frac{7}{8}$
2. (a) $\frac{7}{9}$ (b) $-\frac{5}{18}$
3. (a) $-\frac{6}{11}$ (b) $\frac{11}{13}$
4. (a) $\frac{1}{14}$ (b) $\frac{5}{7}$
5. (a) $-\frac{3}{13}$ (b) $-\frac{10}{11}$

In each of the following exercises, find the rational number represented by the decimal. (Express your answers in the form a/b, where a and b are integers.)

6. (a) 1.27   (b) -2.071
7. (a) 0.1777 . . .   (b) 1.4242 . . .
8. (a) 3.296296 . . .   (b) 0.054545 . . .
9. (a) 0.5333 . . .   (b) 0.63636 . . .
10. (a) -1.8333 . . .   (b) -1.4545 . . .

In each of the following exercises, find the rational number represented by the radical expression.

11. (a) $\sqrt{\frac{1}{4}}$ (b) $\sqrt{\frac{81}{25}}$
12. (a) $\sqrt[3]{125}$ (b) $\sqrt[3]{-216}$
13. (a) $\sqrt{0.36}$ (b) $\sqrt{0.0004}$
14. (a) $\sqrt[3]{0.008}$ (b) $\sqrt[4]{0.0081}$
15. (a) $\sqrt[7]{-128}$ (b) $\sqrt[3]{-\frac{27}{64}}$
16. (a) $\sqrt[3]{-343}$ (b) $\sqrt[3]{-1}$
17. (a) $\sqrt[3]{-8}$ (b) $\sqrt[3]{0.000001}$
18. (a) $\sqrt{\frac{121}{144}}$ (b) $\sqrt{\frac{225}{36}}$
19. (a) $\sqrt[4]{625}$ (b) $\sqrt[5]{1}$
20. (a) $\sqrt[4]{0.0016}$ (b) $\sqrt[5]{0.00001}$

Calculator Activities

Estimating nth roots

Recall, x is an nth root of y if $x^n = y$. The following example shows how to approximate nth roots of a number.

Example Approximate the third root of 10, that is, $\sqrt[3]{10}$. First, by inspection, determine two integers closest to $\sqrt[3]{10}$. Clearly,

$$2 < \sqrt[3]{10} < 3 \quad 2^3 = 8 < 10, \, 3^3 = 27 > 10$$

Now test 2.5. Does $2.5 = \sqrt[3]{10}$ or is it less than $\sqrt[3]{10}$ or more than $\sqrt[3]{10}$? To determine this, compare $(2.5)^3$ and 10. Clearly,

$$(2.5)^3 > 10 \quad (2.5)^3 = 15.625$$

and hence $2 < \sqrt[3]{10} < 2.5$.

✓ CHECK: $(2.2)^3$ and $(2.3)^3$. You see that

$$2.2 < \sqrt[3]{10} < 2.3 \quad (2.2)^3 = 9.261, \, (2.3)^3 = 12.167$$

This gives one-digit accuracy beyond. To obtain two-digit accuracy, first test 2.25. If $(2.25)^3 > 10$, then $2.20 < \sqrt[3]{10} < 2.25$; otherwise, $2.25 < \sqrt[3]{10} < 2.30$. Then test different hundredths digits to trap $\sqrt[3]{10}$. Test yourself on

(a) $\sqrt{20}$ (b) $\sqrt[4]{14}$ (c) $\sqrt{69}$ (d) $\sqrt[3]{31}$ (e) $\sqrt[3]{72}$ (f) $\sqrt{8}$

Give two-digit accuracy.

11-2 SIMPLIFYING ROOTS

From the equations

$$\sqrt{25 \times 4} = \sqrt{100} = 10$$

and

$$\sqrt{25} \times \sqrt{4} = 5 \times 2 = 10$$

you conclude that

$$\sqrt{25 \times 4} = \sqrt{25} \times \sqrt{4}$$

Similarly, you can show that

$$\sqrt{\frac{25}{4}} = \frac{\sqrt{25}}{\sqrt{4}}$$

These examples illustrate the following properties of radicals:

$\sqrt{a \times b} = \sqrt{a} \times \sqrt{b}$ *1st law of radicals*

$\sqrt{\dfrac{a}{b}} = \dfrac{\sqrt{a}}{\sqrt{b}}$ *2nd law of radicals*

Therefore, the product of $\sqrt{a}$ and $\sqrt{b}$ can be written as

$$\sqrt{a} \times \sqrt{b} \quad \text{or} \quad \sqrt{a} \cdot \sqrt{b} \quad \text{or} \quad \sqrt{a}\sqrt{b} \quad \text{or} \quad \sqrt{a \cdot b}$$

The quotient of $\sqrt{a}$ divided by $\sqrt{b}$ can be written as

$$\sqrt{a} \div \sqrt{b} \quad \text{or} \quad \frac{\sqrt{a}}{\sqrt{b}} \quad \text{or} \quad \sqrt{\frac{a}{b}}$$

Analogous laws hold for cube roots, fourth roots, and so on. For example,

$\sqrt{50} = \sqrt{25 \times 2} = \sqrt{25} \times \sqrt{2} = 5\sqrt{2}$ *1st law of radicals*

$\sqrt[3]{56} = \sqrt[3]{8 \times 7} = \sqrt[3]{8} \times \sqrt[3]{7} = 2\sqrt[3]{7}$

$\sqrt{\dfrac{32}{9}} = \dfrac{\sqrt{32}}{\sqrt{9}} = \dfrac{\sqrt{16 \times 2}}{3} = \dfrac{\sqrt{16} \times \sqrt{2}}{3} = \dfrac{4\sqrt{2}}{3}$ *1st and 2nd laws of radicals*

$\sqrt[4]{\dfrac{16}{81}} = \dfrac{\sqrt[4]{16}}{\sqrt[4]{81}} = \dfrac{2}{3}$

Using the laws in the opposite direction,

$\sqrt{2}\sqrt{8} = \sqrt{2 \times 8} = \sqrt{16} = 4$ *1st law of radicals*

$\dfrac{\sqrt{18}}{\sqrt{3}} = \sqrt{\dfrac{18}{3}} = \sqrt{6}$ *2nd law of radicals*

$\sqrt[3]{9}\sqrt[3]{3} = \sqrt[3]{9 \times 3} = \sqrt[3]{27} = 3$ *1st law of radicals*

If N is a positive integer that is not a perfect square, then the square root of N can always be expressed in the form

$$\sqrt{N} = A\sqrt{B}$$

where A and B are positive integers and B has no perfect-square factor other than 1. This is called the **simplified form** of $\sqrt{N}$. For example, you saw that

$\sqrt{50} = 5\sqrt{2}$ *2 has no perfect-square factor other than 1*

This is the simplified form of $\sqrt{50}$. As another example, find the simplified form of $\sqrt{108}$. First of all, 4 is a factor of 108: $108 = 4 \cdot 27$. Therefore,

$$\sqrt{108} = \sqrt{4 \cdot 27} = \sqrt{4}\sqrt{27} = 2\sqrt{27}$$

Is this the simplified form of $\sqrt{108}$? No! The integer 27 has a perfect-square factor of 9, and

$$\sqrt{27} = \sqrt{9 \cdot 3} = \sqrt{9}\sqrt{3} = 3\sqrt{3}$$

Then

$$\sqrt{108} = 2(3\sqrt{3}) = 6\sqrt{3} \quad \blacksquare = A\sqrt{B}, \text{ where } A = 6, B = 3$$

This is the simplified form of $\sqrt{108}$ because 3 has no perfect-square factor other than 1.

If N is a positive rational number that is not a perfect square, then the square root of N can always be expressed in the form

$$\sqrt{N} = A\sqrt{B}$$

where A is a positive rational number and B is a positive integer that has no perfect-square factor other than 1. Again, this is called the *simplified form* of $\sqrt{N}$. For example,

$$\sqrt{\frac{2}{5}} = \sqrt{\frac{2 \cdot 5}{5 \cdot 5}} = \sqrt{\frac{10}{25}} \quad \blacksquare \quad \frac{2}{5} = \frac{2 \cdot 5}{5 \cdot 5}$$

$$= \frac{\sqrt{10}}{\sqrt{25}} = \frac{\sqrt{10}}{5}$$

Thus, in simplified form,

$$\sqrt{\frac{2}{5}} = \frac{1}{5}\sqrt{10} \qquad A = \tfrac{1}{5}, B = 10;\ 10 \text{ has no perfect-square factor other than } 1$$

As this example illustrates, you express the rational number N as a quotient of two integers, with the denominator a perfect square. Then the problem of finding the simplified form of $\sqrt{N}$ reduces to that of finding the simplified form of the square root of N's numerator. For example,

$$\sqrt{\frac{9}{20}} = \sqrt{\frac{9 \cdot 5}{20 \cdot 5}} = \sqrt{\frac{45}{100}} \qquad \textit{Least perfect-square multiple of } 20 \textit{ is } 100,\ \textit{since } 20 = 2^2 \cdot 5$$

$$= \frac{\sqrt{45}}{\sqrt{100}} = \frac{\sqrt{45}}{10}$$

Now express $\sqrt{45}$ in simplified form:

$$\sqrt{45} = \sqrt{9 \cdot 5}$$

$$= \sqrt{9}\sqrt{5}$$

$$= 3\sqrt{5} \qquad\qquad \textit{Simplified form}$$

Thus,

$$\sqrt{\frac{9}{20}} = \frac{3\sqrt{5}}{10}, \text{ or } \frac{3}{10}\sqrt{5} \quad \blacksquare \quad A = \tfrac{3}{10}, B = 5$$

What you have just done for square roots can also be done for cube roots, fourth roots, and so on. That is, for every positive rational number N, the cube root of N can be expressed in the form

$\sqrt[3]{N} = A\sqrt[3]{B}$ $\quad$ *A a rational number, B an integer having no perfect-cube factor other than 1*

A similar statement can be made for fourth roots, fifth roots, and so on. For example, you saw that

$\sqrt[3]{56} = 2\sqrt[3]{7}$

This is the simplified form of $\sqrt[3]{56}$.

As another example, find the simplified form of $\sqrt[3]{\frac{8}{5}}$. First, express $\frac{8}{5}$ as a quotient of two integers such that the denominator is a perfect cube:

$$\frac{8}{5} = \frac{8 \cdot 5^2}{5 \cdot 5^2} = \frac{200}{5^3}$$

Thus,

$$\sqrt[3]{\frac{8}{5}} = \sqrt[3]{\frac{200}{5^3}} = \frac{\sqrt[3]{200}}{\sqrt[3]{5^3}} = \frac{\sqrt[3]{200}}{5}$$

Is $\sqrt[3]{200}$ in simplified form? No, because 8 is a perfect-cube factor of 200:

$200 = 8 \cdot 25$

Thus,

$$\sqrt[3]{200} = \sqrt[3]{8 \cdot 25} = \sqrt[3]{8}\sqrt[3]{25}$$
$$= 2\sqrt[3]{25} \quad \text{25 has no perfect-cube factor other than 1}$$

Putting everything together,

$$\sqrt[3]{\frac{8}{5}} = \frac{\sqrt[3]{200}}{5} = \frac{2\sqrt[3]{25}}{5}$$

That is,

$$\sqrt[3]{\frac{8}{5}} = \frac{2}{5}\sqrt[3]{25} \quad \text{Simplified form}$$

Problem 1 Simplify.

(a) $\sqrt{\dfrac{8}{27}}$ $\qquad$ (b) $\sqrt[3]{\dfrac{3}{4}}$ $\qquad$ (c) $\sqrt{\dfrac{45}{98}}$

SOLUTION:

(a) Multiply numerator and denominator of $\frac{8}{27}$ by the same integer such that the resulting denominator is a perfect square.

$$\frac{8}{27} = \frac{8}{27} \cdot \frac{3}{3} \qquad 27 = 3 \cdot 3 \cdot 3, \; 81 = 27 \cdot 3 = 9^2$$

$$= \frac{24}{81}$$

Then

$$\sqrt{\frac{8}{27}} = \sqrt{\frac{24}{81}} = \frac{\sqrt{24}}{\sqrt{81}}$$

$$= \frac{\sqrt{24}}{9}$$

In turn, $24 = 4 \cdot 6$, where 4 is the largest perfect-square factor of 24. Therefore,

$$\sqrt{\frac{8}{27}} = \frac{\sqrt{4 \cdot 6}}{9}$$

$$= \frac{\sqrt{4} \cdot \sqrt{6}}{9}, \quad \text{or} \quad \frac{2\sqrt{6}}{9}$$

Thus, the simplified form is

$$\sqrt{\frac{8}{27}} = \frac{2}{9}\sqrt{6}$$

(b) Proceed as above, using cubes instead of squares.

$$\frac{3}{4} = \frac{3}{4} \cdot \frac{2}{2} = \frac{6}{8} \qquad \begin{array}{l} 4 = 2 \cdot 2, \\ 8 = 2 \cdot 2 \cdot 2 = 2^3 \end{array}$$

Then

$$\sqrt[3]{\frac{3}{4}} = \sqrt[3]{\frac{6}{8}} = \frac{\sqrt[3]{6}}{\sqrt[3]{8}}, \quad \text{or} \quad \frac{\sqrt[3]{6}}{2}$$

Thus, the simplified form is

$$\sqrt[3]{\frac{3}{4}} = \frac{1}{2}\sqrt[3]{6}$$

(c) $\dfrac{45}{98} = \dfrac{45}{98} \cdot \dfrac{2}{2} = \dfrac{90}{196}$

- $98 = 49 \cdot 2 = 7^2 \cdot 2$
  $98 \cdot 2 = 7^2 \cdot 2^2 = (7 \cdot 2)^2$

$\sqrt{\dfrac{45}{98}} = \sqrt{\dfrac{90}{196}} = \dfrac{\sqrt{90}}{\sqrt{196}}, \quad \text{or} \quad \dfrac{\sqrt{90}}{14}$

- $90 = 9 \cdot 10$; 9 is largest perfect-square factor of 90

$\dfrac{\sqrt{90}}{14} = \dfrac{\sqrt{9 \cdot 10}}{14} = \dfrac{\sqrt{9}\sqrt{10}}{14} = \dfrac{3\sqrt{10}}{14}$

Thus, the simplified form is

$$\sqrt{\frac{45}{98}} = \frac{3}{14}\sqrt{10} \qquad \sqrt{10} \text{ is in reduced form}$$

Radical expressions containing variables can also be simplified. For example,

$$\sqrt{18x^3y^2} = \sqrt{3^2 \cdot 2 \cdot x^2 \cdot x \cdot y^2} \qquad \text{Assume variables have positive values}$$
$$= \sqrt{(3^2x^2y^2) \cdot (2x)} \qquad 3^2x^2y^2 \text{ is a perfect square,}$$
$$= \sqrt{3^2x^2y^2} \cdot \sqrt{2x} \qquad 2x \text{ has no perfect-square factor}$$
$$= 3xy\sqrt{2x}$$

Thus,

$$\sqrt{18x^3y^2} = 3xy\sqrt{2x}$$

To simplify

$$\sqrt{\frac{2a}{3b}}$$

find an equivalent fraction whose denominator is a perfect square.

$$\sqrt{\frac{2a}{3b}} = \sqrt{\frac{2a \cdot 3b}{3b \cdot 3b}} = \sqrt{\frac{6ab}{(3b)^2}}$$
$$= \frac{\sqrt{6ab}}{\sqrt{(3b)^2}} = \frac{\sqrt{6ab}}{3b}$$

Thus,

$$\sqrt{\frac{2a}{3b}} = \frac{1}{3b}\sqrt{6ab} \qquad \frac{1}{3b} \text{ is a rational expression, } 6ab \text{ is a monomial}$$

Observe that in each example, a root of a rational expression was simplified to a rational expression times a root of a monomial that has no perfect-square factor other than 1.

Problem 2 Simplify.

(a) $\sqrt{49x^4y^5}$ (b) $\sqrt{125ab^3c^7}$ (c) $\sqrt{\frac{63x}{4y^5}}$ (d) $\frac{\sqrt{3ab^3}}{\sqrt{4a^4b^2}}$

SOLUTION:

(a) $\sqrt{49x^4y^5} = \sqrt{7^2(x^2)^2(y^2)^2 y}$
$= \sqrt{(7x^2y^2)^2 y}$
$= \sqrt{(7x^2y^2)^2}\sqrt{y}$
$= 7x^2y^2\sqrt{y}$

Thus, in simplified form,
$$\sqrt{49x^4y^5} = 7x^2y^2\sqrt{y}$$

(b) $\sqrt{125ab^3c^7} = \sqrt{5 \cdot 5^2 \cdot a \cdot b \cdot b^2 \cdot c \cdot c^6}$

$\qquad\qquad\quad = \sqrt{5abc \cdot 5^2b^2c^6}$

$\qquad\qquad\quad = \sqrt{5abc}\sqrt{5^2b^2c^6}$

$\qquad\qquad\quad = \sqrt{5abc} \cdot 5bc^3$

Thus,
$$\sqrt{125ab^3c^7} = 5bc^3\sqrt{5abc} \qquad \textit{Simplified form}$$

(c) $\sqrt{\dfrac{63x}{4y^5}} = \sqrt{\dfrac{63x \cdot y}{4y^5 \cdot y}} = \sqrt{\dfrac{63xy}{4y^6}}$ *Denominator $4y^6$ is a perfect square ($2y^3$ squared)*

$\qquad = \dfrac{\sqrt{63xy}}{\sqrt{4y^6}} = \dfrac{\sqrt{9 \cdot 7xy}}{2y^3}$

$\qquad = \dfrac{\sqrt{9}\sqrt{7xy}}{2y^3} = \dfrac{3\sqrt{7xy}}{2y^3}$

Thus,
$$\sqrt{\dfrac{63x}{4y^5}} = \dfrac{3}{2y^3}\sqrt{7xy} \qquad \textit{Simplified form}$$

(d) $\dfrac{\sqrt{3ab^3}}{\sqrt{4a^4b^2}} = \sqrt{\dfrac{3ab^3}{4a^4b^2}} = \sqrt{\dfrac{3b}{4a^3}}$ ■ $\dfrac{a}{a^4} = \dfrac{1}{a^3}, \dfrac{b^3}{b^2} = b$

$\qquad = \sqrt{\dfrac{3b \cdot a}{4a^3 \cdot a}}$ ■ $4a^3 \cdot a = 4a^4$, *a perfect square*

$\qquad = \dfrac{\sqrt{3ba}}{\sqrt{4a^4}} = \dfrac{\sqrt{3ba}}{2a^2}$

Thus,
$$\dfrac{\sqrt{3ab^3}}{\sqrt{4a^4b^2}} = \dfrac{1}{2a^2}\sqrt{3ba} \qquad \textit{Simplified form}$$

Quick Reinforcement

Simplify.

(a) $\sqrt{48}$ (b) $\sqrt[4]{32}$ (c) $\sqrt{\dfrac{7}{50}}$ (d) $\sqrt[3]{24x^4y^3}$ (e) $\sqrt{\dfrac{4}{5b^3}}$

Answers (a) $4\sqrt{3}$ (b) $2\sqrt[4]{2}$ (c) $\dfrac{1}{10}\sqrt{14}$ (d) $2xy\sqrt[3]{3x}$ (e) $\dfrac{2}{5b^2}\sqrt{5b}$

EXERCISES 11-2

Express each of the following numbers in simplified form.

1. (a) $\sqrt{32}$ (b) $\sqrt{51}$
2. (a) $\sqrt[3]{54}$ (b) $\sqrt[3]{192}$
3. (a) $\sqrt{7.2}$ (b) $\sqrt{\dfrac{12}{9}}$
4. (a) $\sqrt{98}$ (b) $\sqrt{216}$
5. (a) $\sqrt{\dfrac{52}{27}}$ (b) $\sqrt{\dfrac{24}{49}}$
6. (a) $\sqrt[3]{2000}$ (b) $\sqrt[3]{-54}$
7. (a) $\sqrt[4]{162}$ (b) $\sqrt[5]{64}$
8. (a) $\sqrt[6]{1}$ (b) $\sqrt{\dfrac{11}{72}}$
9. (a) $\sqrt{\dfrac{75}{20}}$ (b) $\sqrt{\dfrac{242}{720}}$
10. (a) $\sqrt[3]{0.016}$ (b) $\sqrt[4]{0.0001}$
11. (a) $\sqrt[3]{\dfrac{16}{27}}$ (b) $\sqrt[3]{\dfrac{250}{64}}$
12. (a) $\sqrt[3]{-320}$ (b) $\sqrt[5]{96}$

Change each of the following radical expressions into simplified form.

13. (a) $\sqrt{a^2bc}$ (b) $\sqrt{x^3y^2z}$
14. (a) $\sqrt{x^3y^2}$ (b) $\sqrt{25r^2s^2}$
15. (a) $\sqrt{72p^3q}$ (b) $\sqrt{162x^3y^3}$
16. (a) $\sqrt{\dfrac{9}{7m^4}}$ (b) $\sqrt{\dfrac{16x^2}{25y^3}}$
17. (a) $\sqrt{81m^5}$ (b) $\sqrt{125x^3}$
18. (a) $\sqrt{\dfrac{98tv}{11z^5}}$ (b) $\dfrac{\sqrt{5a^2b}}{\sqrt{18ab^3}}$
19. (a) $\sqrt[3]{\dfrac{x^3y^6}{z^9}}$ (b) $\sqrt[3]{\dfrac{m^4n^3}{54p^7}}$
20. (a) $\sqrt[4]{\dfrac{32a^5}{b^8}}$ (b) $\sqrt[3]{-216a^5b^3}$
21. (a) $\sqrt{108a^2}$ (b) $\sqrt[3]{250a^3b}$
22. (a) $\sqrt[3]{-27a^4b^4}$ (b) $\sqrt[4]{16a^4b^5}$

11-3 OPERATIONS WITH SQUARE ROOTS

Certain sums involving roots can be simplified, while others cannot. For example, $5\sqrt{7} + 8\sqrt{7}$ can be simplified as follows:

$$5\sqrt{7} + 8\sqrt{7} = (5+8)\sqrt{7} \quad \text{\textit{Distributive property}}$$
$$= 13\sqrt{7}$$

On the other hand, the sum

$$2\sqrt{5} + 9\sqrt{11}$$

cannot be simplified. There is no way to express such a sum in terms of a single square root of an integer. However, you can approximate its value using a calculator:

$$\sqrt{5} \approx 2.236, \quad \sqrt{11} \approx 3.316 \quad \approx \text{\textit{means "is approximately equal to"}}$$

Therefore,

$$2\sqrt{5} + 9\sqrt{11} \approx 34.316$$

Sometimes, sums that look like they cannot be simplified really can be. That is, terms that look different from all other terms actually may be closely related to one or more other terms. For example, the sum

$$2\sqrt{5} + 6\sqrt{45}$$

can be simplified as follows:

$$2\sqrt{5} + 6\sqrt{45} = 2\sqrt{5} + 6\sqrt{9 \cdot 5} = 2\sqrt{5} + 6\sqrt{9}\sqrt{5}$$
$$= 2\sqrt{5} + 6 \cdot 3\sqrt{5} = 2\sqrt{5} + 18\sqrt{5}$$
$$= (2+18)\sqrt{5} = 20\sqrt{5}$$

Thus,

$$2\sqrt{5} + 6\sqrt{45} = 20\sqrt{5}$$

Problem 1 Simplify.

(a) $3\sqrt{24} + 5\sqrt{54}$ \quad (b) $2\sqrt{20} + 6\sqrt{80} - 5\sqrt{125}$

(c) $(7\sqrt{3} + 2\sqrt{13}) + (2\sqrt{3} - 5\sqrt{13})$

SOLUTION:

(a) $3\sqrt{24}$ and $5\sqrt{54}$ can be simplified.

$3\sqrt{24} = 3\sqrt{4 \cdot 6} = 3 \cdot \sqrt{4} \cdot \sqrt{6} = 3 \cdot 2 \cdot \sqrt{6} = 6\sqrt{6}$ \quad \textit{Simplified form}

$5\sqrt{54} = 5\sqrt{9 \cdot 6} = 5 \cdot \sqrt{9} \cdot \sqrt{6} = 5 \cdot 3 \cdot \sqrt{6} = 15\sqrt{6}$ \quad \textit{Simplified form}

Then
$$3\sqrt{24} + 5\sqrt{54} = 6\sqrt{6} + 15\sqrt{6}$$
$$= (6 + 15)\sqrt{6}$$
$$= 21\sqrt{6}$$

Hence,
$$3\sqrt{24} + 5\sqrt{54} = 21\sqrt{6}$$

(b) Simplify each term, then combine like terms.
$$2\sqrt{20} + 6\sqrt{80} - 5\sqrt{125} = 2\sqrt{4 \cdot 5} + 6\sqrt{16 \cdot 5} - 5\sqrt{25 \cdot 5}$$
$$= 2\sqrt{4}\sqrt{5} + 6\sqrt{16}\sqrt{5} - 5\sqrt{25}\sqrt{5}$$
$$= 2 \cdot 2\sqrt{5} + 6 \cdot 4\sqrt{5} - 5 \cdot 5\sqrt{5}$$
$$= 4\sqrt{5} + 24\sqrt{5} - 25\sqrt{5}$$
$$= (4 + 24 - 25)\sqrt{5}$$
$$= 3\sqrt{5}$$

Hence,
$$2\sqrt{20} + 6\sqrt{80} - 5\sqrt{125} = 3\sqrt{5}$$

(c) The roots are all in simplified form. Thus, using the associative and commutative properties of addition,
$$(7\sqrt{3} + 2\sqrt{13}) + (2\sqrt{3} - 5\sqrt{13}) = (7\sqrt{3} + 2\sqrt{3}) + (2\sqrt{13} - 5\sqrt{13})$$

Collecting like roots
$$= (7 + 2)\sqrt{3} + (2 - 5)\sqrt{13}$$
$$= 9\sqrt{3} - 3\sqrt{13}$$

Simplified form

Products involving roots can also be simplified at times. For example, using the associative and commutative properties of multiplication,
$$3\sqrt{60} \cdot 5\sqrt{50} = (3 \cdot 5) \cdot (\sqrt{60} \cdot \sqrt{50})$$
$$= 15\sqrt{3000}$$

■ $3000 = 100 \cdot 30$; 30 *has no perfect-square factor*

$$= 15\sqrt{100} \cdot \sqrt{30}$$
$$= 15 \cdot 10\sqrt{30}$$
$$= 150\sqrt{30}$$

11-3: OPERATIONS WITH SQUARE ROOTS

Therefore,

$$3\sqrt{60} \cdot 5\sqrt{50} = 150\sqrt{30} \qquad \textit{Simplified form}$$

As another example, find $4\sqrt{14} \cdot 3\sqrt{22}$.

$$\begin{aligned} 4\sqrt{14} \cdot 3\sqrt{22} &= 4 \cdot 3 \cdot \sqrt{14} \cdot \sqrt{22} \\ &= 12\sqrt{14 \cdot 22} \qquad \blacksquare \; 14 \cdot 22 = (2 \cdot 7)(2 \cdot 11) = 4 \cdot 77 \\ &= 12\sqrt{4 \cdot 77} \\ &= 12\sqrt{4}\sqrt{77}, \quad \text{or} \quad 24\sqrt{77} \end{aligned}$$

Therefore,

$$4\sqrt{14} \cdot 3\sqrt{22} = 24\sqrt{77} \qquad \textit{Simplified form}$$

You find a product such as

$$\sqrt{6}(2\sqrt{5} + 3\sqrt{6})$$

by using the distributive property:

$$\begin{aligned} \sqrt{6}(2\sqrt{5} + 3\sqrt{6}) &= 2\sqrt{6}\sqrt{5} + 3\sqrt{6}\sqrt{6} \qquad \blacksquare \; \sqrt{6}\sqrt{6} = 6 \\ &= 2\sqrt{30} + 18 \qquad \textit{Simplified form} \end{aligned}$$

Products such as

$$(2\sqrt{3} + 5\sqrt{2})(\sqrt{3} + 6\sqrt{2})$$

can be simplified as follows, again using the distributive property:

$$(2\sqrt{3} + 5\sqrt{2})(\sqrt{3} + 6\sqrt{2})$$
$$= 2\sqrt{3} \cdot \sqrt{3} + 2\sqrt{3} \cdot 6\sqrt{2} + 5\sqrt{2} \cdot \sqrt{3} + 5\sqrt{2} \cdot 6\sqrt{2}$$

(Multiply each term of one sum by each term of the other sum)

$$\begin{aligned} &= 2\sqrt{9} + 12\sqrt{6} + 5\sqrt{6} + 30\sqrt{4} \\ &= 6 + 17\sqrt{6} + 60 \\ &= 66 + 17\sqrt{6} \end{aligned}$$

Thus,

$$(2\sqrt{3} + 5\sqrt{2})(\sqrt{3} + 6\sqrt{2}) = 66 + 17\sqrt{6} \qquad \textit{Simplified form}$$

Problem 2 Simplify.

(a) $(4\sqrt{2} - 5)(4\sqrt{2} + 5)$ \qquad (b) $(\sqrt{7} - 2\sqrt{10})(3\sqrt{7} + 4\sqrt{10})$

(c) $(\sqrt{2} + \sqrt{3})^2$

SOLUTION:

(a) This is a difference of two squares:
$$(4\sqrt{2} - 5)(4\sqrt{2} + 5) = (4\sqrt{2})^2 - 5^2 \quad \blacksquare \ (A - B)(A + B) = A^2 - B^2$$
$$= 4^2(\sqrt{2})^2 - 5^2$$
$$= 16(2) - 25$$
$$= 32 - 25$$
$$= 7$$

Thus,
$$(4\sqrt{2} - 5)(4\sqrt{2} + 5) = 7$$

(b) $(\sqrt{7} - 2\sqrt{10})(3\sqrt{7} + 4\sqrt{10})$ $\blacksquare \ \sqrt{7} \cdot \sqrt{7} = 7, \ \sqrt{7} \cdot \sqrt{10}$
$$= 3 \cdot 7 + 4\sqrt{70} - 6\sqrt{70} - 8 \cdot 10 \quad = \sqrt{70}, \ \sqrt{10} \cdot \sqrt{10} = 10$$
$$= 21 - 2\sqrt{70} - 80$$
$$= -59 - 2\sqrt{70}$$

Thus,
$$(\sqrt{7} - 2\sqrt{10})(3\sqrt{7} + 4\sqrt{10}) = -59 - 2\sqrt{70} \quad \textit{Simplified form}$$

(c) $(\sqrt{2} + \sqrt{3})^2 = (\sqrt{2} + \sqrt{3})(\sqrt{2} + \sqrt{3})$
$$= (\sqrt{2})^2 + 2\sqrt{2}\sqrt{3} + (\sqrt{3})^2 \quad \blacksquare \ (A + B)^2 = A^2 + 2AB + B^2$$
$$= 2 + 2\sqrt{2 \cdot 3} + 3$$
$$= 5 + 2\sqrt{6}$$

Thus,
$$(\sqrt{2} + \sqrt{3})^2 = 5 + 2\sqrt{6}$$

A quotient of two expressions involving square roots, such as
$$\frac{3 + \sqrt{3}}{2 - \sqrt{3}}$$
can be simplified by a process called **rationalizing the denominator**. In this process, you multiply numerator and denominator by an expression that will eliminate roots from the denominator. How this is done is illustrated below.

$$\frac{3 + \sqrt{3}}{2 - \sqrt{3}} = \frac{(3 + \sqrt{3}) \cdot (2 + \sqrt{3})}{(2 - \sqrt{3}) \cdot (2 + \sqrt{3})} \quad \textit{Multiply numerator and denominator by } 2 + \sqrt{3}, \textit{ which makes the new denominator a difference of two squares}$$

$$= \frac{6 + 3\sqrt{3} + 2\sqrt{3} + 3}{2^2 - (\sqrt{3})^2}$$

$$= \frac{9 + 5\sqrt{3}}{4 - 3}$$

$$= 9 + 5\sqrt{3}$$

Thus,

$$\frac{3 + \sqrt{3}}{2 - \sqrt{3}} = 9 + 5\sqrt{3} \qquad \textit{Simplified form}$$

Problem 3 Simplify by rationalizing the denominator.

(a) $\dfrac{2}{2\sqrt{5} - 3\sqrt{2}}$ \qquad (b) $\dfrac{4 + 2\sqrt{7}}{3 + \sqrt{7}}$

SOLUTION:

(a) $\dfrac{2}{2\sqrt{5} - 3\sqrt{2}} = \dfrac{2}{2\sqrt{5} - 3\sqrt{2}} \cdot \dfrac{2\sqrt{5} + 3\sqrt{2}}{2\sqrt{5} + 3\sqrt{2}}$ \qquad *Make denominator a difference of two squares*

$$= \frac{2(2\sqrt{5} + 3\sqrt{2})}{(2\sqrt{5})^2 - (3\sqrt{2})^2}$$

$$= \frac{2(2\sqrt{5} + 3\sqrt{2})}{4 \cdot 5 - 9 \cdot 2}$$

$$= \frac{2(2\sqrt{5} + 3\sqrt{2})}{2}, \quad \text{or} \quad 2\sqrt{5} + 3\sqrt{2}$$

Thus,

$$\frac{2}{2\sqrt{5} - 3\sqrt{2}} = 2\sqrt{5} + 3\sqrt{2} \qquad \textit{Simplified form}$$

(b) $\dfrac{4 + 2\sqrt{7}}{3 + \sqrt{7}} = \dfrac{4 + 2\sqrt{7}}{3 + \sqrt{7}} \cdot \dfrac{3 - \sqrt{7}}{3 - \sqrt{7}}$ \qquad *Make denominator a difference of two squares*

$$= \frac{(4 + 2\sqrt{7})(3 - \sqrt{7})}{(3 + \sqrt{7})(3 - \sqrt{7})}$$

$$= \frac{12 - 4\sqrt{7} + 6\sqrt{7} - 2 \cdot 7}{3^2 - (\sqrt{7})^2}$$

$$= \frac{-2 + 2\sqrt{7}}{2}, \quad \text{or} \quad -1 + \sqrt{7}$$

Thus,

$$\frac{4 + 2\sqrt{7}}{3 + \sqrt{7}} = -1 + \sqrt{7} \qquad \textit{Simplified form}$$

Quick Reinforcement

Perform the indicated operations and simplify.

(a) $-4\sqrt{2} + \sqrt{18} - 5\sqrt{8}$ (b) $\sqrt{6} \cdot \sqrt{18}$

(c) $(\sqrt{5} - \sqrt{2})(2\sqrt{5} + 3\sqrt{2})$ (d) $\dfrac{5}{\sqrt{3} - 1}$

Answers (a) $-11\sqrt{2}$ (b) $6\sqrt{3}$ (c) $4 + \sqrt{10}$ (d) $\dfrac{5(\sqrt{3}+1)}{2}$

EXERCISES 11-3

Simplify each expression by combining terms where possible.

1. (a) $\sqrt{2} + \sqrt{50}$ (b) $\sqrt{5} + \sqrt{45}$
2. (a) $\sqrt{45} - \sqrt{20}$ (b) $\sqrt{9} + \sqrt{16}$
3. (a) $9\sqrt{18y} - 3\sqrt{72y}$ (b) $2\sqrt{27x} - \sqrt{3x}$
4. (a) $5\sqrt{72} - 2\sqrt{50}$ (b) $(\sqrt{75} - \sqrt{27}) + (4\sqrt{3} - 3\sqrt{12})$
5. (a) $4\sqrt{3} - 3\sqrt{12}$ (b) $7\sqrt{2} - 2\sqrt{128}$
6. (a) $-5\sqrt{32a} + \sqrt{98a}$ (b) $\sqrt{72} + 6\sqrt{98}$
7. (a) $9\sqrt{24} - 2\sqrt{54} + 3\sqrt{20}$ (b) $-4\sqrt{12x} + 12\sqrt{2y} - 5\sqrt{27x}$
8. (a) $10\sqrt{72a^3} + 4\sqrt{128a^3} - 3\sqrt{48b}$ (b) $4\sqrt{48x^2} + 18\sqrt{28y^3} - 6\sqrt{18x}$

Simplify each product below.

9. (a) $\sqrt{0.5} \cdot \sqrt{200}$ (b) $\sqrt{6} \cdot \sqrt{54}$
10. (a) $\sqrt{0.18} \cdot \sqrt{2}$ (b) $\sqrt{3} \cdot \sqrt{27}$
11. (a) $\sqrt{6} \cdot \sqrt{24}$ (b) $(2\sqrt{3}) \cdot (3\sqrt{5})$
12. (a) $(5\sqrt{3}) \cdot (-2\sqrt{6})$ (b) $(4\sqrt{2}) \cdot (2\sqrt{18})$
13. (a) $(3\sqrt{7}) \cdot (4\sqrt{14})$ (b) $\sqrt{98} \cdot \sqrt{8}$
14. (a) $\sqrt{60} \cdot \sqrt{1.2}$ (b) $(-4\sqrt{2}) \cdot (5\sqrt{0.5})$

In each of the following exercises, find the indicated product and simplify.

15. (a) $\sqrt{10}(\sqrt{5} + 2\sqrt{10})$ (b) $\sqrt{3}(2\sqrt{27} - 5\sqrt{6})$
16. (a) $\sqrt{3}(\sqrt{3} + \sqrt{0.2})$ (b) $\sqrt{5}(\sqrt{20} - 3\sqrt{125})$
17. (a) $\sqrt{6}(4\sqrt{6} - 2\sqrt{3})$ (b) $(1 + \sqrt{2})(2 + \sqrt{2})$
18. (a) $(2\sqrt{7} - 1)(\sqrt{7} + 2.2)$ (b) $(\sqrt{5} - \sqrt{3})(\sqrt{5} + \sqrt{3})$
19. (a) $(\sqrt{8} - 5)(3\sqrt{8} + 7)$ (b) $(2\sqrt{3} + 4)(2\sqrt{3} - 4)$
20. (a) $(\sqrt{10} + 5)(\sqrt{10} + 3)$ (b) $(4\sqrt{2} - 1.3)^2$
21. (a) $(5\sqrt{7} - 1)^2$ (b) $(3\sqrt{15} + \sqrt{5})(\sqrt{5} - 2\sqrt{10})$
22. (a) $(4\sqrt{12} + 3\sqrt{3})(2\sqrt{12} - \sqrt{3})$ (b) $(\sqrt{8} - 2\sqrt{12})(3\sqrt{8} + \sqrt{12})$

In each of the following exercises, rationalize the denominator and simplify.

23. (a) $\dfrac{1}{\sqrt{2}-1}$ (b) $\dfrac{4}{\sqrt{3}+1}$

24. (a) $\dfrac{6}{\sqrt{5}-2}$ (b) $\dfrac{2}{2\sqrt{7}-5}$

25. (a) $\dfrac{5}{2\sqrt{3}-4}$ (b) $\dfrac{7}{4\sqrt{2}+3}$

26. (a) $\dfrac{3}{8-2\sqrt{7}}$ (b) $\dfrac{11}{12+4\sqrt{3}}$

27. (a) $\dfrac{\sqrt{2}-4}{3\sqrt{3}-5}$ (b) $\dfrac{8+\sqrt{5}}{6\sqrt{7}-1}$

28. (a) $\dfrac{\sqrt{6}+2\sqrt{2}}{\sqrt{5}-4\sqrt{7}}$ (b) $\dfrac{2\sqrt{10}-3\sqrt{15}}{2\sqrt{10}+3\sqrt{15}}$

11-4 ALGEBRAIC EXPRESSIONS

An expression consisting of sums, differences, products, quotients, and roots of numbers and variables is called an **algebraic expression.** It becomes a number when values are given to the variables. Some examples of algebraic expressions are

$$\dfrac{3\sqrt{x}+\sqrt{y}}{x^2+y^2}, \quad \dfrac{\sqrt[3]{a}+3b}{\sqrt[3]{a}+\sqrt[3]{b}}, \quad \sqrt{3x}\cdot\sqrt{6x^3}$$

You are restricted to giving values to the variables such that all expressions make sense. For example, x and y are limited to positive values in the algebraic expression $(3\sqrt{x}+\sqrt{y})/(x^2+y^2)$.

You can simplify certain algebraic expressions just as you could simplify similar numerical expressions in the preceding section. For example, $3\sqrt{y}-5\sqrt{y}$ can be simplified as follows:

$$3\sqrt{y}-5\sqrt{y}=(3-5)\sqrt{y} \quad \textit{Distributive property}$$
$$=-2\sqrt{y}$$

Thus,

$$3\sqrt{y}-5\sqrt{y}=-2\sqrt{y}$$

Similarly,

$$(4\sqrt{x}+2\sqrt{z})+(5\sqrt{x}-6\sqrt{z}) \quad \textit{Associative and}$$
$$=4\sqrt{x}+5\sqrt{x}+2\sqrt{z}-6\sqrt{z} \quad \textit{commutative properties}$$

$$= (4+5)\sqrt{x} + (2-6)\sqrt{z}$$
$$= 9\sqrt{x} + -4\sqrt{z}$$

Hence,
$$(4\sqrt{x} + 2\sqrt{z}) + (5\sqrt{x} - 6\sqrt{z}) = 9\sqrt{x} - 4\sqrt{z}$$

As another example,
$$2\sqrt{x} \cdot 3\sqrt{x} = 2 \cdot 3(\sqrt{x})^2 = 6x$$

Hence,
$$2\sqrt{x} \cdot 3\sqrt{x} = 6x$$

A slightly more complicated example follows:

$$(5\sqrt{a} - 7\sqrt{b})(5\sqrt{a} + 7\sqrt{b})$$
$$= (5\sqrt{a})^2 - (7\sqrt{b})^2$$
$$= 25(\sqrt{a})^2 - 49(\sqrt{b})^2$$
$$= 25a - 49b$$

■ $(A-B)(A+B) = A^2 - B^2$; difference of two squares

Thus,
$$(5\sqrt{a} - 7\sqrt{b})(5\sqrt{a} + 7\sqrt{b}) = 25a - 49b$$

Similarly,
$$(3\sqrt{a} - 8\sqrt{b})^2$$
$$= (3\sqrt{a})^2 - 2 \cdot 3 \cdot 8\sqrt{a}\sqrt{b} + (8\sqrt{b})^2$$
$$= 9(\sqrt{a})^2 - 48\sqrt{ab} + 64(\sqrt{b})^2$$
$$= 9a - 48\sqrt{ab} + 64b$$

■ $(A-B)^2 = A^2 - 2AB + B^2$

Hence,
$$(3\sqrt{a} - 8\sqrt{b})^2 = 9a - 48\sqrt{ab} + 64b$$

Problem 1 Simplify.

(a) $(\sqrt{x} + 3)(2\sqrt{x} - 5)$ (b) $(3\sqrt{x} + \sqrt{y})(2\sqrt{x} + 5\sqrt{y})$

SOLUTION:

(a) $(\sqrt{x} + 3)(2\sqrt{x} - 5) = 2\sqrt{x} \cdot \sqrt{x} - 5 \cdot \sqrt{x} + 3 \cdot 2\sqrt{x} - 3 \cdot 5$
$$= 2x - 5\sqrt{x} + 6\sqrt{x} - 15 \quad ■ \ \sqrt{x} \cdot \sqrt{x} = x$$
$$= 2x + \sqrt{x} - 15 \quad ■ \ -5\sqrt{x} + 6\sqrt{x} = (-5+6)\sqrt{x}$$

Thus,

$$(\sqrt{x} + 3)(2\sqrt{x} - 5) = 2x - 15 + \sqrt{x}$$

(b) $(3\sqrt{x} + \sqrt{y})(2\sqrt{x} + 5\sqrt{y})$
$$= 3\sqrt{x} \cdot 2\sqrt{x} + 3\sqrt{x} \cdot 5\sqrt{y} + \sqrt{y} \cdot 2\sqrt{x} + \sqrt{y} \cdot 5\sqrt{y}$$
$$= 6x + 15\sqrt{x}\sqrt{y} + 2\sqrt{y}\sqrt{x} + 5y$$
$$= 6x + 5y + 17\sqrt{xy}$$

Thus,

$$(3\sqrt{x} + \sqrt{y})(2\sqrt{x} + 5\sqrt{y}) = 6x + 5y + 17\sqrt{xy}$$

Denominators of certain algebraic expressions can be rationalized as they were in the preceding section. For example,

$$\frac{1}{3\sqrt{a} - 2} = \frac{1}{3\sqrt{a} - 2} \cdot \frac{3\sqrt{a} + 2}{3\sqrt{a} + 2}$$ *Make the denominator a difference of two squares*

$$= \frac{3\sqrt{a} + 2}{(3\sqrt{a})^2 - 2^2}, \quad \text{or} \quad \frac{3\sqrt{a} + 2}{9a - 4}$$

Thus,

$$\frac{1}{3\sqrt{a} - 2} = \frac{3\sqrt{a} + 2}{9a - 4}$$ *The root is now in the numerator; the denominator is a polynomial*

Problem 2 Rationalize the denominator of the algebraic expression

$$\frac{4\sqrt{x} + 5\sqrt{y}}{2\sqrt{x} - 3\sqrt{y}}$$

SOLUTION:

$$\frac{4\sqrt{x} + 5\sqrt{y}}{2\sqrt{x} - 3\sqrt{y}} = \frac{4\sqrt{x} + 5\sqrt{y}}{2\sqrt{x} - 3\sqrt{y}} \cdot \frac{2\sqrt{x} + 3\sqrt{y}}{2\sqrt{x} + 3\sqrt{y}}$$ *Make the denominator a difference of two squares*

$$= \frac{(4\sqrt{x} + 5\sqrt{y})(2\sqrt{x} + 3\sqrt{y})}{(2\sqrt{x})^2 - (3\sqrt{y})^2}$$

$$= \frac{8x + 12\sqrt{x}\sqrt{y} + 10\sqrt{y}\sqrt{x} + 15y}{4x - 9y}$$

$$= \frac{8x + 15y + 22\sqrt{xy}}{4x - 9y}$$

Thus,

$$\frac{4\sqrt{x} + 5\sqrt{y}}{2\sqrt{x} - 3\sqrt{y}} = \frac{8x + 15y + 22\sqrt{xy}}{4x - 9y}$$ *The root is in the numerator; the denominator is a polynomial*

Quick Reinforcement

Perform the indicated operations.

(a) $(3\sqrt{x} - \sqrt{2})(\sqrt{x} + \sqrt{2})$

(b) $\dfrac{1}{5\sqrt{y} - \sqrt{x}}$

Answers (a) $3x - 2 + 2\sqrt{2x}$ (b) $\dfrac{5\sqrt{y} + \sqrt{x}}{25y - x}$

EXERCISES 11-4

In each of the following exercises, find the indicated product and simplify.

1. (a) $(\sqrt{y} + 3)(\sqrt{y} - 5)$ (b) $(2\sqrt{x} - 1)(3\sqrt{x} + 2)$
2. (a) $(2\sqrt{a} - 7)(5\sqrt{a} + 1)$ (b) $(\sqrt{xy} + 8)(3\sqrt{xy} - 5)$
3. (a) $(4\sqrt{x} + 3\sqrt{y})(7\sqrt{x} - \sqrt{y})$ (b) $(\sqrt{6a} + 3)(4\sqrt{6a} - 1)$
4. (a) $(\sqrt{3y} - 1)^2$ (b) $(5\sqrt{x} - 7)(5\sqrt{x} + 7)$
5. (a) $(5\sqrt{x} + 4)(2\sqrt{x} - 8)$ (b) $(\sqrt{7r} + 3)(2\sqrt{7r} - 4)$
6. (a) $(9\sqrt{a} + 3)(9\sqrt{a} - 3)$ (b) $(\sqrt{6a} + \sqrt{2b})(\sqrt{3a} - 2\sqrt{2b})$
7. (a) $(10\sqrt{2x} + \sqrt{5})(3\sqrt{2x} - 4\sqrt{5})$ (b) $(2\sqrt{y} + 4)^2$
8. (a) $(\sqrt{3xy} + 2)(2\sqrt{3xy} - 6)$ (b) $(5\sqrt{ab} + 2\sqrt{c})(5\sqrt{ab} - 2\sqrt{c})$

In each of the following exercises, rationalize the denominator and simplify.

9. (a) $\dfrac{1}{\sqrt{x} - 1}$ (b) $\dfrac{2}{2\sqrt{y} + 3}$
10. (a) $\dfrac{4}{3\sqrt{a} + 7}$ (b) $\dfrac{8}{6\sqrt{x} - 5}$
11. (a) $\dfrac{2\sqrt{x} + 5\sqrt{y}}{\sqrt{x}}$ (b) $\dfrac{6}{2\sqrt{x} - 3\sqrt{y}}$
12. (a) $\dfrac{5}{-2\sqrt{a} - 4\sqrt{5}}$ (b) $\dfrac{5\sqrt{a} - 7}{\sqrt{a}}$
13. (a) $\dfrac{12}{3\sqrt{2x} - 5}$ (b) $\dfrac{9}{2\sqrt{5a} + 7}$
14. (a) $\dfrac{\sqrt{2x} + 1}{\sqrt{2x} - 1}$ (b) $\dfrac{\sqrt{x} - 3}{\sqrt{x} + 4}$
15. (a) $\dfrac{3\sqrt{y} + 7\sqrt{z}}{\sqrt{y} - \sqrt{z}}$ (b) $\dfrac{\sqrt{3x} - \sqrt{5y}}{\sqrt{3x} + \sqrt{5y}}$
16. (a) $\dfrac{\sqrt{a} - \sqrt{b}}{\sqrt{a} + \sqrt{b}}$ (b) $\dfrac{6\sqrt{x} + 2\sqrt{y}}{5\sqrt{x} - 7\sqrt{y}}$

Calculator Activities

Evaluate each of the following. Estimate your answer first.

1. $\dfrac{45\sqrt{x} - 2x\sqrt{y}}{3xy}$
   (a) for $x = 18$, $y = 6$
   (b) for $x = 20$, $y = 10$

2. $\dfrac{0.2a\sqrt{bc} - 31\sqrt{c}}{\sqrt{a} + \sqrt{b}}$
   (a) for $a = 5$, $b = 101$, $c = 12$
   (b) for $a = 17$, $b = 50$, $c = 6$

11-5 RADICAL EQUATIONS, FORMULAS, AND APPLICATIONS

A **radical equation** is an equation whose sides involve radicals and which has a variable under a radical sign. Examples of radical equations are

$$\sqrt{x} = 4, \quad \sqrt{x - 3} - 3 = 2, \quad \sqrt{x - 1} + x = 3$$

To solve a radical equation involving only square roots, as in the equations above, isolate one of the radicals on one side of the equation and then square both sides. Repeat this process, if necessary, until you eliminate the radicals. Then solve the resulting equation. For example,

$\sqrt{x - 3} - 3 = 2$	*Given equation*
$\sqrt{x - 3} = 5$	*Add 3 to each side, to isolate $\sqrt{x - 3}$*
$(\sqrt{x - 3})^2 = 5^2$	*Square each side*
$x - 3 = 25$	
$x = 28$	*Possible solution*

√ CHECK:

Replace x by 28 in the given equation.

$$\sqrt{28 - 3} - 3 = 2$$
$$\sqrt{25} - 3 = 2$$
$$5 - 3 = 2 \;\checkmark$$

Thus, $x = 28$ is a solution of the radical equation $\sqrt{x - 3} - 3 = 2$.

Any solution of an equation is also a solution of the equation obtained by squaring each side. However, the new equation may have solutions that are not solutions of the given equation. Such solutions are called **extraneous solutions** of the given equation. As a simple example, consider

$x = 1$ 1 *is the only solution*
$x^2 = 1^2$ *Square each side*
$x^2 = 1$ *1 and -1 are solutions; -1 is extraneous*

Problem 1 Solve each radical equation.
(a) $\sqrt{7-x} = \sqrt{2}$ (b) $6 + \sqrt{2x+3} = 7$
(c) $\sqrt{x-1} + x = 3$

SOLUTION:

(a) $\sqrt{7-x} = \sqrt{2}$ *Given equation*
   $(\sqrt{7-x})^2 = (\sqrt{2})^2$ *Square each side*
            $7 - x = 2$
                $x = 5$

✓ CHECK:

Let $x = 5$
$$\sqrt{7-x} = \sqrt{2}$$
$$\sqrt{7-5} = \sqrt{2}$$
$$\sqrt{2} = \sqrt{2} \checkmark$$

Thus, $x = 5$ is the solution.

(b) $6 + \sqrt{2x+3} = 7$ *Given equation*
          $\sqrt{2x+3} = 1$ *Add -6 to each side to isolate $\sqrt{2x+3}$ on one side*
       $(\sqrt{2x+3})^2 = 1^2$ *Square each side*
              $2x + 3 = 1$
                  $2x = -2$
                   $x = -1$

✓ CHECK:

Let $x = -1$
$$6 + \sqrt{2x+3} = 7$$
$$6 + \sqrt{2 \cdot (-1) + 3} = 7$$
$$6 + \sqrt{-2+3} = 7$$
$$6 + \sqrt{1} = 7$$
$$6 + 1 = 7 \checkmark$$

Thus, $x = -1$ is the solution.

(c) $\sqrt{x-1} + x = 3$ *Given equation*
$\sqrt{x-1} = 3 - x$ *Isolate radical on left side*
$(\sqrt{x-1})^2 = (3-x)^2$ *Square each side*
$x - 1 = 9 - 6x + x^2$
$0 = x^2 - 7x + 10$ *Add $1 - x$ to each side*
$0 = (x-5)(x-2)$
$x - 5 = 0$ or $x - 2 = 0$
$x = 5$ or $x = 2$ *Possible solutions*

√ CHECK:

$\sqrt{x-1} + x = 3$

Let $x = 5$ Let $x = 2$
$\sqrt{5-1} + 5 = 3$ $\sqrt{2-1} + 2 = 3$
$\sqrt{4} + 5 = 3$ $\sqrt{1} + 2 = 3$
$2 + 5 = 3$ $1 + 2 = 3$ √

$2 + 5 \ne 3$; therefore 5 is *not* a solution.

Thus, $x = 2$ is the only solution of the radical equation $\sqrt{x-1} + x = 3$. The other possible solution, $x = 5$, is extraneous.

The lengths of the sides of a right triangle are related by the equation

$a^2 + b^2 = c^2$

according to the Pythagorean theorem. As shown in Figure 11-1, c is the length of the hypotenuse (the longest side), and a and b are the lengths of the legs. You can solve the equation above for c by taking square roots of both sides, obtaining

$c = \sqrt{a^2 + b^2}$

Since $a^2 = c^2 - b^2$,

$a = \sqrt{c^2 - b^2}$

and, similarly,

$b = \sqrt{c^2 - a^2}$

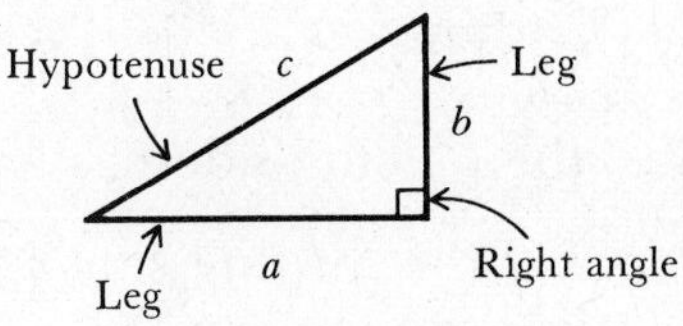

FIGURE 11-1

Problem 2 Find the length of the missing side of each right triangle.

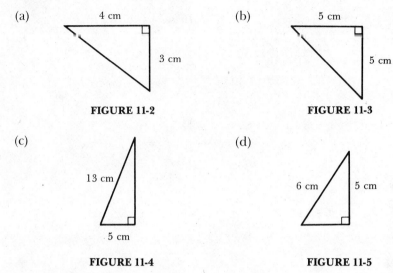

FIGURE 11-2

FIGURE 11-3

FIGURE 11-4

FIGURE 11-5

SOLUTION:

(a) Label the missing side c. Then

$$c = \sqrt{3^2 + 4^2}$$
$$= \sqrt{25}, \text{ or } 5$$

Thus, the missing side has length 5 centimeters.

(b) Label the missing side c. Then

$$c = \sqrt{5^2 + 5^2}$$
$$= \sqrt{50} = \sqrt{25 \cdot 2}$$

Thus, the missing side has length $5\sqrt{2}$ (≈ 7.071) centimeters.

(c) Label the missing side b. Then

$$b = \sqrt{13^2 - 5^2}$$
$$= \sqrt{169 - 25}$$
$$= \sqrt{144}, \text{ or } 12$$

Thus, the missing side has length 12 centimeters.

(d) Label the missing side a. Then

$$a = \sqrt{6^2 - 5^2}$$
$$= \sqrt{36 - 25}, \text{ or } \sqrt{11}$$

Thus, the missing side has length $\sqrt{11}$ (≈ 3.32) centimeters.

Radicals also appear in many formulas. For example, a circle of area A has diameter d given by

$$d = 2\sqrt{\frac{A}{\pi}}$$

If $A = 10$ square meters, then

$$d = 2\sqrt{\frac{10}{\pi}} \quad d \approx 3.568 \text{ on a calculator}$$

If you are h meters above the earth (see Figure 11-6), then you can see at most a distance of V kilometers to the horizon, where

$$V = 3.6\sqrt{h}$$

For example, if you are 10 meters above the earth, then

$$V = 3.6\sqrt{10} \quad \approx 11.384 \text{ kilometers}$$

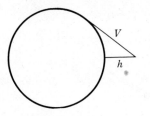

FIGURE 11-6

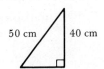

FIGURE 11-7

Problem 3

(a) Find the missing side of the right triangle in Figure 11-7.

(b) Find the diameter of a circular room whose area is 48 square meters.

(c) Find the distance you can see to the horizon if you are in an airplane 10,000 meters above the earth.

SOLUTION:

(a) If a is the length of the missing side, then

$$a = \sqrt{50^2 - 40^2}$$
$$= \sqrt{2500 - 1600}$$
$$= \sqrt{900}, \text{ or } 30$$

The missing side has a length of 30 centimeters.

(b) By the formula,

$$d = 2\sqrt{\frac{48}{\pi}} \quad \blacksquare \ \sqrt{48} = \sqrt{16 \cdot 3} = 4\sqrt{3}$$

$$= 8\sqrt{\frac{3}{\pi}}$$

Using a calculator, $d \approx 7.818$ meters.

(c) By the formula,

$$V = 3.6\sqrt{10{,}000}$$
$$= 3.6 \cdot 100$$
$$= 360$$

The distance you can see is approximately 360 kilometers.

Quick Reinforcement

(a) Solve the equation $\sqrt{x+1} - 5 = 0$.

(b) Solve for the missing side of the triangle in the figure.

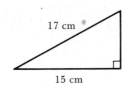
17 cm

15 cm

(c) If $A = 12{,}500$ and $P = 10{,}000$, determine r, given

$$r = \sqrt{\frac{A}{P}} - 1$$

Use a calculator.

Answers: (a) $x = 24$ (b) 8 (c) 0.118

EXERCISES 11-5

Solve each of the following radical equations. Check your solutions.

1. (a) $\sqrt{5x} = 2$ (b) $\sqrt{12y} = 4$
2. (a) $4 + \sqrt{2x} = 12$ (b) $\sqrt{6x - 2} = 3$
3. (a) $\sqrt{3 + x} = 4$ (b) $\sqrt{5 - y} = 2$
4. (a) $\sqrt{2 - 3x} = 6$ (b) $\sqrt{5x + 1} = 1$
5. (a) $7 - \sqrt{2x + 1} = 5$ (b) $4 + \sqrt{x - 3} = 7$

6. (a) $7 - 2\sqrt{5y} = 0$ (b) $4\sqrt{3z} = 6$
7. (a) $9 + 3\sqrt{2a} = 11$ (b) $\sqrt{7b} - 4 = -3$
8. (a) $4 - 3\sqrt{x} = 2 - \sqrt{x}$ (b) $2 + 5\sqrt{3y} = 3\sqrt{3y} + 12$
9. (a) $\sqrt{x-1} = \sqrt{2x+1}$ (b) $\sqrt{4x-3} = \sqrt{7x}$
10. (a) $\sqrt{2x+3} = \sqrt{4x+1}$ (b) $\sqrt{12x-5} = \sqrt{11x+2}$
11. (a) $x - \sqrt{x^2+6} = -2$ (b) $x + \sqrt{x^2-7} = 7$
12. (a) $\sqrt{x^2+1} + 2x = 3x$ (b) $4x - \sqrt{x^2+2} = 3 + 5x$

Find the length of the missing side of the triangle, given the following pairs of sides.

13. (a) $a = 6$ meters and $b = 8$ meters
    (b) $a = 9$ meters and $b = 12$ meters
14. (a) $a = 3$ centimeters and $c = 5$ centimeters
    (b) $a = 12$ centimeters and $c = 20$ centimeters
15. (a) $b = 1.2$ meters and $c = 5$ meters
    (b) $b = 35$ meters and $c = 50$ meters
16. (a) $a = 12$ centimeters and $b = 2$ centimeters
    (b) $a = 2.5$ meters and $c = 9.2$ meters

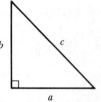

17. Use the formula $V = 3.6\sqrt{h}$ to find how far to the horizon you can see if
    (a) you are in an airplane at the height of 16,000 meters.
    (b) you are at the top of a tall building 150 meters high.
18. Use the formula $d = 2\sqrt{A/\pi}$ to find the diameter of a circle when
    (a) you are on a circular stage whose area is 128 square meters.
    (b) you have bought a round bed whose area is 4 square meters.
19. For a simple pendulum of length l, in centimeters, the time of the period, T, in seconds is given by

$$T = 2\pi\sqrt{\frac{l}{g}}$$

where g is given in centimeters per second squared. Find the time T if
(a) $l = 12$ cm, $g = 3$ cm/sec² (b) $l = 200$ cm, $g = 25$ cm/sec²

20. A balance that has unequal arms with weights W_1 and W_2, respectively, will have a true mass W according to $W = \sqrt{W_1 \cdot W_2}$. What is the true mass of a balance with
    (a) $W_1 = 25$ kilograms, $W_2 = 8$ kilograms?
    (b) $W_1 = 10$ grams, $W_2 = 5$ grams?
21. Using the formula for exercise 20 above, answer the following questions.
    (a) If the true mass of a balance were 30 kilograms and one of the unequal arms were 10 kilograms, what would the mass of the other arm be?
    (b) If the true mass of a balance were 15 grams and one of the unequal arms were 5 grams, what would the mass of the other arm be?

KEY TERMS

Finite decimals
Infinite repeating decimals
Infinite nonrepeating decimals
Radical sign
Index
Radicand

Simplified form
Rationalizing the denominator
Algebraic expression
Radical equation
Extraneous solutions

REVIEW EXERCISES

Find the decimal representation of the rational number. Check with a calculator.

1. (a) $\dfrac{6}{11}$ (b) $\dfrac{-4}{7}$
2. (a) $\dfrac{-4}{13}$ (b) $\dfrac{11}{19}$

Find the rational number represented by the given decimal.

3. (a) 3.025 (b) -0.893
4. (a) 0.5333 . . . (b) 1.148148 . . .

Find the rational number represented by each radical expression.

5. (a) $\sqrt{\dfrac{49}{625}}$ (b) $\sqrt{\dfrac{121}{36}}$
6. (a) $\sqrt{0.0144}$ (b) $\sqrt[3]{-0.216}$
7. (a) $\sqrt[6]{64}$ (b) $\sqrt[5]{-1}$

Express in simplified form.

8. (a) $\sqrt{243}$ (b) $\sqrt{28}$
9. (a) $\sqrt[3]{5000}$ (b) $\sqrt[3]{-250}$
10. (a) $\sqrt{\dfrac{75}{36}}$ (b) $\sqrt{\dfrac{405}{49}}$
11. (a) $\sqrt{289x^3}$ (b) $\sqrt[3]{-8a^3b^4}$
12. (a) $\sqrt[3]{27y^4z^5}$ (b) $\sqrt{175x^3y^2}$
13. (a) $\sqrt{\dfrac{48m^5n^4}{p^3r^6}}$ (b) $\dfrac{\sqrt[3]{7a^5b^4}}{\sqrt[3]{c^6}}$

Simplify each expression where possible.

14. (a) $3\sqrt{3} - 2\sqrt{48}$ (b) $\sqrt{180} + 5\sqrt{5}$
15. (a) $2\sqrt{63} + 4\sqrt{112}$ (b) $7\sqrt{27} - 4\sqrt{75}$

16. (a) $(6\sqrt{8})(5\sqrt{2})$ (b) $(7\sqrt{12})(8\sqrt{3})$
17. (a) $(-5.2\sqrt{6})(9.1\sqrt{8})$ (b) $(0.2\sqrt{7})(1.4\sqrt{21})$
18. (a) $\sqrt{3}(2\sqrt{3}-6)$ (b) $\sqrt{7}(4\sqrt{2}+3\sqrt{10})$
19. (a) $(3\sqrt{5}-2\sqrt{8})(6\sqrt{5}+3\sqrt{8})$ (b) $(8\sqrt{11}+3\sqrt{3})(-\sqrt{11}-2\sqrt{3})$

Rationalize the denominators and simplify.

20. (a) $\dfrac{3}{\sqrt{7}-2}$ (b) $\dfrac{5}{\sqrt{11}+6}$
21. (a) $\dfrac{\sqrt{8}+1}{\sqrt{8}-3}$ (b) $\dfrac{\sqrt{5}-\sqrt{2}}{2\sqrt{5}+\sqrt{2}}$

Find the indicated products and simplify.

22. (a) $(2\sqrt{x}-6)(3\sqrt{x}-5)$ (b) $(\sqrt{3a}-\sqrt{5})(\sqrt{3a}+\sqrt{5})$
23. (a) $(\sqrt{5x}-\sqrt{2y})(\sqrt{5x}-3\sqrt{2y})$ (b) $(2\sqrt{r}-5\sqrt{s})(2\sqrt{r}+5\sqrt{s})$

Rationalize the denominators and simplify.

24. (a) $\dfrac{5}{2\sqrt{x}-1}$ (b) $\dfrac{7}{3\sqrt{y}+4}$
25. (a) $\dfrac{2\sqrt{a}-3}{5\sqrt{a}+7}$ (b) $\dfrac{4\sqrt{r}-2\sqrt{s}}{4\sqrt{r}+2\sqrt{s}}$

Solve each of the radical equations. Check your solutions.

26. (a) $\sqrt{7x}=3$ (b) $\sqrt{9y-4}=5$
27. (a) $\sqrt{4-x}=-2$ (b) $\sqrt{2x+1}=11$
28. (a) $6+4\sqrt{y}=12$ (b) $\sqrt{5a+3}=2$
29. (a) $\sqrt{x-3}=\sqrt{4x+7}$ (b) $\sqrt{3x+2}-\sqrt{5x-1}=0$
30. (a) $\sqrt{x^2-3}=x-5$ (b) $x+\sqrt{x^2-7}=4$

Find the length of the missing side of the right triangle, given two sides.

31. (a) $a=0.2$ centimeters, $b=1.3$ centimeters
    (b) $a=12$ meters, $b=4$ meters
32. (a) $a=35$ centimeters, $c=70$ centimeters
    (b) $b=40$ kilometers, $c=90$ kilometers

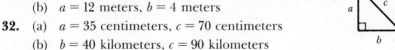

33. The circumference of an ellipse whose semiaxes are a and b is

$$C = 2\pi\sqrt{\dfrac{a^2+b^2}{2}}$$

(a) Find the circumference when $a=3$ centimeters and $b=4$ centimeters.
(b) If the circumference is 12.56 meters and $a=b$, find a. (What has happened to the ellipse?)

Calculator Activities

1. Estimate the following roots to two-place accuracy:
   (a) $\sqrt{15}$
   (b) $\sqrt[3]{30}$

2. Evaluate the following, estimating your answer first.

   $$\frac{22\sqrt{a} - 4a\sqrt{b}}{5ab}$$

   (a) for $a = 6$, $b = 12$
   (b) for $a = 72$, $b = 75$

12 Quadratic Equations

Susan Newman, an editor with The Benjamin/Cummings Publishing Company, is trying to figure out the breakeven point for a computer science textbook she has recently acquired. She knows that the revenue plots a quadratic equation, while her costs are linear. She has drawn a graph of the two equations to more clearly show the situation.

12-1 SOLVING QUADRATIC EQUATIONS BY FACTORING

You recall that an equation is called a **quadratic equation** in the variable x if it is equivalent to an equation of the form

$ax^2 + bx + c = 0$ Where a, b, c represent numbers with $a \neq 0$

For example, the equation $x(x + 2) = 7x + 6$ is equivalent to each of the following equations.

$$x(x + 2) = 7x + 6$$
$$x^2 + 2x = 7x + 6$$
$$x^2 + 2x - 7x - 6 = 0 \quad \text{Add } -7x - 6 \text{ to each side}$$
$$x^2 - 5x - 6 = 0 \quad \text{Of the form } ax^2 + bx + c = 0, a = 1$$

(equivalent)

Therefore, $x(x + 2) = 7x + 6$ is a quadratic equation.

If the polynomial $ax^2 + bx + c$ on the left side of the quadratic equation

$$ax^2 + bx + c = 0$$

can be factored, then the equation can be solved. How we solve such an equation was shown in Section 9-6. The following problems illustrate this method.

Problem 1 Solve the quadratic equation $x(x + 2) = 7x + 6$.

SOLUTION:

You saw above that the given equation is equivalent to the equation

$x^2 - 5x - 6 = 0$ *If $x^2 - 5x - 6 = (x + A)(x + B)$, then $AB = -6$, $A + B = -5$; by testing various possibilities, you see that $A = -6$, $B = 1$ works.*

or, in factored form,

$(x - 6)(x + 1) = 0$

Thus,

$x - 6 = 0$ or $x + 1 = 0$
$x = 6$ or $x = -1$

The solution set is $\{6, -1\}$.

✓ CHECK:

$x(x + 2) = 7x + 6$

Let $x = 6$ | Let $x = -1$
$6 \cdot (6 + 2) = 7 \cdot 6 + 6$ | $(-1) \cdot (-1 + 2) = 7 \cdot (-1) + 6$
$6 \cdot 8 = 42 + 6$ | $(-1) \cdot 1 = -7 + 6$
$48 = 48$ ✓ | $-1 = -1$ ✓
$x(x + 2) = 7x + 6$

This is the method of solving a quadratic equation by factoring introduced in Chapter 9.

Problem 2 The sum of twice a number and nine times its reciprocal equals nine. What is the number?

SOLUTION:

Let x designate the number, where $x \neq 0$. Then

$2x$ Twice the number

$\dfrac{9 \cdot \dfrac{1}{x}}{9}$ Nine times the reciprocal of the number

     Sum

$$2x + \dfrac{9}{x} = 9 \quad \blacksquare \ 9 \cdot \dfrac{1}{x} = \dfrac{9}{x}$$

$$2x^2 + 9 = 9x \quad \text{Multiply each side by } x$$

$$2x^2 - 9x + 9 = 0$$

$$(2x - 3)(x - 3) = 0$$

$$2x - 3 = 0 \quad \text{or} \quad x - 3 = 0$$

$$x = \dfrac{3}{2} \quad \text{or} \quad x = 3$$

The solution set is $\{\tfrac{3}{2}, 3\}$.

✓ CHECK:
$$2x + \dfrac{9}{x} = 9$$

Let $x = \tfrac{3}{2}$ Let $x = 3$

$2 \cdot \dfrac{3}{2} + 9 \cdot \left(\dfrac{2}{3}\right) = 9$ $2 \cdot 3 + \dfrac{9}{3} = 9$

$3 + 6 = 9$ ✓ $6 + 3 = 9$ ✓

Quick Reinforcement

Solve each of the following quadratic equations by factoring.

(a) $12y^2 = 17y - 6$ (b) $x(5x - 1) - 1 = -5x^2 + 8x$

Answers (a) $\{\tfrac{4}{3}, \tfrac{3}{2}\}$ (b) $\{1, -\tfrac{1}{10}\}$

EXERCISES 12-1

Solve each of the following quadratic equations by factoring. Check your solutions.

1. (a) $x(x + 3) = 4$ (b) $x(x - 6) = 3x - 20$
2. (a) $x(x + 17) = 14(x + 2)$ (b) $2(y^2 - 66) = -13y$
3. (a) $4m^2 + 2m - 3 = 2(m + 3)$ (b) $3(3a^2 + 4a) = -4$

4. (a) $p^2 - 4p + 4 = -(p-2)^2$ (b) $(2x+3)(x+1) = 3x(x+1)$
5. (a) $6a(a+1) + 3 = 5(a+1)$ (b) $4p(p+1) = 12(p+1)$
6. (a) $y^2 - 1 = 3(y-1)$ (b) $m^2 + 10m - 5 = 7(m+4) - 5$
7. (a) $3x(x-1) = 4x - 2$ (b) $a^2 + 12a - 3 = 7(3+a)$
8. (a) $a^2 + 9a + 5 = 5(2a+5)$ (b) $3x^2 - x + 56 = (x+6)^2$
9. (a) $5y^2 - y + 6 = 4y(y+1)$ (b) $6a(2a+1) = a+2$
10. (a) $4b^2 + 7(b+1) = 3b^2 - 5$ (b) $2c^2 + 18(c+2) - 3 = c^2 + 5c - 3$

Solve the following applied problems using quadratic equations.

11. (a) Jasper built a large rectangular sandbox for his triplets. The length of the sandbox is two meters greater than its width. If the width were increased by seven meters and the length by two meters, the area of the new sandbox would be 91 square meters greater than the original. What were the original dimensions?

    (b) Fred and Shirley were playing number games. Fred asked Shirley if she could find three consecutive integers such that the product of the first two exceeded the third integer by 47. Assuming she was a good algebra student, what would be her answer?

12. (a) In another number puzzle, Shirley asked Fred if he could find two numbers such that, first, one number is four more than the other, and, second, if the smaller one is squared and added to three times the larger, the result is 66. For what numbers was Fred looking?

    (b) The quilting club was cutting material into triangles for a new quilt. The pattern showed that the height of each triangle was one-half the base. They decided to add two centimeters to the height and subtract two from the base. This would give them an area of 18 square centimeters. What were the dimensions of the original pattern? (Area of a triangle $= \frac{1}{2} \times$ base $\times$ height.)

13. (a) A square fountain in Maseo's back yard has a 3-foot paved walk surrounding it. If Masco decided to tear out the walk and plant flowers, the combined area of the fountain and the flower bed would have four times the area of the fountain. What is the length of one side of the fountain?

    (b) A swimming pool slide is to be built at the recreation center in town. The bottom of the slide will be at a point four feet from the bottom of the pool. The ladder built vertically from the bottom of the pool will be sixteen feet high. If the ladder and the end of the slide are to be sixteen feet apart, what will be the length of the slide?

𝔑ostalgia 𝔓roblems

1. (a) A person purchased a number of horses for 240 dollars. If he had obtained 3 more for the same money, each horse would have cost

him 4 dollars less. Find the number of horses. (Hint: Let x = original number of horses, and $240/x$ = cost per horse.)

(b) In a certain number of hours a man traveled 36 miles; if he had traveled one more mile per hour, it would have taken him 3 hours less to perform his journey; how many miles did he travel per hour?

Calculator Activities

1. If x denotes the number of items produced in hundreds, and if your company's profits in thousands are given by

$$-(x - 3.5)(x - 7.5)$$

then

(a) what are your profits if you poduce five hundred items?
(b) what are your break-even production rates?
(c) will you ever lose money? Explain.

12-2 SOLVING QUADRATIC EQUATIONS BY TAKING SQUARE ROOTS

If A is a number such that $A^2 = 16$, then either $A = 4$ or $A = -4$. Stated more generally, if $A^2 = B$, then $A = \sqrt{B}$ or $A = -\sqrt{B}$ (where $B \geq 0$, of course). This property of numbers allows you to solve certain quadratic equations, as follows.

Problem 1 Solve the quadratic equation $x^2 = 9$.

SOLUTION:

By the property above,

$x = 3$ or $x = -3$

The solution set of the equation is $\{3, -3\}$.

Problem 2 Solve the quadratic equation $(x - 2)^2 = 25$.

SOLUTION:

By the property above,

$x - 2 = \sqrt{25}$ or $x - 2 = -\sqrt{25}$

Thus,

$x - 2 = 5$ or $x - 2 = -5$

$x = 7$ or $x = -3$ *Add 2 to each side*

The solution set is $\{7, -3\}$.

✓ CHECK:

Let $x = 7$

$(x - 2)^2 = 25$
$(7 - 2)^2 = 25$
$5^2 = 25$ ✓

Let $x = -3$

$(x - 2)^2 = 25$
$(-3 - 2)^2 = 25$
$(-5)^2 = 25$ ✓

Problem 3 Solve the quadratic equation $(x - 9)^2 = 0$.

SOLUTION:

By the property above,

$x - 9 = \sqrt{0}$ or $x - 9 = -\sqrt{0}$ *If $A^2 = 0$, then $A = 0$*

However, $\sqrt{0} = -\sqrt{0} = 0$. Thus,

$x - 9 = 0$ and $x = 9$

Thus, the quadratic equation $(x - 9)^2 = 0$ has only one solution, $x = 9$.

Problem 4 Solve the quadratic equation $(2y - 1)^2 = 12$.

SOLUTION:

By the property discussed earlier,

$2y - 1 = \sqrt{12}$ or $2y - 1 = -\sqrt{12}$
$2y = 1 + 2\sqrt{3}$ or $2y = 1 - 2\sqrt{3}$ ■ $\sqrt{12} = \sqrt{4 \cdot 3} = 2\sqrt{3}$
$y = \frac{1}{2} + \sqrt{3}$ or $y = \frac{1}{2} - \sqrt{3}$ *Divide by 2*

The solution set is $\{\frac{1}{2} + \sqrt{3}, \frac{1}{2} - \sqrt{3}\}$.

✓ CHECK:

One solution is checked below. You check the other one. Let $y = \frac{1}{2} + \sqrt{3}$

$(2y - 1)^2 = 12$
$\left(2\left(\frac{1}{2} + \sqrt{3}\right) - 1\right)^2 = 12$
$(1 + 2\sqrt{3} - 1)^2 = 12$
$(2\sqrt{3})^2 = 12$
$4 \cdot 3 = 12$ ✓

Some quadratic equations have no solutions. For example, the equation

$$(x - 7)^2 = -4$$

has no solution because the square of a real number, $(x - 7)^2$, can never be a negative number, -4. From the previous problems, you see that a quadratic equation might have one solution, two solutions, or no solution.

Quick Reinforcement

Solve each of the following quadratic equations by taking square roots.

(a) $5x^2 = 125$ (b) $(y + 1)^2 - 49 = 0$

Answers (a) $\{5, -5\}$ (b) $\{6, -8\}$

EXERCISES 12-2

Solve each of the following quadratic equations by taking square roots.

1. (a) $x^2 = 49$ (b) $x^2 = 81$
2. (a) $x^2 - 121 = 0$ (b) $x^2 - 144 = 0$
3. (a) $4y^2 = 3$ (b) $x^2 - 16 = 20$
4. (a) $36x^2 = 169$ (b) $25y^2 - 196 = 0$
5. (a) $x^2 + 1 = \dfrac{x^2}{4} + 4$ (b) $5a^2 = 3a^2 + 1$
6. (a) $(3x - 1)^2 = 18$ (b) $(7a - 10)^2 = 144$
7. (a) $r^2 = \dfrac{2r^2}{3} + 2$ (b) $(2a - 1)^2 = 9$
8. (a) $(y - 2)^2 = 48$ (b) $(y + 3)^2 = \tfrac{1}{2}$
9. (a) $2x^2 - 5 = 4x^2 - 55$ (b) $\dfrac{2x^2}{3} + 4 = 9 - x^2$
10. (a) $5t^2 + \tfrac{3}{5} = 2t^2 + \tfrac{9}{10}$ (b) $(6m - 5)^2 - 11 = 0$

Calculator Activities

Solve each of the following quadratic equations by taking square roots.

1. (a) $48x^2 - 142 = 0$ (b) $2.11x^2 = 48.3$
2. (a) $5(x + 0.2)^2 = 5.25$ (b) $-31(4x - 5)^2 = -289$

3. A state's population is calculated each month (t). The initial population is 850,000 and the population at each month thereafter is $t^2 + 850{,}000$.
   (a) After 24 months, what is the population?
   (b) About how many months are required to boost the population to one million?

12-3 COMPLETING THE SQUARE

Starting with a binomial such as $x^2 + 6x$, you can always find a number to add to it such that the resulting trinomial is a perfect square.

$x^2 + 6x + 9$ Add 9 in this example

$(x+3)^2$

Starting with

$x^2 + bx$ Coefficient of x^2 is 1, $b \neq 0$

what do you add? Look at the pattern for a perfect binomial square:

$x^2 + 2Ax + A^2 = (x + A)^2$ $b = 2A,\ A = b/2,\ A^2 = b^2/4$

$x^2 + bx + \dfrac{b^2}{4} = \left(x + \dfrac{b}{2}\right)^2$ Add $b^2/4$ to $x^2 + bx$ to make $x^2 + bx$ into a perfect square

The process of adding a number to a binomial of the form $x^2 + bx$ to obtain a perfect-square trinomial is called **completing the square.**

Problem 1 Complete the square of each of the following.

(a) $x^2 + 14x$ (b) $x^2 - 22x$ (c) $x^2 + x$

SOLUTION:

(a) Take half of 14 and then square. Add this number to the binomial.

$x^2 + 14x + 49$ ■ $\dfrac{14}{2} = 7,\ 7^2 = 49$

Thus,

$x^2 + 14x + 49 = (x + 7)^2$ Perfect square

(b) $x^2 - 22x + 121$ ■ $-\dfrac{22}{2} = -11,\ (-11)^2 = 121$

Then

$$x^2 - 22x + 121 = (x - 11)^2 \quad \text{Perfect square}$$

(c) $x^2 + x + \dfrac{1}{4}$ The coefficient of x is 1, $\dfrac{1}{2} \cdot 1 = \dfrac{1}{2}$, $\left(\dfrac{1}{2}\right)^2 = \dfrac{1}{4}$

Then

$$x^2 + x + \frac{1}{4} = \left(x + \frac{1}{2}\right)^2 \quad \text{Perfect square}$$

Completing the square is an aid in solving quadratic equations, as you will see in the next section.

Quick Reinforcement

Complete the square of each binomial.

(a) $z^2 - 18z$ \hspace{2cm} (b) $x^2 + 7x$

Answers (a) $z^2 - 18z + 81 = (z - 9)^2$ (b) $x^2 + 7x + \dfrac{49}{4} = \left(x + \dfrac{7}{2}\right)^2$

EXERCISES 12-3

Complete the square of each of the following binomials.

1. (a) $x^2 + 18x$ \hspace{2cm} (b) $x^2 + 40x$
2. (a) $a^2 - 10a$ \hspace{2cm} (b) $b^2 + 24b$
3. (a) $x^2 - x$ \hspace{2.5cm} (b) $y^2 + 12y$
4. (a) $y^2 + 3y$ \hspace{2.2cm} (b) $x^2 - 13x$
5. (a) $b^2 + 9b$ \hspace{2.2cm} (b) $s^2 + 15s$
6. (a) $x^2 - 3x$ \hspace{2.2cm} (b) $r^2 - 5r$
7. (a) $y^2 - 17y$ \hspace{2cm} (b) $x^2 + 25x$
8. (a) $y^2 - 7y$ \hspace{2.2cm} (b) $y^2 + 11y$
9. (a) $a^2 + 19a$ \hspace{2cm} (b) $t^2 - 28t$
10. (a) $x^2 - 98x$ \hspace{2cm} (b) $y^2 + 0.5y$
11. (a) $m^2 - \tfrac{3}{2}m$ \hspace{2cm} (b) $x^2 + \tfrac{3}{4}x$
12. (a) $b^2 + \tfrac{1}{3}b$ \hspace{2.2cm} (b) $a^2 - \tfrac{2}{5}a$
13. (a) $x^2 - 1.2x$ \hspace{2cm} (b) $y^2 - 2.2y$
14. (a) $m^2 + 0.6m$ \hspace{2cm} (b) $n^2 + 0.8n$

12-4 SOLVING QUADRATIC EQUATIONS BY COMPLETING THE SQUARE

Every quadratic equation can be solved by completing the square. The examples below illustrate this process.

Problem 1 Solve the equation $x^2 + 6x + 8 = 0$ by completing the square.

SOLUTION:

$x^2 + 6x + 8 = 0$	*Given equation*
$x^2 + 6x = -8$	*Add -8 to each side so that the x terms are on the left side and the constant term is on the right side*
$x^2 + 6x + 9 = -8 + 9$	*Add $(\frac{6}{2})^2$ to each side to make the left side a perfect-square trinomial*
$(x + 3)^2 = 1$	*Solve by taking square roots*
$x + 3 = 1$ or $x + 3 = -1$	$\sqrt{1} = 1, -\sqrt{1} = -1$
$x = -2$ or $x = -4$	

The solution set is $\{-2, -4\}$. You could have solved this equation by factoring: $x^2 + 6x + 8 = (x + 2)(x + 4)$.

Problem 2 Solve the equation $x(x + 3) = -7(x + 1)$.

SOLUTION:

$x(x + 3) = -7(x + 1)$	*Given equation*
$x^2 + 3x = -7x - 7$	*Multiply as indicated*
$x^2 + 10x = -7$	*Add $7x$ to each side; now complete the square on the left side*
$x^2 + 10x + \left(\frac{10}{2}\right)^2 = -7 + \left(\frac{10}{2}\right)^2$	*Add $(\frac{10}{2})^2 = 5^2$, or 25, to each side*
$x^2 + 10x + 25 = -7 + 25$	
$(x + 5)^2 = 18$	*Solve as before*
$x + 5 = \sqrt{18}$ or $x + 5 = -\sqrt{18}$	
$x = -5 + 3\sqrt{2}$ or $x = -5 - 3\sqrt{2}$	∎ $\sqrt{18} = \sqrt{9 \cdot 2} = \sqrt{9} \cdot \sqrt{2} = 3\sqrt{2}$

The solution set is $\{-5 + 3\sqrt{2}, -5 - 3\sqrt{2}\}$.

12-4: SOLVING QUADRATIC EQUATIONS BY COMPLETING THE SQUARE

√ CHECK:

One solution is checked below. You check the other solution. Let $x = -5 + 3\sqrt{2}$

$$x(x + 3) = -7(x + 1)$$
$$(-5 + 3\sqrt{2})(-5 + 3\sqrt{2} + 3) = -7(-5 + 3\sqrt{2} + 1)$$
$$(-5 + 3\sqrt{2})(-2 + 3\sqrt{2}) = -7(-4 + 3\sqrt{2})$$
$$10 - 15\sqrt{2} - 6\sqrt{2} + 9 \cdot 2 = 28 - 21\sqrt{2}$$
$$(10 + 18) - 15\sqrt{2} - 6\sqrt{2} = 28 - 21\sqrt{2} \quad \checkmark$$

Problem 3 Solve the equation $8x(x - 1) = 4x - 3$ by completing the square.

SOLUTION:

$8x(x - 1) = 4x - 3$	*Given equation*
$8x^2 - 8x = 4x - 3$	
$8x^2 - 12x = -3$	*Add $-4x$ to each side*
$x^2 - \dfrac{12}{8}x = -\dfrac{3}{8}$	*Multiply each side by $\tfrac{1}{8}$ so that x^2 has a coefficient of 1*
$x^2 - \dfrac{3}{2}x + \left(\dfrac{3}{4}\right)^2 = -\dfrac{3}{8} + \left(\dfrac{3}{4}\right)^2$	*Take half of $-\tfrac{3}{2}$, square it, and add to each side*
$\left(x - \dfrac{3}{4}\right)^2 = -\dfrac{3}{8} + \dfrac{9}{16} \qquad -\dfrac{3}{8} = \dfrac{-6}{16}$	
$\left(x - \dfrac{3}{4}\right)^2 = \dfrac{3}{16}$	*Solve as before*

$$x - \dfrac{3}{4} = \sqrt{\dfrac{3}{16}} \quad \text{or} \quad x - \dfrac{3}{4} = -\sqrt{\dfrac{3}{16}} \qquad \blacksquare \ \sqrt{\dfrac{3}{16}} = \dfrac{\sqrt{3}}{\sqrt{16}} = \dfrac{\sqrt{3}}{4}$$

$$x = \dfrac{3}{4} + \dfrac{\sqrt{3}}{4} \quad \text{or} \quad x = \dfrac{3}{4} - \dfrac{\sqrt{3}}{4}$$

The solution set is $\left\{\dfrac{3}{4} + \dfrac{\sqrt{3}}{4}, \dfrac{3}{4} - \dfrac{\sqrt{3}}{4}\right\}$

√ CHECK:

One solution is checked below. You check the other solution. Let $x = 3/4 - \sqrt{3}/4$.

$$8x(x - 1) = 4x - 3$$
$$8\left(\dfrac{3}{4} - \dfrac{\sqrt{3}}{4}\right)\left(\dfrac{3}{4} - \dfrac{\sqrt{3}}{4} - 1\right) = 4\left(\dfrac{3}{4} - \dfrac{\sqrt{3}}{4}\right) - 3$$

$$(6 - 2\sqrt{3})\left(-\frac{1}{4} - \frac{\sqrt{3}}{4}\right) = 3 - \sqrt{3} - 3$$

$$-\frac{6}{4} - \frac{6\sqrt{3}}{4} + \frac{2}{4}\sqrt{3} + \frac{2 \cdot 3}{4} = -\sqrt{3}$$

$$-\frac{3}{2} - \frac{3}{2}\sqrt{3} + \frac{1}{2}\sqrt{3} + \frac{3}{2} = -\sqrt{3}$$

$$\left(-\frac{3}{2} + \frac{1}{2}\right)\sqrt{3} = -\sqrt{3}$$

$$(-1)\sqrt{3} = -\sqrt{3} \quad \checkmark$$

Quick Reinforcement

Solve each quadratic equation by completing the square.

(a) $x^2 - 5x = 4$

(b) $2y^2 = 3y - 1$

Answers: (a) $\left\{\frac{\sqrt{41}+5}{2}, -\frac{\sqrt{41}+5}{2}\right\}$ (b) $\left\{1, \frac{1}{2}\right\}$

EXERCISES 12-4

Solve each of the following quadratic equations by completing the square. Check one solution.

1. (a) $x^2 + 10x + 9 = 0$ (b) $x^2 + 12x + 32 = 0$
2. (a) $x^2 - 4x - 45 = 0$ (b) $x^2 + 2x - 15 = 0$
3. (a) $6a - 300 = 204 - 3a^2$ (b) $5y^2 + 80 = -505 - 110y$
4. (a) $9 - 3b^2 = 2b - 7b$
   (b) $(2 - x)(x + 2) = 10(x + 2) - 72$
5. (a) $3y^2 - 180 = 8y - 177$ (b) $2z^2 + 5z - 6 = 0$
6. (a) $a^2 - 14a - 38 = -10$ (b) $3x^2 - 7x = 7x - 8$
7. (a) $2x^2 = 8x + 5 - 4x^2$ (b) $4m - 3 = m^2 + 2m$
8. (a) $\frac{1}{2}x^2 - \frac{3}{4}x = 1$ (b) $3a^2 + a - 1 = 0$
9. (a) $2m^2 + 3m - 1 = 0$ (b) $y^2 - \frac{4 - 32y}{3} = \frac{8(1 + y)}{3} + 16$
10. (a) $4x(x - 1) = 2x - 1$ (b) $x^2 - \frac{5}{6}x = \frac{1}{6}$
11. (a) $20(t - 3) = t^2$ (b) $12(m - 1) - m^2 = 0$
12. (a) $p^2 + 16(p + 2) = 0$ (b) $12(n + 5) + 2n^2 = 0$

13. (a) $2a(3a - 5) = -3$ (b) $5x(x - 2) = 2$
14. (a) $(2y - 5)^2 = (y + 3)^2$ (b) $(2a + 3)^2 = (4a - 1)^2$

12-5 THE QUADRATIC FORMULA

Some problems are easily solved by using a formula. Such is the case for solving quadratic equations. There is a formula that allows you to quickly write down the solutions of any quadratic equation. Called the **quadratic formula,** it is developed below.

Every quadratic equation is equivalent to one in the form

$$ax^2 + bx + c = 0 \qquad a, b, c \text{ represent numbers with } a > 0$$

To find the formula, solve the equation above by completing the square.

$$x^2 + \frac{b}{a}x + \frac{c}{a} = 0 \qquad \text{Multiply each side by } 1/a \text{ so that } x^2 \text{ has a coefficient of } 1$$

$$x^2 + \frac{b}{a}x = -\frac{c}{a} \qquad \text{Add } -\frac{c}{a} \text{ to each side}$$

$$x^2 + \frac{b}{a}x + \left(\frac{b}{2a}\right)^2 = -\frac{c}{a} + \left(\frac{b}{2a}\right)^2 \qquad \text{Take half of } \frac{b}{a}, \text{ square it, and add the result to each side}$$

$$\left(x + \frac{b}{2a}\right)^2 = -\frac{c}{a} + \frac{b^2}{4a^2} \qquad \text{The left side is a perfect-square trinomial}$$

$$\left(x + \frac{b}{2a}\right)^2 = \frac{b^2 - 4ac}{4a^2} \qquad -\frac{c}{a} + \frac{b^2}{4a^2} = -\frac{c \cdot 4a}{a \cdot 4a} + \frac{b^2}{4a^2} = \frac{b^2}{4a^2} - \frac{4ac}{4a^2}$$

$$x + \frac{b}{2a} = \frac{\sqrt{b^2 - 4ac}}{2a} \qquad \blacksquare \sqrt{\frac{b^2 - 4ac}{4a^2}} = \frac{\sqrt{b^2 - 4ac}}{\sqrt{4a^2}}$$

or

$$x + \frac{b}{2a} = -\frac{\sqrt{b^2 - 4ac}}{2a}$$

$$x = -\frac{b + \sqrt{b^2 - 4ac}}{2a}$$

or

$$x = -\frac{b - \sqrt{b^2 - 4ac}}{2a} \qquad \blacksquare \sqrt{4a^2} = 2a \text{ because } a > 0$$

The Quadratic Formula

The quadratic equation

$$ax^2 + bx + c = 0 \qquad a,b,c \text{ represent numbers with } a > 0$$

has solutions

$$x = \frac{-b \pm \sqrt{b^2 - 4ac}}{2a} \qquad \text{Use the plus sign in } \pm \text{ for one solution,} \\ \text{the minus sign for the other}$$

provided $b^2 - 4ac \geq 0$. If $b^2 - 4ac < 0$, the equation has no real-number solution.

The number

$$b^2 - 4ac$$

appearing in the formula is called the **discriminant** of the equation $ax^2 + bx + c = 0$. It is the first expression you evaluate in using the quadratic formula.

Problem 1 Use the quadratic formula to solve the equation

$$2x^2 - 3x - 5 = 0$$

SOLUTION:

$$\begin{array}{c} 2x^2 - 3x - 5 = 0 \\ \uparrow \quad \uparrow \quad \uparrow \\ ax^2 + bx + c = 0 \end{array} \qquad \begin{array}{l} = 2x^2 + (-3)x + (-5) \\ a = 2, b = -3, c = -5 \end{array}$$

Therefore,

$$b^2 - 4ac = (-3)^2 - 4 \cdot 2 \cdot (-5)$$
$$= 9 + 40$$
$$= 49 \qquad\qquad \blacksquare\; b^2 - 4ac = 49 \geq 0$$

By the formula,

$$x = \frac{-(-3) \pm \sqrt{49}}{2 \cdot 2}$$

$$x = \frac{3 \pm 7}{4}$$

The solutions are

$$x = \frac{3 + 7}{4} = \frac{5}{2}, \qquad x = \frac{3 - 7}{4} = -1$$

The solution set is $\{\frac{5}{2}, -1\}$.

Problem 2 Use the quadratic formula to solve $4x^2 + 7x + 2 = 0$.

SOLUTION:

For the equation $4x^2 + 7x + 2 = 0$, $a = 4$, $b = 7$, and $c = 2$. Hence,

$$b^2 - 4ac = 7^2 - 4 \cdot 4 \cdot 2 = 49 - 32 = 17$$

By the formula,

$$x = \frac{-7 \pm \sqrt{17}}{2 \cdot 4}$$

The solution set is $\left\{ \dfrac{-7 + \sqrt{17}}{8}, \dfrac{-7 - \sqrt{17}}{8} \right\}$.

Problem 3 Use the quadratic formula to solve $4y^2 = 12y - 9$.

SOLUTION:

First, put the equation in the form $ay^2 + by + c = 0$.

$4y^2 = 12y - 9$ *Given equation*

$4y^2 - 12y + 9 = 0$ *Add $-12y + 9$ to each side*

For this equation, $a = 4$, $b = -12$, and $c = 9$. Thus,

$$b^2 - 4ac = (-12)^2 - 4 \cdot 4 \cdot 9 = 0$$

and

$$y = \frac{-(-12) \pm \sqrt{0}}{2 \cdot 4} = \frac{12}{8} = \frac{3}{2}$$

The equation has one solution, $y = \frac{3}{2}$. It has only one solution because $4y^2 - 12y + 9 = (2y - 3)^2$, a perfect-square trinomial.

Problem 4 Use the quadratic formula to solve $x^2 + 4x + 8 = 0$.

SOLUTION:

For this equation, $a = 1$, $b = 4$, and $c = 8$. Hence,

$$b^2 - 4ac = 4^2 - 4 \cdot 1 \cdot 8$$
$$= 16 - 32$$
$$= -16$$

Since $b^2 - 4ac = -16 < 0$, the equation

$$x^2 + 4x + 8 = 0$$

has no real-number solution.

Quick Reinforcement

Solve, using the quadratic formula.

(a) $5y^2 - 2y - 4 = 0$ (b) $2y^2 = -3y + 3$

Answers: (a) $\dfrac{1 \mp \sqrt{21}}{5}$ (b) $\dfrac{-3 \mp \sqrt{33}}{4}$

EXERCISES 12-5

Use the quadratic formula to solve each of the following quadratic equations. Check one solution.

1. (a) $x^2 + 2x - 2 = 0$ (b) $3y^2 - 5y + 1 = 0$
2. (a) $y^2 - 6y - 3 = 0$ (b) $y^2 - 10y - 3 = 0$
3. (a) $2p^2 - 4p + 1 = 0$ (b) $2x^2 = 3x + 14$
4. (a) $6a^2 + 13a + 5 = 0$ (b) $2a^2 = 6a - 1$
5. (a) $2y^2 = -3(y + 1)$ (b) $2z^2 + 3 = 5z$
6. (a) $2a = 3 + \dfrac{3}{a}$ (b) $x^2 = 2x - 3$
7. (a) $x^2 - 10x + 30 = 0$ (b) $2y^2 + 2y + 4 = 4 - 2y$
8. (a) $2x^2 - 5x + 3 = 0$ (b) $2m^2 = 1 - 3m$
9. (a) $2m^2 = 3m + 5$ (b) $b^2 = -19 + 20b$
10. (a) $(4y - 3)^2 = 5(y - 1)$ (b) $3x(x + 1) = 2$
11. (a) $5m^2 - 4m = 2$ (b) $6x^2 - 5 = 5x(x + 2)$
12. (a) $(x - 3)^2 = 2(x + 4)$ (b) $(m + 2)^2 = 10 - m$

Calculator Activities

Use the quadratic formula to solve each of the following equations.

1. (a) $14x^2 + 21x - 4 = 0$ (b) $23x^2 + 10 = 52 - 19x$
2. (a) $44(x - 15) = 69x^2 - 182x$
   (b) $15(14x^2 - 21x) = 4(143 - 29x) + 140$
3. The vertical height at t seconds obtained by an object thrust with initial velocity 154 ft/sec is $154t - 16t^2$.
   (a) When will the object be at 300 feet?
   (b) When will the object be at 150 feet?

12-6 APPLIED PROBLEMS

The following problems can be solved by solving a related quadratic equation.

Problem 1 Is it possible for a rectangular picture to have a perimeter of 2 meters and an area of only 0.24 square meter?

SOLUTION:

From Figure 12-1,

$$2x + 2y = 2 \quad \text{Perimeter}$$
$$xy = 0.24 \quad \text{Area}$$

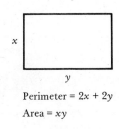

Perimeter = $2x + 2y$
Area = xy

FIGURE 12-1

Thus,

$$x + y = 1 \quad \blacksquare \; \frac{1}{2}(2x + 2y) = \frac{1}{2} \cdot 2$$

$$y = 1 - x \quad \text{Add } -x \text{ to each side}$$
$$x(1 - x) = 0.24 \quad \text{Substitute } 1 - x \text{ for } y \text{ in } xy = 0.24$$
$$x - x^2 = 0.24$$
$$x^2 - x + 0.24 = 0 \quad \text{Add } -x + x^2 \text{ to each side}$$

By the quadratic formula, with $a = 1$, $b = -1$, and $c = 0.24$,

$$b^2 - 4ac = (-1)^2 - 4 \cdot 1 \cdot (0.24)$$
$$= 1 - 0.96$$
$$= 0.04$$

and

$$x = \frac{1 \pm \sqrt{0.04}}{2} = \frac{1 \pm 0.2}{2} \quad \blacksquare \; 0.04 = (0.2)^2, \; \sqrt{0.04} = 0.2$$

$$x = 0.6 \quad \text{or} \quad x = 0.4 \quad \text{Meters}$$

Thus, $x = 60$ centimeters or $x = 40$ centimeters. If $x = 60$ centimeters, then $y = 40$ centimeters; if $x = 40$ centimeters, then $y = 60$ centimeters. In either case, the picture is 40 centimeters × 60 centimeters. It checks, as you should see.

Problem 2 A company's profits P (in thousands of dollars) when x items are produced (in hundreds) has been shown to be $P = -x^2 + 8x - 12$. How many items must be produced to yield a profit of \$4000?

SOLUTION:

You are asked to find a value of x that gives P a value of 4. In other words, you are asked to solve the equation

$$4 = -x^2 + 8x - 12$$

$x^2 - 8x + 12 + 4 = 0$ *Add $x^2 - 8x + 12$ to each side*

$x^2 - 8x + 16 = 0$

$(x - 4)^2 = 0$

$x - 4 = 0$ *If $A^2 = 0$, then $A = 0$*

$x = 4$

Thus, 400 items must be produced to yield a profit of \$4000.

√ CHECK:

If 400 items are produced, then $x = 4$ and

$P = -4^2 + 8 \cdot 4 - 12$

$ = -16 + 32 - 12$

$ = 4$

That is, profits equal \$4000. √

Problem 3 All but 10 members of the chess club contributed a total of \$60 to buy a trophy for the winner of next week's tournament. If all the members had contributed, each contribution would have been \$1 less than it was. How much did each contributing member give?

SOLUTION:

Let x = number of contributing members

Then

$x + 10 =$ number of members of chess club

$\dfrac{60}{x} =$ number of dollars each contributing member gave

$\dfrac{60}{x + 10} = $ number of dollars each member would have contributed if all members had given

$\dfrac{60}{x} - \dfrac{60}{x + 10} = 1 \quad$ *Difference in contribution (if all had given) would have been $1*

Solve this equation:

$\dfrac{60(x + 10) - 60x}{x(x + 10)} = 1 \quad$ *Put fractions over common denominator*

$60(x + 10) - 60x = x(x + 10) \quad$ *Multiply each side by $x(x + 10)$*

$60x + 600 - 60x = x^2 + 10x$

$0 = x^2 + 10x - 600 \quad$ *Add -600 to each side*

Use the quadratic formula, with $a = 1$, $b = 10$, and $c = -600$. Then

$b^2 - 4ac = 10^2 - 4 \cdot 1 \cdot (-600)$

$= 100 + 2400$

$= 2500$

Hence, by the formula,

$x = \dfrac{-10 \pm \sqrt{2500}}{2}$

$= \dfrac{-10 \pm 50}{2}$

$= 20 \quad \text{or} \quad -30 \quad$ *You can't have -30 contributing members!*

Thus, $x = 20$.

✓ CHECK:

There are 20 contributing members, and each gives $\frac{60}{20}$, or $3. There are 30 members of the chess club. If each contributed, he or she would give $\frac{60}{30}$, or $2. The difference in the contributions is $3 - 2$, or $1. ✓

Problem 4 Tom lives 30 kilometers from his office. The round trip takes him 1 hour to drive. His average speed going to work is 10

km/hr faster than his speed going home. What is his average speed each way?

SOLUTION:

You may follow the flowchart in Figure 6-6 if you wish.

Tom drives 30 kilometers to his office and
30 kilometers to his home
The round trip takes 1 hour   *Bits of information*
His speed to work is 10 km/hr faster
than his speed home

Let $x =$ speed to home, in kilometers per hour

$x + 10 =$ speed to office, in kilometers per hour

$\dfrac{30}{x} =$ time to drive home, in hours   *Distance ÷ rate = time*

$\dfrac{30}{x + 10} =$ time to drive to work, in hours

$\dfrac{30}{x + 10} + \dfrac{30}{x} = 1$ *Time to drive to office +
time to drive home = 1 hour*

Solve this equation:

$\dfrac{30x + 30(x + 10)}{x(x + 10)} = 1$ *Put fractions over common denominator*

$30x + 30x + 300 = x^2 + 10x$ *Multiply each side by $x(x + 10)$*

$x^2 - 50x - 300 = 0$ *Add $-60x - 300$ to each side*

Solve by the quadratic formula, with $a = 1$, $b = -50$, and $c = -300$.

$b^2 - 4ac = (-50)^2 - 4 \cdot 1 \cdot (-300)$
$= 2500 + 1200$
$= 3700$

Then

$x = \dfrac{50 \pm 10\sqrt{37}}{2}$ $\sqrt{3700} = \sqrt{100 \cdot 37} = 10\sqrt{37}$

$x = 25 \pm 5\sqrt{37}$ $25 - 5\sqrt{37}$ *is negative. You can't have a negative answer!*

Thus,

$x = 25 + 5\sqrt{37} \approx 25 + 5 \cdot 6.08$
≈ 55.4

Therefore, Tom travels at a speed of 65.4 km/hr to his office and 55.4 km/hr home.

✓ CHECK:

$$\frac{30}{65.4} + \frac{30}{55.4} \approx 0.46 + 0.54 \approx 1 \quad \checkmark \quad 0.46 \text{ hours to office}, 0.54 \text{ hours to home}$$

Quick Reinforcement

For every x thousand items the Acme Company produces, it has costs of $x^2 - 3x + 7$ thousands of dollars and revenue of $3x$ thousands of dollars. How many items must the Acme Company produce to break even?

Answer: 4414 or 1586

EXERCISES 12-6

1. (a) Kelly is cutting tiles for a large floor. If each tile is a rectangle with the width two centimeters less than the length, what are the dimensions of the tiles if they have an area of 12 square centimeters?
   (b) Hans and Julio have agreed to mow their neighbor's back yard. If the yard is a rectangle and has a perimeter of 94 meters and an area of 496 square meters, what are the length and width of the yard?
2. (a) Shana and Tammy have to write some number puzzles that they hope will stump the other students in an "algebra bee." Shana's puzzle is, "I am thinking of a number that I will triple and then reduce by 1. If I multiply that number by the original number increased by 1, the result will be 20. What is the number?"
   (b) Tammy's number puzzle is, "What are all possible numbers such that if I add the original number to itself it will be the same as if I multiplied the number by itself?"
3. (a) A given right triangle has a hypotenuse of 6 centimeters in length. If one of the legs is double the other leg, what are the lengths of the legs?
   (b) Another right triangle has one leg 8 centimeters in length, and the length of the other leg is 1 less than half the length of the hypotenuse. What are the lengths of the second leg and the hypotenuse?
4. (a) The Masters family took an 8-hour hike. The trip out was uphill while the trip back was downhill. If they walked exactly ten kilometers each way and their speed downhill was three kilometers per hour faster than their speed uphill, how fast did they walk uphill?

416 CHAPTER 12: QUADRATIC EQUATIONS

(b) Melody's power boat has an average speed of ten kilometers per hour in still water. On an outing on the American River, she noticed that it took two hours longer to go 48 kilometers up the river than it did to go down. What was the average speed of the current?

5. (a) The Steak and Taters Restaurant went to a wholesale meat firm to buy their steaks. Given a choice of club steaks or rib steaks costing $4 less per box, they decided to buy a number of boxes of club steaks at a total cost of $240. Had they bought rib steaks, they could have purchased three more boxes for the same total cost. How many boxes of club steaks did they buy?

(b) While training for the Olympics, André ran 48 kilometers one afternoon. The next day he ran the same distance at an average speed of 4 km/hr faster than the day before. If it took him one hour less running time the second day than the first, what was his average speed the first day?

6. (a) Jasper and Jenny own stock in two companies. The stock in Company X is worth $6000 and the stock in Company Y is worth $6250. They have five more shares of stock in Company Y than in company X, but each share of Company Y stock is worth 50% less than a share of Company X stock. How many shares of stock do they own?

(b) Gustave loves fine expensive cheeses from Europe. Last week he paid $5.76 for his order of cheese. This week he spent the same amount but received two more ounces because his cheese was on sale for four cents less per ounce than normal. How many ounces of cheese did he buy last week?

7. (a) What number is it that when you add it to its square you will have twenty?

(b) Find two negative numbers such that one is seven less than the other and their product is thirty.

8. (a) The Togetherness Club is building a clubhouse. The floor plan shows that the clubhouse will be a rectangle of eight square meters. If the width of the clubhouse is to be $\frac{2}{3}$ of a meter less than $\frac{2}{3}$ of the length, what will be the length and width of the clubhouse?

(b) Two friends on bicycles leave Troy on roads going at right angles to each other. They leave at the same time and one bicyclist travels seven kilometers per hour faster than his friend. After one hour, they are thirteen kilometers apart. What is the speed of each bicyclist?

Nostalgia Problems

1. (a) A company dining at a house of entertainments had to pay 3 dollars and 50 cents; but before the bill was presented, two of them left, in consequence of which, those who remained had to each pay 20 cents more than if all had been present. How many persons dined?

2. (b) A company at a tavern had 1 dollar and 75 cents to pay; but before the bill was paid, two of them left, when those who remained each had 10 cents more to pay. How many were in the company at first?

2. (a) It is required to divide the number 14 into two such parts that the quotient of the greater divided by the less, divided by the quotient of the less divided by the greater equals $\frac{16}{9}$. Find the two parts.

   (b) If four times the square of a certain number be diminished by twice the number, it will leave a remainder of 30; what is the number?

3. (a) A square courtyard has a rectangular gravel walk round it. The side of the court is two yards less than six times the width of the gravel walk, and the number of square yards in the walk exceeds the number of yards in the periphery of the court by 164; required the area of the court exclusive of the walk.

   (b) A fisherman being asked how many fish he had caught, replied, "If you add 11 to the square of the number, 9 times the square root of the sum, diminished by 4, will equal 50." How many had he caught?

4. (a) A and B distributed 1200 dollars each among a certain number of persons. A relieved 40 persons more than B, and B gave to each individual 5 dollars more than A; how many were relieved by A and B?

   (b) What number is that, whose square plus 18 is equal to half its square plus $30\frac{1}{2}$?

Calculator Activities

1. Use the formula $h = 50t - 16t^2$ to find the time it takes an object thrown vertically upward to reach a height of
   (a) 30 feet   (b) 10 feet

2. Find the number of items x (in thousands) that a company must produce to break even if revenue is $5.5x$ and cost is
   (a) $x^2 - 0.9x + 1.5$ (b) $x^2 - 0.7x + 2.2$

3. Two cyclists leave the same place at the same time. How far apart are they after the first cyclist travels 11.5 kilometers east and the second travels
   (a) 7.3 kilometers south?   (b) 8.2 kilometers north?

12-7 GRAPHS OF QUADRATIC EQUATIONS

An equation in two variables of the form

$$y = ax^2 + bx + c \quad a, b, c \text{ numbers with } a \neq 0$$

is called a **quadratic equation in two variables.** Such equations occur in many applications of mathematics, as in the study of freely falling objects in physics and in break-even analysis in economics. You can get some idea of the relationship between the two variables x and y by studying the graph of their equation.

The simplest quadratic equation is

$$y = x^2 \quad a = 1, b = 0, c = 0 \text{ in the equation above}$$

Using a table of values, such as the following, you can plot some points on the graph, as shown in Figure 12-2.

x	−10	−3	−2.5	−2	−1.5	−1	−0.5	0	0.5	1	1.5	2	2.5	3	10
y	100	9	6.25	4	2.25	1	0.25	0	0.25	1	2.25	4	6.25	9	100

The points (±10, 100) are not plotted — they are way off the graph paper!

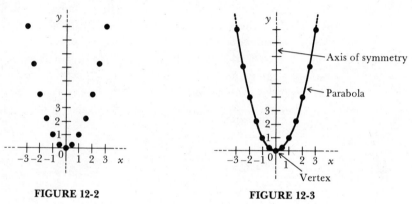

FIGURE 12-2　　　　　　　　FIGURE 12-3

There is a point (x, y) on the graph for every value of x. Thus, the graph consists of a curve that you can visualize by connecting points in order of increasing values of x with a smooth curve. This is done in Figure 12-3. The curve opens upward and extends infinitely far in both directions, as indicated in the figure.

The graph of the equation

$$y = x^2$$

is called a **parabola.** A mirror in a reflecting telescope has parabolic cross sections. Such a mirror has the property that light rays from space reflect off the mirror to a single point, called the *focus,* where the observations are made. The y-axis is an **axis of symmetry** of the parabola. That is, for every point (x, y) on the parabola, there is a corresponding point $(-x, y)$ on the parabola the same distance on the other side of the y-axis (as is evident from the figure above). The point $(0, 0)$ on the parabola shown is called the **vertex** of the parabola. The vertex is the only point on both the parabola and the axis of symmetry.

Every quadratic equation $y = ax^2 + bx + c$ has a parabola as its graph. What effect the numbers a, b, and c have on the graph is shown below.

Effect of a. The graphs of the three equations

$y = \frac{1}{2}x^2 \qquad a = \frac{1}{2}, b = 0, c = 0$

$y = x^2 \qquad a = 1, b = 0, c = 0$

$y = 2x^2 \qquad a = 2, b = 0, c = 0$

are sketched in Figure 12-4 from the three tables of values below.

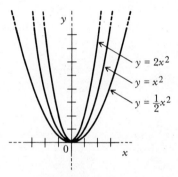

FIGURE 12-4

$y = \frac{1}{2}x^2$

x	0	±1	±2	±3
y	0	0.5	2	4.5

$y = x^2$

x	0	±1	±2	±3
y	0	1	4	9

$y = 2x^2$

x	0	±1	±2	±3
y	0	2	8	18

As these three examples illustrate, the parabola $y = ax^2$ has the y-axis as an axis of symmetry and the origin $(0, 0)$ as a vertex for every positive number a. If $a < 1$, the parabola $y = ax^2$ is *flatter* than $y = x^2$; if $a > 1$, the parabola $y = ax^2$ is *thinner* than $y = x^2$. The smaller a is, the flatter the parabola is; the larger a is, the thinner the parabola is.

If a is a negative number, then the y-values of the parabola $y = ax^2$ are negative. This means that the parabola opens downward. The parabolas $y = ax^2$ with $a = -\frac{1}{2}$, $a = -1$, and $a = -2$ are sketched in Figure 12-5. These three parabolas are mirror images (in the x-axis) of the three parabolas in the earlier figure. In terms of absolute values, the parabola $y = ax^2$ is flatter than the parabola $y = x^2$ if $0 < |a| < 1$, and thinner if $|a| > 1$.

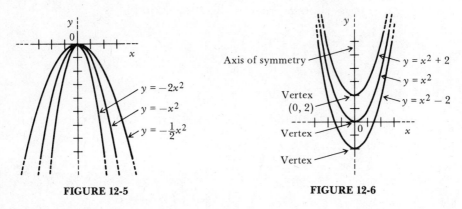

FIGURE 12-5   FIGURE 12-6

Effect of c. The graphs of

$$y = x^2, \quad y = x^2 + 2, \quad y = x^2 - 2$$

are similar parabolas. For each point (x, y) on the first parabola, there corresponds a point $(x, y + 2)$ on the second one and $(x, y - 2)$ on the third. Evidently, one parabola is two units above $y = x^2$ and the other two units below. Thus, if you were to push up the parabola $y = x^2$ two units, it would coincide with the parabola $y = x^2 + 2$; and if you were to pull down the parabola $y = x^2$ two units, it would coincide with $y = x^2 - 2$ (see Figure 12-6). More generally, for any number c, the graph of

$$y = x^2 + c$$

is a parabola having vertex $(0, c)$ and the y-axis as axis of symmetry. It is c units above the parabola $y = x^2$ if $c > 0$ and $|c|$ units below if $c < 0$.

Effect of b. Just as the number c in the equation

$$y = ax^2 + bx + c$$

influences the up and down movement of the parabola $y = x^2$, so does the number b help describe the right and left movement. As a simple example, the equation

$$y = x^2 - 2x + 1 \qquad a = 1, b = -2, c = 1$$

is equivalent to

$$y = (x - 1)^2$$

Its graph is sketched in Figure 12-7 from the table of values below.

x	−2	−1	0	1	2	3	4
y	9	4	1	0	1	4	9

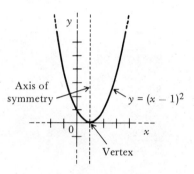

FIGURE 12-7

Clearly, it is the parabola $y = x^2$ moved one unit to the right! Similarly, the graph of

$$y = (x - h)^2 \qquad y = x^2 - 2hx + h^2, b = -2h$$

is the parabola $y = x^2$ moved h units to the right if $h > 0$ and h units to the left if $h < 0$ (Figure 12-8).

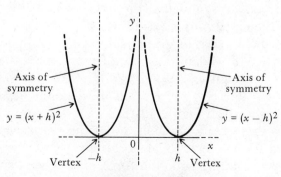

FIGURE 12-8

Every quadratic equation

$$y = ax^2 + bx + c$$

is equivalent to one of the form

$$y = a(x - h)^2 + k$$

for some numbers h and k. This is shown by completing the square on the polynomial $ax^2 + bx$. The following example illustrates how this is done.

$y = 3x^2 - 18x + 23$	*Given equation*
$= 3(x^2 - 6x) + 23$	*Factor out 3 from x-terms to give x^2 a coefficient of 1*
$= 3(x^2 - 6x + 9 - 9) + 23$	
$= 3(x^2 - 6x + 9) - 27 + 23$	*Add 9 inside parentheses to complete square; to compensate, add -27 to right side*

$$\underbrace{27}_{} \quad \underbrace{-27}_{}$$
$$0$$

Thus,

$$y = 3(x - 3)^2 + 4 \quad \textit{Equivalent to given equation}$$

When you express a quadratic equation in the form

$$y = a(x - h)^2 + k$$

the three numbers a, h, and k help you describe the graph. The first thing you know is that the graph is a parabola. As shown above, the number a describes the direction the parabola opens (upward if $a > 0$, downward if $a < 0$) and its flatness (flat if $|a| < 1$, thin if $|a| > 1$). The numbers h and k describe the vertex (the point (h, k)) and the axis of symmetry (the line $x = h$) of the parabola.

For the equation $y = 3x^2 - 18x + 23$, $a = 3$, $h = 3$, and $k = 4$. Thus, the parabola that is its graph is thin ($a > 1$), opens upward ($a > 0$), has vertex $(3, 4)$, and has axis of symmetry $x = 3$. The parabola is sketched in Figure 12-9.

The parabola $y = -x^2 + 4$ cuts the x-axis at -2 and 2 and the y-axis at 4, as shown in Figure 12-10. You call the numbers -2 and 2 the **x-intercepts** of the parabola and 4 the **y-intercept** of the parabola. To find the x-intercepts algebraically, let $y = 0$ in the equation and solve the resulting equation for x.

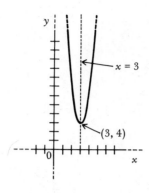

FIGURE 12-9   FIGURE 12-10

$0 = -x^2 + 4$ *Let $y = 0$*

$x^2 = 4$

$x = \pm 2$ *x-intercepts*

Similarly, let $x = 0$ to find the y-intercept:

$y = -0^2 + 4$

$y = 4$ *y-intercept*

Problem 1 Find the intercepts and describe the graph of each equation.

(a) $y = 2x^2 + 20x + 18$ (b) $2y = -x^2 + 6x - 5$

SOLUTION:

(a)   Complete the square as follows:

$y = 2x^2 + 20x + 18$ *Given equation*

$ = 2(x^2 + 10x) + 18$ *Complete the square on $x^2 + 10x$*

$ = 2(x^2 + 10x + 25) - 50 + 18$ *Added $2 \cdot 25$, so compensate by adding -50*

Thus,

$y = 2(x + 5)^2 - 32$

The graph is a thin parabola ($a = 2$), opens upward with vertex $(-5, -32)$, and has an axis of symmetry $x = -5$. Let $y = 0$ to find the x-intercept:

$$0 = 2(x+5)^2 - 32$$
$$(x+5)^2 = 16$$
$$x + 5 = \pm 4$$
$$x = -9 \quad \text{or} \quad x = -1 \quad \textit{x-intercepts}$$

If you let $x = 0$ in the given equation, you see that

$$y = 18 \quad \textit{y-intercept}$$

(b) Complete the square as follows:

$$2y = -x^2 + 6x - 5 \quad \textit{Given equation}$$
$$= -(x^2 - 6x + 9) + 9 - 5$$
$$= -(x-3)^2 + 4$$

Thus,

$$y = -\frac{1}{2}(x-3)^2 + 2$$

The graph is a parabola that opens downward ($a < 0$), is flat ($|a| < 1$), has a vertex $(3, 4)$, and has an axis of symmetry $x = 3$. Letting $y = 0$, you get

$$(x-3)^2 = 4$$
$$x - 3 = \pm 2$$
$$x = 5 \quad \text{or} \quad x = 1 \quad \textit{x-intercepts}$$

Letting $x = 0$, you get

$$y = -\frac{5}{2} \quad \textit{y-intercept}$$

Quick Reinforcement

Find the intercepts and describe the graph of

$$y = -3x^2 - 6x - 4$$

Answer: Parabola, thin, opens downward, vertex $(-1, -1)$, axis of symmetry $x = -1$, no x-intercept, y-intercept -4

EXERCISES 12-7

Make a table of values and graph each of the following equations. Determine the intercepts for each. Use your calculator where necessary.

1. (a) $y = 4x^2$ (b) $y = 6x^2$
2. (a) $y = -3x^2$ (b) $y = -2x^2$
3. (a) $y = \frac{1}{2}x^2$ (b) $y = -\frac{1}{4}x^2$
4. (a) $y = -2x^2 + 1$ (b) $y = \frac{3}{4}x^2 - 2$
5. (a) $y = 0.5x^2 + 3$ (b) $y = 0.8x^2 + 2$
6. (a) $y = 2(x-1)^2 - 1$ (b) $y = -\frac{1}{2}(x+1)^2 + 1$
7. (a) $y = -3(x+2)^2 + 2$ (b) $y = 4(x-2)^2 - 3$
8. (a) $y = 2x^2 + x - 1$ (b) $y = -3x^2 + 2x + 1$

Sketch the graph of each of the following equations. Clearly label vertex, axis of symmetry, and intercepts.

9. (a) $y = 2x^2 - 3$ (b) $y = \frac{1}{3}x^2 - 4$
10. (a) $y = -\frac{1}{2}x^2 + 2$ (b) $y = -5x^2 + 1$
11. (a) $y = (x+2)^2 - 3$ (b) $y = (x+4)^2 - 2$
12. (a) $y = -(x-1)^2 - 4$ (b) $y = -(x+3)^2 - 5$
13. (a) $y = -\frac{1}{3}(x-1)^2 - 2$ (b) $y = 3(x+1)^2 + 2$
14. (a) $y = 1.5(x - \frac{4}{3})^2 + 2.6$ (b) $y = -4(x + 1.75)^2 - 0.5$
15. Your company's profits are given by the equation $y = -2(x-3)^2 + 4$, where y denotes the profit and x denotes the number of items produced (both in units of a thousand).

    (a) Sketch the profit equation.

    (b) Determine the break-even points. Show these on the graph.
    (c) Determine the production rate that gives the maximum profit.
    (d) What is the maximum profit? Show it on the graph.
    (e) If you produce 4.9 thousand items, what are your profits? Show this on the graph.
    (f) When do you lose money? Show this on the graph.
    (g) What is the number of items you must produce to yield $3000 in profit?
16. Your company's costs are given by the equation $y = 4(x-2)^2 + 10$, where y denotes costs and x denotes the number of items produced (both in units of a thousand).
    (a) Sketch the cost equation.
    (b) With no production, what are your costs? Show this on the graph.
    (c) What is the production rate that gives minimum cost?

(d) What is your minimum cost? Show this on the graph.
(e) How many items must you produce to keep costs at $14,000? Show this on your graph.
(f) Are your costs ever zero? Explain, using the graph.

17. The owner of a residential hotel with 60 units has found that he can rent all his studios at $200 a month. However, for each $5 increase in rent per unit he loses one tenant. His revenue is $y = -5(x - 10)^2 + 17,000$, where y denotes the revenue and x is the number of $5 increases.
   (a) Sketch the revenue equation.
   (b) What happens if the owner raises the rent on all 60 units? Show this on the graph.
   (c) How many rent raises will give him his maximum revenue? Show this on the graph.
   (d) What is his maximum revenue? Find this on the graph.
   (e) What would his revenue be if he chose not to raise any rents?
   (f) What would his revenue be if he chose to decrease rents by $5?
   (g) What two situations would cause his revenue to go to zero? Show this on your graph and explain.

KEY TERMS

Quadratic equation
Completing the square
Quadratic formula
Discriminant
Quadratic equation
  in two variables

Parabola
Axis of symmetry
Vertex
x-intercept
y-intercept

REVIEW EXERCISES

Solve by factoring.

1. (a) $y^2 = 4(y + 3)$ (b) $x^2 - 8x + 15 = 0$
2. (a) $2m(m - 1) = 3(m + 1)$ (b) $3p(p - 2) = 2(p - 2)$

Solve by taking square roots.

3. (a) $(x + 3)^2 = 25$ (b) $x^2 - 9 = 0$
4. (a) $7x^2 - 5 = 0$ (b) $(p - \frac{1}{3})^2 = \frac{7}{9}$

Complete the square of each of the following binomials.

5. (a) $x^2 + 7x$ (b) $y^2 - \dfrac{26y}{3}$

6. (a) $m^2 - \dfrac{15m}{2}$ (b) $x^2 + 19x$

Solve by completing the square.

7. (a) $x^2 + 8x - 1 = 0$ (b) $x^2 = 3(2x + 1)$
8. (a) $4m^2 + 2m - 9 = 0$ (b) $2x^2 = 2x + 3$

Solve by the quadratic formula.

9. (a) $5y^2 = y + 5$ (b) $6x + 3 = 2x^2$
10. (a) $3x^2 + 1 = 5x$ (b) $x(x + 4) = 3x(x - 1)$

Solve using quadratic equations.

11. (a) Melinda was water-skiing and wanted to know how fast the boat was pulling her. Michael, who was piloting the boat, estimated that the current in the river was $2\frac{1}{2}$ km/hr. Michael and Melinda took the boat 7.5 kilometers up the river and then returned back to their dock. They found that the round trip in the boat took a half hour. How fast would the boat be going in still water?

(b) Given a triangle where the height exceeds the base by three, if three times the length of the base (numerically speaking) is subtracted from the area of the triangle, the result is two. What is the area of the triangle?

Make a table of values and graph each of the following equations.

12. (a) $y = -5x^2 + 4$ (b) $y = \frac{2}{3}x^2 - 3$
13. (a) $y = 4(x - 3)^2 - 2$ (b) $y = -(x + 6)^2 + 1$

Sketch the graph of each of the following equations. Clearly label vertex, axis of symmetry, and intercepts.

14. (a) $y = -\frac{3}{4}x^2 - \frac{1}{2}$ (b) $y = 0.5x^2 + 2.5$
15. (a) $y = 3(x - 6)^2 - 5$ (b) $y = 4(x + 3)^2 - 3$
16. A driver determines that the cost in cents per kilometer of driving his car at speed x km/hr is given by $y = 0.018x^2 - 1.44x + 40$, where y denotes the cost per kilometer, and x the speed.

(a) Sketch the graph.

(b) Find the most efficient speed for this car. (Hint: Since the most efficient speed is that which costs the least per mile, you are looking for the minimum y.)

Nostalgia Problems

1. (a) A and B start at the same time to travel 150 miles; A travels 3 miles an hour faster than B, and finishes his journey $8\frac{1}{3}$ hours before him; at what rate per hour does each travel?

    (b) The plate of a mirror, 18 inches by 12, is to be set in a frame of uniform width, and the area of the frame is to be equal to that of the glass; required the width of the frame.

Calculator Activities

1. If x denotes the number of items produced in hundreds and if your company's profits are given by

    $$-(x - 4.2)(x - 5.3)$$

    in thousands of dollars, then

    (a) what are your profits if you produce 400 items?

    (b) what are your break-even production rates?

2. Solve by taking square roots.

    (a) $96x^2 - 300 = 0$ (b) $6.2 - 8.4x^2 = 0$

3. Use the quadratic formula to solve.

    (a) $2.3x^2 - 5.4x - 4.1 = 0$ (b) $56x^2 = 34x + 63$

Answers to "(a)" Exercises

Answers to chapter-opening photo problems appear on page 492.

Exercises 1-1

1. (a) $\frac{7}{8}$ is a rational number because it can be expressed as a quotient of two integers.
2. (a) Irrational
3. (a) Integer
4. (a) Rational
5. (a) Integer
6. (a) Irrational
7. (a) Integer
8. (a) Irrational
9. (a) Rational
10. (a) Rational
11. (a) Both the numbers 37 and 29 are integers
12. (a) Irrational
13. (a) Rational
14. (a) Integer
15. (a) Rational

Exercises 1-2

1. (a) $A \cup B$ means the union of the sets A and B. That is the collection of all the elements in A or B, but using each only once.

    use only once

    $A = \{2, 4, 6, 8, 10\}$ $B = \{0, 3, 6, 9, 12\}$
    $A \cup B = \{0, 2, 3, 4, 6, 8, 9, 10, 12\}$

2. (a) $\{0, 12\}$
3. (a) $\{4, 6, 8\}$
4. (a) $\{0, 2, 4, 6, 8, 10, 12\}$
5. (a) $\{0, 3, 4, 6, 8, 9, 12\}$
6. (a) $\{2, 3, 4, 5, 6, 8, 10\}$
7. (a) This is true, since all elements in O are also in M.
8. (a) F
9. (a) T
10. (a) F (no 9!)
11. (a) T
12. (a) F
13. (a) $(6 + 5) \cdot 3 + 9$
    $11 \cdot 3 + 9$
    $33 + 9$
    42
14. (a) 1
15. (a) 33
16. (a) 56
17. (a) 64
18. (a) 4
19. (a) 1
20. (a) 0
21. (a) 5
22. (a) 2
23. (a) 5
24. (a) $x(3 + 7)$, given x is 2
    $2(3 + 7)$
    $2(10)$
    20
25. (a) 10
26. (a) 10
27. (a) 12
28. (a) 54
29. (a) 44
30. (a) There are two possible choices here. $4 + (2 \cdot 3)$, which is *not* 18, or $(4 + 2) \cdot 3$, which evaluates to 18.
31. (a) $9 \div (5 - 2) = 3$
32. (a) $14 - 3 \cdot (2 - 2) = 14$
33. (a) $(15 \div 3) \cdot 2 = 10$
34. (a) $8 \cdot (4 - 2) \cdot 3 = 48$
35. (a) $(2 \cdot 3) + 5 = 11$
    $2 \cdot (3 + 5) = 16$
36. (a) $(18 - 6) \div 2 + 1 = 7$
    $18 - (6 \div 2) + 1 = 16$
    $18 - [6 \div (2 + 1)] = 16$
    More possible

Calculator Activities 1-2

1. (a) 5
2. (a) 782.7
3. (a) 0.61
4. (a) 23.2
5. (a) 0.7
6. (a) 5.260
7. (a) 236.5

ANSWERS TO "(a)" EXERCISES

Exercises 1-3

1. (a) $p = 2l + 2w$, given $p = 10, l = 2, w = 3$
   Substitute values, giving
   $10 = 2 \cdot 2 + 2 \cdot 3$
   $10 = 4 + 6$
   $10 = 10$, which is true
2. (a) T
3. (a) F
4. (a) F
5. (a) Six less than a number means subtract six, from some number which we can call "x": $x - 6$
6. (a) $x + 2x$
7. (a) $x - 7$ or $7 - x$
8. (a) $10 - 3x$
9. (a) $(x + 3)/x$
10. (a) 24 units
11. (a) 6 square units
12. (a) $-25/9°C$
13. (a) $60.8°F$
14. (a) 113.04 cubic units
15. (a) $600
16. (a) 20 square units
17. (a) 6.1
18. (a) If we let $x =$ number of biographies and y the number of mysteries, then $y = x + 3$.
19. (a) $x =$ son's age, $y =$ Martha's age, then $y = 4x$
20. (a) $a =$ number of paper clips in small box, b in large box, then $b = 2.5a$
21. (a) $x =$ large bulb wattage, $y =$ small bulb wattage, then $y = \frac{1}{3}x$
22. (a) $x =$ Juan's pay, $y =$ Lucia's pay, then $x + y = 121$
23. (a) $x =$ milliliters of sugar, $y =$ milliliters of flour, then $y = x + 750$
24. (a) $x =$ old computer, $y =$ new computer, then $y = 3x$
25. (a) $x =$ burners last year, $y =$ burners now, then $y = 6 + x$
26. (a) $x =$ first payment, $y =$ second payment, then $x + y = 835$
27. (a) One-quarter less than five
28. (a) Twice a number
29. (a) The difference between the sum of three and a number, and four
30. (a) The quotient of the difference between a number and one, and three
31. (a) The sum of a number and two is five.
32. (a) The sum of four times a number and one is nine.
33. (a) The difference between four and the sum of a number and two is twice the number.

Calculator Activities 1-3

1. (a) 67.1 centimeters
2. (a) 6.0 cubic centimeters
3. (a) $33,732.63
4. (a) 10.45 years
5. (a) 63.73°

Exercises 1-4

1. (a) The commutative property allows the order of the addends to be rearranged, while the associative property allows the addends to be regrouped. In this problem, the
2. (a) Additive identity
3. (a) Commutative property of multiplication
4. (a) Reflexive property of equality
5. (a) Associative property of multiplication
6. (a) Distributive
7. (a) Multiplicative inverse
8. (a) Distributive
9. (a) Additive inverse
10. (a) Additive identity
11. (a) Symmetric property of equality
12. (a) Additive property of equality
13. (a) Multiplicative property of equality
14. (a) Transitive property of equality
15. (a) Reflexive property of equality
16. (a) Multiplicative inverse
17. (a) Commutative of addition
18. (a) Symmetric property of equality

ANSWERS TO "(a)" EXERCISES

Exercises 1-5

1. (a) Associative property of addition,
By arithmetic
By arithmetic

2. (a) Associative property of multiplication
By arithmetic
By arithmetic

3. (a) Distributive property
By arithmetic
By arithmetic

4. (a) Associative of multiplication
Commutative of multiplication
Associative of multiplication
Multiplicative inverse
Multiplicative identity

5. (a) Distributive
Additive inverse
Multiplication by 0

6. (a) Commutative of addition
Associative of addition
Additive inverse
Additive identity
Additive inverse

7. (a) Associative of addition
Associative of addition
Commutative of addition
Associative of addition
Associative of addition
Distributive
By arithmetic

8. (a) Distributive
Commutative of addition
Associative of addition
By arithmetic

9. (a) Distributive
Distributive
Additive inverse
Multiplication by 0

10. (a) Distributive
By definition

Exercises 1-6

1. (a) $\frac{1}{3}$ 2. (a) $\frac{3}{5}$ 3. (a) $\frac{2}{3}$ 4. (a) 9
5. (a) $\frac{8}{5} \div \frac{30}{5} = \frac{8}{5} \cdot \frac{5}{30} = \frac{40}{150} = \frac{4}{15}$ 6. (a) 50
7. (a) 4 8. (a) $\frac{4}{7}$ 9. a) Prime 10. (a) $3 \cdot 3$
11. (a) $2 \cdot 2 \cdot 2 \cdot 2 \cdot 2$ 12. (a) $2 \cdot 2 \cdot 5 \cdot 5$
13. (a) $2 \cdot 2$ 14. (a) Prime 15. (a) $2 \cdot 2 \cdot 5$
16. (a) 3 is prime, 5 is prime, $9 = 3 \cdot 3$, so LCM $= 3 \cdot 3 \cdot 5 = 45$
17. (a) 180 18. (a) 24 19. (a) 315 20. (a) 90 21. (a) $\frac{5}{4}$ 22. (a) $\frac{5}{8}$
23. (a) $\frac{11}{10}$ 24. (a) $\frac{41}{48}$

Review Exercises Chapter 1

1. (a) Rational 2. (a) Rational 3. (a) Integer 4. (a) Integer
5. (a) Rational 6. (a) {2, 3, 5, 7}
7. (a) {3, 5, 7} 8. (a) T
9. (a) F 10. (a) T 11. (a) 43 12. (a) 44 13. (a) 30 14. (a) 36
15. (a) 36 16. (a) $(5 \cdot 2) - (3 \cdot 3) = 1$
17. (a) $(9 + 7) \div 2 = 8$ 18. (a) $(5 \cdot 2) - (3 \cdot 3) = 1$
19. (a) $3x + 4$ 20. (a) $(x + 3)/4$
21. (a) $\frac{5}{9}$°C 22. (a) 1.4 cubic units
23. (a) $56 24. (a) $x + y = 530$
25. (a) $7x = y$ 26. (a) The quotient of 5 and x
27. (a) Two less than five times a number is eight.
28. (a) Distributive 29. (a) Commutative of multiplication

ANSWERS TO "(a)" EXERCISES

30. (a) Associative of addition
Additive inverse
Additive identity

31. (a) Commutative of addition
Associative of addition
Associative of addition
Distributive
By definition

32. (a) 2 **33.** (a) $\frac{8}{5}$ **34.** (a) $3 \cdot 3 \cdot 3$ **35.** (a) Prime **36.** (a) 36
37. (a) 126 **38.** (a) $\frac{1}{3}$ **39.** (a) $\frac{31}{66}$

Calculator Activities

1. (a) 113,417 2. (a) 938.87 3. (a) 14,892.24 4. (a) 25.8 5. (a) 12.91
6. (a) −119.7   7. (a) $38.96
8. (a) −27.65 m/sec   9. (a) 35.07°
10. (a) 15.7 years

Exercises 2-1

1. (a) number line with points at −2, 0, 3, 5, 8

2. (a) number line with points at −5, −1, 0, 8

3. (a) number line with points at 0, $\frac{1}{3}$, $1\frac{1}{4}$, $\frac{7}{2}$, 5

4. (a) number line with points at −1.5, 0, 1.3, 2.1, 3

5. (a) number line with points at $-\frac{2}{3}$, 0, $\frac{2}{3}$, .75

6. (a) number line with points at −1.2, −1, 0, $2\frac{3}{4}$, 3

7. (a) number line with points at $-\frac{1}{2}$, 0, $1\frac{1}{2}$, 2.7, 4.2

8. (a) number line with points at 8.2, 9.3, 10.5, 11.9

9. (a) $\{-\frac{3}{2}, 3, \frac{1}{2}\}$ 10. (a) $\{-\frac{3}{2}, -2, -\frac{1}{2}, 3, \frac{3}{2}\}$
11. (a) $\{\frac{1}{2}, 1, -2, -\frac{1}{2}\}$ 12. (a) $\{-\frac{3}{2}, -2, -1, \frac{1}{2}, 2, 1\}$
13. (a) $\{-\frac{3}{2}, 3, \frac{1}{2}, \frac{3}{2}, \frac{5}{2}\}$
14. (a) number line with points at −3, 0, 3 --- $3 > -3$

15. (a) number line with points at 0, $\frac{1}{3}$, $\frac{1}{2}$, 1 --- $\frac{1}{2} > \frac{1}{3}$

ANSWERS TO "(a)" EXERCISES   **433**

16. (a) [number line showing −2, −3/2, −1, 0] $-1 > -\frac{3}{2}$

17. (a) [number line showing 1, 1.2, 2] $2 > 1.2$

18. (a) [number line showing 5, 7] $7 > 5$

19. (a) [number line showing −10.5, −7, −6] $-7 > -10.5$

20. (a) [number line showing 1, 2, 5/2, 3] $3 > \frac{5}{2}$

21. (a) [number line showing 3, 3.75, 4] $4 > 3.75$

22. (a) [number line showing −9, 8] $8 > -9$

23. (a) [number line showing 2, 2⅔, 2⅘, 3] $2\frac{4}{5} > 2\frac{2}{3}$

24. (a) [number line showing 3, 10/3, 4] $\frac{10}{3} > 3$

25. (a) [number line showing −4, −3.2, −3, −2.7, −2, 0] $-2.7 > -3.2$

Exercises 2-2

1. (a) By definition, this is the transitive property of >.
2. (a) Transitive property of <
3. (a) Additive property of >
4. (a) Transitive property of >
5. (a) Trichotomy property
6. (a) Multiplicative property of >
7. (a) Multiplicative property of <
8. (a) Transitive property of >
9. (a) Multiplicative property of <
10. (a) Transitive property of <
11. (a) Transitive property of >
12. (a) Multiplicative property of <

Exercises 2-3

1. (a) 3   2. (a) 9   3. (a) 4   4. (a) 2.2   5. (a) −1   6. (a) −5.7
7. (a) 12   8. (a) 5   9. (a) ¾
10. (a) $(-6.5) + (3.2)$
       $(3.2) + (-6.5)$
       $3.2 - 6.5$
       -3.3
11. (a) −13
12. (a) −6.7
13. (a) −3

434 ANSWERS TO "(a)" EXERCISES

14. (a) 6.5 15. (a) $-2\frac{1}{8}$ 16. (a) -13 17. (a) -3.6 18. (a) 4
19. (a) -11 20. (a) $-3\frac{5}{7}$ 21. (a) -31 22. (a) -5 23. (a) -11.8
24. (a) $-\frac{3}{8}$ 25. (a) -17 26. (a) -14 27. (a) $\frac{1}{4}$ 28. (a) -1
29. (a) 20 30. (a) -1 31. (a) 3 32. (a) -1131 33. (a) $1\frac{1}{2}$
34. (a) -0.1 35. (a) -5 36. (a) 5
37. (a) $[3.2 + (-2.1 - 4.3)] + (-5.6)$ 38. (a) -6
    $3.2 + (-6.4) + (-5.6)$ 39. (a) 17
    $3.2 - 6.4 - 5.6$ 40. (a) $6 + 4 = 10$
    -8.8
41. (a) $6 + 8 = 14$ 42. (a) $22 - 16 = 6$
43. (a) $-56 - 21 - 75 = -152$

Calculator Activities 2-3

1. (a) -359 2. (a) -783 3. (a) 6.221 4. (a) -12.956 5. (a) -2.545
6. (a) $\$652.21 + 43.25 - 33.67 - 102.99 - 1.55 + 23.40 - 14.50 - 12.20 = \$553.95$

Exercises 2-4

1. (a) 42 2. (a) -12 3. (a) 16 4. (a) -35 5. (a) -48
6. (a) $-[(-4) \times 7]$ 7. (a) 15
    $-[-(4 \times 7)]$ 8. (a) -6
    28
9. (a) 8 10. (a) -5 11. (a) -6 12. (a) 8 13. (a) $\frac{1}{10}$ 14. (a) $-\frac{1}{6}$
15. (a) $\frac{1}{11}$ 16. (a) $\frac{6}{5}$ 17. (a) $\frac{4}{3}$ 18. (a) $-\frac{2}{3}$ 19. (a) 24.15
20. (a) -20 21. (a) 5 22. (a) 4.75 23. (a) -18 24. (a) $-\frac{3}{5}$
25. (a) $\frac{2}{15}$ 26. (a) 8 27. (a) -35 28. (a) -57
29. (a) $12 \times (4)$ 30. (a) $5 \times (-5) = -25$
    $\$48$
31. (a) $1 - \frac{1}{4} = \frac{3}{4}$ 32. (a) $9000/3 = \$3000$
33. (a) $5 \times 24 \times \$0.30 = \$36$

Calculator Activities 2-4

1. (a) $-250{,}958$ 2. (a) -125.754 3. (a) -2.663 4. (a) -83.65
5. (a) -0.33 6. (a) -743.56

Exercises 2-5

1. (a) 93 has a 9 in the 10s place, a 3 in the units place, giving $9 \cdot 10^1 + 3$
2. (a) $1 \cdot 10^2 + 4$ 3. (a) $2 \cdot 10^2 + 6 \cdot 10^1 + 7$
4. (a) $9 \cdot 10^3 + 1 \cdot 10^1$ 5. (a) $3 \cdot 10^3 + 1 \cdot 10^2 + 5$
6. (a) $1 \cdot 10^5 + 1 \cdot 10^4 + 4 \cdot 10^3$ 7. (a) $6 \cdot 10^4 + 3 \cdot 10^3 + 9 \cdot 10^2 + 8 \cdot 10^1$
8. (a) $9 \cdot 10^5 + 3 \cdot 10^4 + 8 \cdot 10^3 + 3 \cdot 10^2 + 2 \cdot 10^1 + 5$
9. (a) 0.201 has a 2 in the tenths place and a 1 in the thousandths place, giving $2 \cdot 10^{-1} + 1 \cdot 10^{-3}$
10. (a) $6 \cdot 10^0 + 3 \cdot 10^{-2}$ 11. (a) $4 \cdot 10^0 + 7 \cdot 10^{-1} + 9 \cdot 10^{-2} + 1 \cdot 10^{-3}$
12. (a) $1 \cdot 10^1 + 1 \cdot 10^0 + 5 \cdot 10^{-1} + 2 \cdot 10^{-2}$
13. (a) $8 \cdot 10^1 + 1 \cdot 10^0 + 9 \cdot 10^{-1} + 6 \cdot 10^{-2} + 3 \cdot 10^{-3}$
14. (a) $2 \cdot 10^1 + 7 \cdot 10^0 + 1 \cdot 10^{-3}$ 15. (a) $2 \cdot 10^4 + 7 \cdot 10^3 + 2 \cdot 10^{-2}$
16. (a) $3 \cdot 10^2 + 8 \cdot 10^1 + 7 \cdot 10^{-2}$
17. (a) 9,800,000: To place the decimal point between the 9 and 8 means 6 moves, giving $9.8 \cdot 10^6$

18. (a) $1.2 \cdot 10^3$
19. (a) $8.02 \cdot 10^5$
20. (a) $5.3 \cdot 10^{-4}$
21. (a) $8.3 \cdot 10^{-8}$
22. (a) $7.7 \cdot 10^{-7}$
23. (a) $1.26 \cdot 10^{11}$
24. (a) $1.0 \cdot 10^{11}$ or 10^{11}
25. (a) $6.4 \cdot 10^3$ kilometers
26. (a) $3.0 \cdot 10^{10}$ centimeters per second
27. (a) $1.09 \cdot 10^8$ kilometers

Calculator Activities 2-5

1. (a) $9.261 \cdot 10^3$
2. (a) $8.318 \cdot 10^{-3}$
3. (a) $3.1 \cdot 10^7$
4. (a) $1.33 \cdot 10^5$
5. (a) $9.8 \cdot 10^{-5}$
6. (a) $2.97 \cdot 10^{-4}$
7. (a) $6.7 \cdot 10^9$
8. (a) $1.13 \cdot 10^1$
9. (a) $4.066 \cdot 10^1$
10. (a) $6.64 \cdot 10^1$

Exercises 2-6

1. (a) $\sqrt{28}$
$5 < \sqrt{28} < 6$
$5.2^2 < 28 < 5.3^2$; therefore, $\sqrt{28}$ is closer to 5 than 6
2. (a) 7
3. (a) 2
4. (a) 10
5. (a) 12
6. (a) 6
7. (a) 15
8. (a) 1
9. (a) $\sqrt{58}$
$7 < \sqrt{58} < 8$
$7.6^2 < 58 < 7.7^2$; thus, 7.6 or 7.7 is a two-digit approximation
10. (a) 3.7
11. (a) 8.8
12. (a) 9.7
13. (a) 11
14. (a) 4.5
15. (a) 7.9
16. (a) 2.6
17. (a) 7.1
18. (a) 10
19. (a) 9.5
20. (a) 12
21. (a) 9.43 meters
22. (a) 14.32 meters

Review Exercises Chapter 2

1. (a) [number line with points at -6, -2, 0, 5, 9]

2. (a) [number line with points at -2, $-.6$, 0, $\frac{4}{5}$, $\frac{7}{5}$]

3. (a) $\{-2.5, 0.5, -2, 0\}$
4. (a) $\{0.5, -1, 1.5, -2.5\}$

5. (a) [number line with points at -7, -3, 0] $-3 > -7$

6. (a) [number line with points at -2, $-\frac{4}{3}$, $-\frac{5}{4}$, -1, 0] $-\frac{5}{4} > -\frac{4}{3}$

7. (a) Transitive property of $>$
8. (a) Multiplicative property of $>$
9. (a) 3
10. (a) -13.3
11. (a) 6.8
12. (a) -11
13. (a) 2
14. (a) -84
15. (a) 72
16. (a) -3
17. (a) $4 - 6 + 18 = 16$
18. (a) $320 - 27 - 63 - 12 - 5.50 - 72 + 35 + 90 = \$265.50$
19. (a) $40 \cdot \frac{5}{8} = 25$
20. (a) $2 \cdot 10^2 + 5$
21. (a) $4 \cdot 10^4 + 5 \cdot 10^3 + 8 \cdot 10^2$
22. (a) $2 \cdot 10^0 + 3 \cdot 10^{-1} + 4 \cdot 10^{-2}$

436 ANSWERS TO "(a)" EXERCISES

23. (a) $7 \cdot 10^0 + 9 \cdot 10^{-1} + 3 \cdot 10^{-3} + 4 \cdot 10^{-4}$ **24.** (a) $5.08 \cdot 10^8$
25. (a) $9.2 \cdot 10^{-9}$ **26.** (a) 11
27. (a) 9   **28.** (a) 14   **29.** (a) 10
30. (a) 13.04 centimeters × 13.04 centimeters

Calculator Activities Chapter 2 Review

1. (a) 3096   **2.** (a) -0.77 **3.** (a) 433,472   **4.** (a) 0.3632
5. (a) $8.3521 \cdot 10^4$ **6.** (a) $1.775 \cdot 10^{-1}$
7. (a) $9.313 \cdot 10^{-10}$ **8.** (a) $2.340 \cdot 10^1$

Exercises 3-1

1. (a) $x + 3 = 12$
Let $x = 7$, $7 + 3 = 12$ is false
Let $x = 8$, $8 + 3 = 12$ is false
Let $x = 9$, $9 + 3 = 12$ is *true*
Let $x = 10$, $10 + 3 = 12$ is false

2. (a) 1   **3.** (a) -5 **4.** (a) 16   **5.** (a) $-4, 3$ **6.** (a) $\frac{1}{2}$ **7.** (a) 4
8. (a) 14   **9.** (a) $\frac{7}{2}$ **10.** (a) $\frac{1}{4}$

Calculator Activities 3-1

1. (a) 2.61   **2.** (a) $\frac{1}{4}$ **3.** (a) 51   **4.** (a) 3.2 (approx.)   **5.** (a) -4

Exercises 3-2

1. (a)
$$x + 11 = 12 \qquad \checkmark \; x + 11 = 12$$
$$x + 11 + (-11) = 12 + (-11) \qquad 1 + 11 = 12$$
$$x + 0 = 12 - 11 \qquad 12 = 12$$
$$x = 1$$

2. (a) $t = 7$ **3.** (a) $x = 2$
4. (a) $x = 8$ **5.** (a) $t = 31$
6. (a)
$$2x = 10 \qquad \checkmark \; 2x = 10$$
$$\tfrac{1}{2} \cdot 2x = \tfrac{1}{2} \cdot 10 \qquad 2(5) = 10$$
$$\frac{2x}{2} = \frac{10}{2} \qquad 10 = 10$$
$$x = 5$$

7. (a) $y = 6$ **8.** (a) $x = 8$
9. (a) $t = \frac{5}{4}$ **10.** (a) $x = 49$
11. (a)
$$4 = 12 + 2x \qquad \checkmark \; 4 = 12 + 2x$$
$$4 + (-12) = 12 + (-12) + 2x \qquad 4 = 12 + 2(-4)$$
$$4 - 12 = 0 + 2x \qquad 4 = 12 - 8$$
$$-8 = 2x \qquad 4 = 4$$
$$-8 \cdot \tfrac{1}{2} = \tfrac{1}{2} \cdot 2x$$
$$\frac{-8}{2} = \frac{2x}{2}$$
$$-4 = x$$

12. (a) $y = 2$ **13.** (a) $m = 12$
14. (a) $x = -\frac{4}{5}$ **15.** (a) $x = -3$
16. (a) $y = \frac{11}{2}$ **17.** (a) $x = -\frac{13}{7}$
18. (a) $x = -6$

ANSWERS TO "(a)" EXERCISES   437

Calculator Activities 3-2

1. (a) $x = 21.109$
2. (a) $x = 150.662$
3. (a) $x = 51$
4. (a) $x = -5.8$
5. (a) $t = -10.92$

Exercises 3-3

1. (a) Let $x =$ son's age, then father's age $= 3x$
   Father's age $-$ son's age $= 24$, therefore
   $$3x - x = 24$$
   $$2x = 24$$
   $$x = 12 \text{ is son's age}$$
   Then $3x = 36$ is father's age

2. (a) $10 and $76
3. (a) 9.5 meters and 15.5 meters
4. (a) Let $x =$ number of quarters, then $18 - x =$ number of dimes.

		Value per	Total value
Quarters	x	0.25	$0.25x$
Dimes	$18-x$	0.10	$0.1(18-x)$
Total	18		$3.60

   $$0.25x + 0.1(18 - x) = 3.60$$
   $$25x + 10(18 - x) = 360$$
   $$25x + 180 - 10x = 360$$
   $$15x + 180 = 360$$
   $$15x = 180$$
   $$x = 12 \text{ quarters}$$
   $$18 - x = 6 \text{ dimes}$$

5. (a) 5275 units
6. (a) 8900 items
7. (a) 175 each of small and extra-large, 350 each of medium and large
8. (a) 1.6 meters by 3.2 meters

Nostalgia Problems 3-3

1. (a) Saddle is $12.50, horse is $87.50
2. (a) 15 needles, 105 pins

Calculator Activities 3-3

1. (a) 11.2%
2. (a) $1,075,000
3. (a) $6811.05 at 9%, $6311.05 at 9.5%, $12,622.11 at 10.2%

Exercises 3-4

1. (a) $3x - 2x + 4 = 3$ √ $3x - 2x + 4 = 3$
        $x + 4 = 3$ $3(-1) - 2(-1) + 4 = 3$
        $x = -1$ $-3 + 2 + 4 = 3$
                                   $3 = 3$

2. (a) $y = 3$
3. (a) $x = -16$
4. (a) $x = 3$
5. (a) $x = 12$
6. (a) $y = 4$
7. (a) $x = 33\frac{1}{3}$
8. (a) $x = 16$
9. (a) $x = 3$
10. (a) $t = 3$
11. (a) $2(x - 11) = 20$ √ $2(x - 11) = 20$
         $2x - 22 = 20$ $2(21 - 11) = 20$
         $2x = 42$ $2(10) = 20$
         $x = 21$ $20 = 20$

ANSWERS TO "(a)" EXERCISES

12. (a) $x = 3$
14. (a) $y = 16$
16. (a) No solution
18. (a) $x = 8$
19. (a) $\frac{x}{2} - 3 + \frac{x}{3} = 2$ ✓ $\frac{x}{2} - 3 + \frac{x}{3} - 2$
    $\frac{x}{2} + \frac{x}{3} = 5$ $\frac{6}{2} - 3 + \frac{6}{3} = 2$
    $6\left(\frac{x}{2} + \frac{x}{3}\right) = 6(5)$ $3 - 3 + 2 = 2$
    $3x + 2x = 30$ $2 = 2$
    $5x = 30$
    $x = 6$

13. (a) $x = 4$
15. (a) $m = -\frac{10}{7}$
17. (a) $m = 5$

20. (a) $x = 12$
22. (a) $x = 8$
24. (a) $z = 3$
26. (a) All real numbers

21. (a) $x = 36$
23. (a) $y = 0$
25. (a) $x = \frac{1}{2}$

Calculator Activities 3-4

1. (a) $x = 16.6$
3. (a) $x = 1.260$
5. (a) $y = -92$

2. (a) $x = -15.1$
4. (a) $x = -0.657$
6. (a) $x = 24.278$

Exercises 3-5

1. Let x = original constant speed
   Then $\frac{2}{3}x$ = speed for last hour

	Rate	× Time =	Distance
First part	x	4	$4x$
Second part	$\frac{2}{3}x$	1	$\frac{2}{3}x$
Total			350

   $4x + \frac{2}{3}x = 350$
   $12x + 2x = 1050$
   $14x = 1050$
   $x = 75$ km/hr

2. (a) 4 centimeters
4. (a) Joe is 36 years old, Martha is 12 years old.
5. (a) 10 nickels, 10 dimes, 40 quarters
7. (a) length is 56 meters, width is 28 meters
9. (a) 194 predators, 147 preys

3. (a) $2\frac{1}{2}$ hours
6. (a) $50, $80, $160
8. (a) $2,000,000
10. (a) $7500 in term, $500 in passbook

Nostalgia Problems 3-5

1. (a) 2400 trees
3. (a) 47 sheep

2. (a) $11,360

Calculator Activities 3-5

1. (a) $3.67

2. (a) $622

Exercises 3-6

1. (a) $t = \dfrac{I}{p \cdot r}$

2. (a) $P = 2(l+w)$ or $P = 2l + 2w$
$\dfrac{P}{2} = l + w$ $P - 2l = 2w$
$\dfrac{P}{2} - l = w$ $\dfrac{P-2l}{2} = w$

(Make sure you understand that these two answers are equivalent!)

3. (a) $b = \dfrac{2A}{h}$
4. (a) $C = S - P$
5. (a) $a = 2A - b$
6. (a) $b_2 = \dfrac{2A}{h} - b_1$
7. (a) $r = \dfrac{C}{2\pi}$
8. (a) $t = \dfrac{k-v}{32}$
9. (a) $G = \dfrac{20(L-1)}{3}$
10. (a) $r = \dfrac{s-a}{s}$
11. (a) $v = \dfrac{m}{d}$
12. (a) (i) 12 hours, (ii) 6 years, (iii) $A = 34 - 2H$

Calculator Activities 3-6

1. (a) $b = \dfrac{P+d}{0.83} - 1235$
2. (a) $r = \dfrac{A-P}{P}$, $33\frac{1}{3}\%$
3. (a) $t = \dfrac{w - 3000}{9000}$

Review Exercises Chapter 3

1. (a) 6
2. (a) 3, −3
3. (a) $x = 20$
4. (a) $x = 6$
5. (a) $x = 2$
6. (a) $x = \frac{3}{2}$
7. (a) $y = 7$
8. (a) $x = -2$
9. (a) $x = -6$
10. (a) $m = -2$
11. (a) $x = 3$
12. (a) All real numbers
13. (a) $y = 12$
14. (a) $n = -15$
15. (a) Isabel is 4 years old, Irwin is 15 years old
16. (a) 19 dimes, 19 quarters, 57 half-dollars
17. (a) $51.95
18. (a) 15 liters
19. (a) $t = \dfrac{w_1 - w_2}{a}$
20. (a) $A = \dfrac{4\pi Cd}{K}$
21. (a) (i) $E = I(R+r)$, (ii) 6 volts

Calculator Activities Chapter 3 Review

1. (a) $x = 52.910$
2. (a) $y = 0.966$
3. (a) $x = 0.021$
4. (a) 14,627 items

Exercises 4-1

1. (a) True
2. (a) $14 - 3 \leq 9 + 1$
$11 \leq 10$ is false
3. (a) False
4. (a) False
5. (a) False
6. (a) $x > -3$ says that any number greater than −3 will satisfy the inequality. Two possible numbers are 0 and −1.

440 ANSWERS TO "(a)" EXERCISES

7. (a) Any $z < 7$, such as 4, -10
8. (a) Any $x < \frac{3}{2}$, such as 1, 0
9. (a) Any $t > -14$, such as -13, 4
10. (a) Any $x \leq 3$, such as 3, 2
11. (a) Any $y < -2$, such as -3, -12
12. (a) Any $x > \frac{3}{2}$, such as 2, 4
13. (a) $D < 2P$
14. (a) $s \geq 300$
15. (a) $s_1 + s_2 \geq 2300$
16. (a) $S > 2C + 10$
17. (a) $S \geq 0.04(250,000) + 9000$
18. (a) $x + 20 < \frac{1}{2}x$

Calculator Activities 4-1

1. (a) True
2. (a) True
3. (a) Any $x < 7.7$, such as 7, 0
4. (a) Any $x < -13.8$, such as -14, -25
5. (a) Any $x \leq 0.545$, such as 0.545, 0

Exercises 4-2

1. (a) $$\begin{aligned} x + 3 &< 11 \\ x + 3 + (-3) &< 11 + (-3) \\ x + 0 &< 11 - 3 \\ x &< 8 \end{aligned}$$
2. (a) $y \geq 41$
3. (a) $z > 5$
4. (a) $w < 4$
5. (a) $x > 7$
6. (a) $x \leq 2$
7. (a) $y > -7$
8. (a) $$\begin{aligned} -x &< -2 \\ (-1)(-x) &> (-1)(-2) \\ x &> 2 \end{aligned}$$
9. (a) $t \leq -\frac{3}{2}$
10. (a) $x > -3$
11. (a) $y < 3$
12. (a) $y \leq -4$
13. (a) $x < -\frac{7}{2}$
14. (a) $x > -14$
15. (a) $p \geq 10$
16. (a) $x < 2$
17. (a) $w < 20$
18. (a) $x > 1$
19. (a) $x \geq -8$
20. (a) $x < 40$
21. (a) Let x = number of men. $x + 14 < 30$, therefore $x < 16$, or there are less than 16 men in the class.
22. (a) Let x = profit margin
    $x + 12 < 19$, therefore profit margin must be less than 7
23. (a) Let x = distance; since $r \cdot t = d$, $r = 50$, and t is at most 4,
    $$\begin{aligned} t &\leq 4 \\ 50t &\leq 200 \end{aligned}$$
    So d is at most 200 kilometers
24. (a) Let x = amount he can lose
    $\frac{x}{7} \leq 6$, therefore $x \leq 42$
25. (a) Let w = width; $14w \geq 154$, therefore $w \geq 11$ meters

Calculator Activities 4-2

1. (a) $x < 5985$
2. (a) $y > -105.325$
3. (a) $x < 22\frac{2}{3}$
4. (a) $x \leq 32.45$

Exercises 4-3

1. (a) $x > 21$

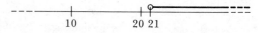

ANSWERS TO "(a)" EXERCISES 441

2. (a) $y \leq -20$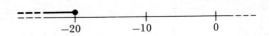

3. (a) $3 - \frac{3}{4}x > \frac{1}{2}(x + 5)$
$3 - \frac{3}{4}x > \frac{1}{2}x + \frac{5}{2}$
$3 - \frac{5}{4}x > \frac{5}{2}$
$-\frac{5}{4}x > -\frac{1}{2}$
$x < \frac{2}{5}$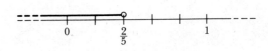

4. (a) $k \leq 1$

5. (a) $t < 5$

6. (a) $x < 3$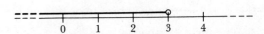

7. (a) $y > 24$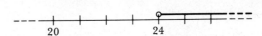

8. (a) $x \leq 9$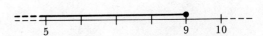

9. (a) $z \geq -2$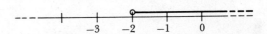

10. (a) $x < 2$

11. (a) $x \geq 7$

12. (a) $x \geq 10$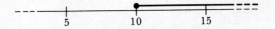

13. (a) $y \geq -9$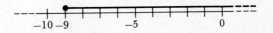

14. (a) $y > \frac{14}{3}$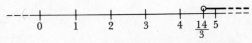

15. (a) $x > 0$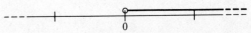

16. (a) $5 - 6x < 23$ and $3x + 5 \leq 7 - 2x$
$-6x < 18$ $\qquad$ $5x \leq 2$
$x > -3$ and $\qquad$ $x \leq \frac{2}{5}$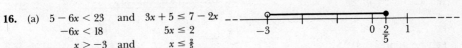

17. (a) $x > 2$ or $x < -2$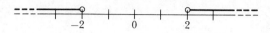

18. (a) $x > 5$ or $x > -4$

19. (a) $x \leq 1250$ units

20. (a) Let $x =$ number of cartons, $x \geq 5$ cartons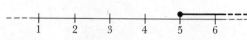

21. (a) Let $x =$ number of trees, $3x - 22 < 68$ and $x \geq 4$ $x < 30$ and $x \geq 4$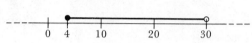

Calculator Activities 4-3

1. (a) $x > 0.276$

2. (a) $x \geq -1.426$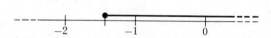

3. (a) $x > 0.348$

4. (a) $x > 0.276$ and $x < 1.655$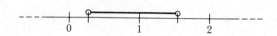

Review Exercises Chapter 4

1. (a) True
2. (a) False
3. (a) Any $x > 3$, such as 4, 10
4. (a) Any $x \geq -\frac{3}{2}$, such as -1, 0
5. (a) Let $x =$ tenors, then sopranos $\geq \frac{3}{2}x$
6. (a) Let $s =$ selling price, then profit $\geq 0.15s$
7. (a) $x > 2$

8. (a) $x \geq -11$

9. (a) $y \leq 8$

10. (a) $x < 7$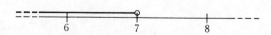

11. (a) $x > 5$

ANSWERS TO "(a)" EXERCISES 443

12. (a) True for all values of x

13. (a) $x > 3$ or $x < -3$

14. (a) $x > -5$ and $x < 3$

15. (a) Greater than 8 years old
16. (a) Between 75 and 90

Calculator Activities Chapter 4 Review

1. (a) True
2. (a) False
3. (a) $x < 5.55$

4. (a) $x \leq -2.10\frac{1}{3}$

5. (a) $x < 853.875$

6. (a) $x < -11.441$

7. (a) $x > 4$ and $x < 5.2$

Exercises 5-1

1. (a) $x + 3y - 3 = 0$
   $x = -3y + 3$
   $\{(-3y + 3, y) \mid y \text{ any real number}\}$ is the solution set
   If $y = 0$, $-3(0) + 3 = 3$ $(3, 0)$
   If $y = 1$, $-3(1) + 3 = 0$ $(0, 1)$
   If $y = 2$, $-3(2) + 3 = -3$ $(-3, 2)$
2. (a) $\{(\frac{2}{3}y - 1, y) \mid y \text{ any real number}\}$ $(-1, 0), (1, 3), (-3, -3)$
3. (a) $\{(4y - 2, y) \mid y \text{ any real number}\}$ $(-2, 0), (6, 2), (-10, -2)$
4. (a) $\{(-5y + \frac{2}{3}, y) \mid y \text{ any real number}\}$ $(\frac{2}{3}, 0), (-1, \frac{1}{3}), (4, -\frac{2}{3})$
5. (a) $\{(3y^2 - 1, y) \mid y \text{ any real number}\}$ $(2, 1), (2, -1), (-1, 0)$
6. (a) $\{(3y^2 - \frac{3}{2}, y) \mid y \text{ any real number}\}$ $(-\frac{3}{2}, 0), (\frac{3}{2}, 1), (\frac{3}{2}, -1)$
7. (a) $\{(3y - 4, y) \mid y \text{ any real number}\}$ $(0, -\frac{4}{3}), (4, 0), (-2, -2)$
8. (a) $\{(-4y^2 - 3, y) \mid y \text{ any real number}\}$ $(-3, 0), (-7, 1), (-7, -1)$
9. (a) $\{(2y + 3, y) \mid y \text{ any real number}\}$ $(3, 0), (5, 1), (1, -1)$
10. (a) $\{(-\frac{1}{2}y^2 + 2, y) \mid y \text{ any real number}\}$ $(2, 0), (0, 2), (0, -2)$
11. (a) $3x + y - 5 = 0$
    $y = -3x + 5$
    $\{(x, -3x + 5) \mid x \text{ any real number}\}$ is the solution set
    If $x = 0$, $-3(0) + 5 = 5$ $(0, 5)$
    If $x = 1$, $-3(1) + 5 = 2$ $(1, 2)$
    If $x = -1$, $-3(-1) + 5 = 8$ $(-1, 8)$

444 ANSWERS TO "(a)" EXERCISES

12. (a) $\{(x, 4x - 7) \mid x \text{ any real number}\}$ $(0, -7)$, $(1, -3)$, $(2, 1)$
13. (a) $\{(x, -0.4x + 2) \mid x \text{ any real number}\}$ $(0, 2)$, $(5, 0)$, $(10, -2)$
14. (a) $\{(x, -2x + 3) \mid x \text{ any real number}\}$ $(0, 3)$, $(1, 1)$, $(-1, 5)$
15. (a) $\left\{\left(x, \dfrac{1}{x}\right) \mid x \text{ any real number except } 0\right\}$ $(1, 1)$, $(2, \tfrac{1}{2})$, $(\tfrac{1}{2}, 2)$
16. (a) $\left\{\left(x, -\dfrac{2}{3x}\right) \mid x \text{ any real number except } 0\right\}$ $(1, -\tfrac{2}{3})$, $(2, -\tfrac{1}{3})$
17. (a) $\left\{\left(x, -\dfrac{2}{3}x^2 + 2\right) \mid x \text{ any real number}\right\}$ $(0, 2)$, $(3, -4)$, $(-3, -4)$
18. (a) $\{(x, -2.5x^2 + 3) \mid x \text{ any real number}\}$ $(0, 3)$, $(1, 0.5)$, $(-1, 0.5)$
19. (a) $\left\{\left(x, -\dfrac{1}{x}\right) \mid x \text{ any real number except } 0\right\}$ $(1, -1)$, $(2, -\tfrac{1}{2})$, $(\tfrac{1}{2}, -2)$
20. (a) $\{(x, -\tfrac{2}{7}x + \tfrac{16}{7}) \mid x \text{ any real number}\}$ $(0, \tfrac{16}{7})$, $(1, 2)$, $(-2, \tfrac{20}{7})$
21. (a) $H = \dfrac{-A + 34}{2}$
22. (a) $t = \dfrac{w - 100}{20}$
23. (a) $P = 1.25t + 15$
24. (a) $F = \dfrac{D - 115}{2}$

Exercises 5-2

1. (a)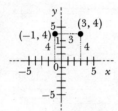

2. (a)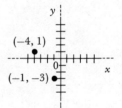

3. (a)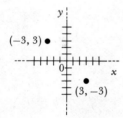

4. (a)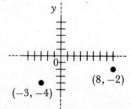

5. (a)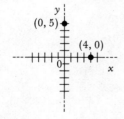

6. (a)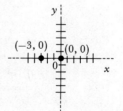

7. (a) $(-4, 2)$
8. (a) $(2, 3)$
9. (a) $(-2, 0)$
10. (a) $(5, 1)$

ANSWERS TO "(a)" EXERCISES 445

11. (a)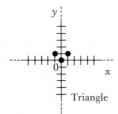
Triangle

12. (a)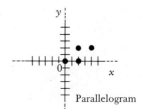
Parallelogram

13. (a)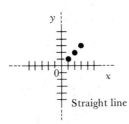
Straight line

14. (a)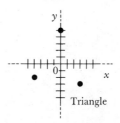
Triangle

15. (a)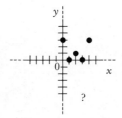
?

16. (a) {(0, 2), (1, 0), (1, 1), (1, 2), (2, 0), (2, 3), (3, 0), (3, 1), (3, 2), (4, 2)}
17. (a) {(0, 0), (1, 0), (1, 1), (2, 0), (2, 2), (3, 0), (3, 3), (4, 1), (4, 3), (5, 2), (5, 3), (6, 3)}
18. (a) (6, 0), on February 6 the temperature was 0°
(4, −5), on February 4 the temperature was −5°
(28, 5), on February 28 the temperature was 5°
19. (a) (1, 100%), the student earned a grade of 100% on quiz #1
(2, 80%), the student earned a grade of 80% on quiz #2
(5, 60%), the student earned a grade of 60% on quiz #5
20. (a) (1979, 130,000), the population in 1979 was 130,000
(1975, 125,000), the population in 1975 was 125,000
(1974, 120,000), the population in 1974 was 120,000

Calculator Activities 5-3

1. (a)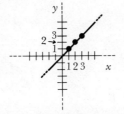

2. (a)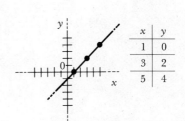

446 ANSWERS TO "(a)" EXERCISES

3. (a)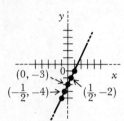

4. (a)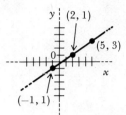

5. (a)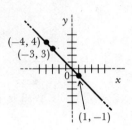

6. (a)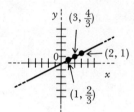

7. (a)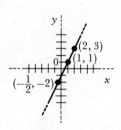

8. (a)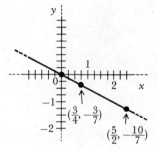

9. (a)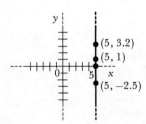

10. (a)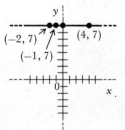

11. (a)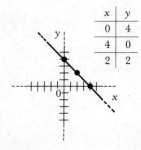

12. (a)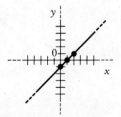

ANSWERS TO "(a)" EXERCISES **447**

13. (a)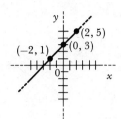

14. (a)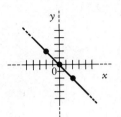

15. (a)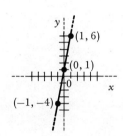

16. (a)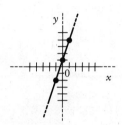

17. (a)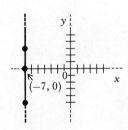

18. (a)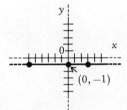

19. (a)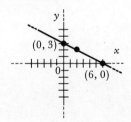

20. (a)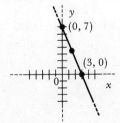

21. (a)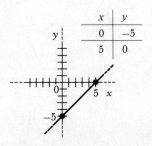

22. (a)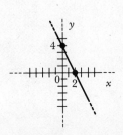

448 ANSWERS TO "(a)" EXERCISES

23. (a)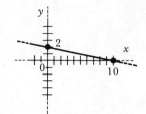

24. (a)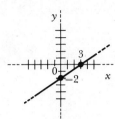

25. (a)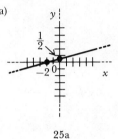

25a

26. (a)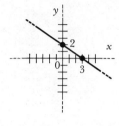

27. (a)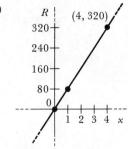

28. (a)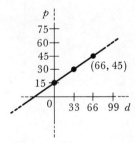

29. (a)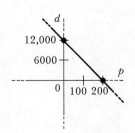

30. (a)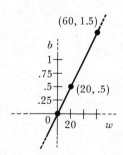

ANSWERS TO "(a)" EXERCISES

Calculator Activities 5-3

1. (a) Pounds

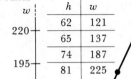

h	w
62	121
65	137
74	187
81	225

2. (a)

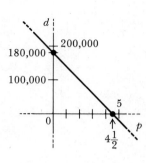

3. (a)

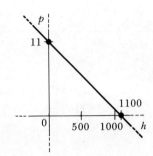

Exercises 5-4

1. (a) $(2, 1)$ and $(6, 3)$; $\dfrac{3-1}{6-2} = \dfrac{1}{2}$ is the slope   2. (a) $\dfrac{1}{5}$
3. (a) $\dfrac{1}{7}$ 4. (a) -1 5. (a) $\dfrac{3}{2}$ 6. (a) $-\dfrac{5}{2}$ 7. (a) 1 8. (a) $\dfrac{1}{2}$
9. (a) $x + y = 3$; two points that satisfy this equation are $(1, 2)$ and $(2, 1)$; $(1 - 2)/(2 - 1) = -1$ is the slope
10. (a) 2 11. (a) 1 12. (a) 1 13. (a) 1 14. (a) 3 15. (a) 5
16. (a) 1 17. (a) 2 18. (a) $-\dfrac{7}{2}$ 19. (a) 5 20. (a) $-\dfrac{2}{11}$ 21. (a) 4
22. (a) $\dfrac{15}{2}$

Exercises 5-5

1. (a) Slope $= 5$, y-intercept $= 4$ 2. (a) Slope $= -2$, y-intercept $= 6$

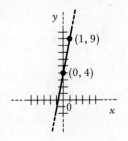
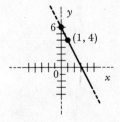

450 ANSWERS TO "(a)" EXERCISES

3. (a) Slope = $\frac{2}{3}$, y-intercept = -2

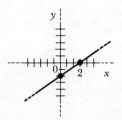

4. (a) Slope = $-\frac{3}{4}$, y-intercept = -5

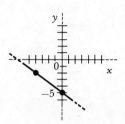

5. (a) Slope = $\frac{1}{5}$, y-intercept = 3

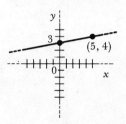

6. (a) Slope = 3, y-intercept = 6

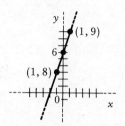

7. (a) Slope = -7, y-intercept = -5

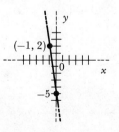

8. (a) Slope = $-\frac{5}{3}$, y-intercept = 5

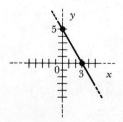

9. (a) Slope = 6, y-intercept = 0

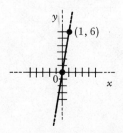

10. (a) Slope = 2, y-intercept = 0

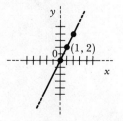

11. (a) $y = mx + b$, $m = 3$, $b = 4$, $y = 3x + 4$ **12.** (a) $y = -x + 3$
13. (a) $y = \frac{1}{2}x$ **14.** (a) $y = \frac{5}{2}x - 3$
15. (a) $y = 0.4x + 1.2$ **16.** (a) $y = -\frac{3}{4}x + 3$
17. (a) $y = \frac{7}{4}x + 7$ **18.** (a) $y = x$
19. (a) $y = \frac{2}{5}x + \frac{2}{5}$ **20.** (a) $y = -4$
21. (a) x-intercept of 5 gives the point $(5, 0)$; y-intercept of -1 gives the point $(0, -1)$; then
$$\frac{-1 - 0}{0 - 5} = \frac{-1}{-5} = \frac{1}{5}; y = mx + b, \text{ where } m = \frac{1}{5}, b = -1; y = \frac{1}{5}x - 1$$
22. (a) $y = -\frac{2}{7}x + 2$ **23.** (a) $y = 0$
24. (a) $y = -\frac{10}{3}x + 1$ **25.** (a) $(5, 2)$ and $(3, 0)$
$$\text{Slope} = \frac{0 - 2}{3 - 5} = \frac{-2}{-2} = 1$$
$y = mx + b$
$y = (1)x + b$, using the point $(3, 0)$
$0 = (1)(3) + b$
$0 = 3 + b$
$-3 = b$, therefore the equation for the line is $y = x - 3$

26. (a) $y = \frac{2}{3}x + \frac{16}{3}$ **27.** (a) $y = 6$
28. (a) $y = \frac{1}{4}x - 7$ **29.** (a) $y = -\frac{1}{10}x + 70$

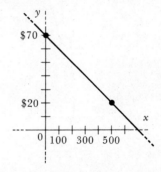

30. (a) $y = 4x - 128$ **31.** (a) $y = \frac{13.75}{250}x$ or $y = 0.055x$

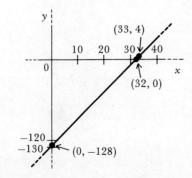
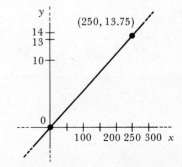

Calculator Activities 5-5

1. (a) Slope = $\frac{2}{5}$, y-intercept = 8.2

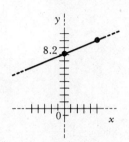

2. (a) Slope = $-\frac{3}{2}$, y-intercept = 4.5

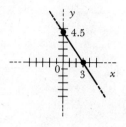

3. (a) $-\$1545.50$
4. (a) 475.54 or 476 items to break even
5. (a)

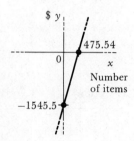

Exercises 5-6

1. (a)

x	y
−.5	−.25
−1	−1
−2	−4

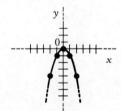

x	y
0	0
.5	−.25
1	−1
2	−4

2. (a)

x	y
−.5	1.25
−1	2
−2	5

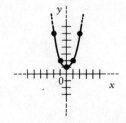

x	y
0	1
.5	1.25
1	2
2	5

ANSWERS TO "(a)" EXERCISES   **453**

3. (a)

x	y
−.5	1.75
−1	1
−2	−2
−3	7

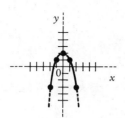

x	y
0	2
.5	1.75
1	1
2	−2
3	.7

4. (a)

x	y
−1	$\frac{1}{3}$
−2	$\frac{1}{9}$
−3	$\frac{1}{27}$

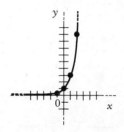

x	y
0	1
1	3
2	9
3	27

5. (a)

x	y
−1	1
−2	2
−3	3

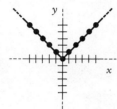

x	y
0	0
1	1
2	2
3	3

6. (a)

x	y
−1	−1
−2	−8
−3	−27

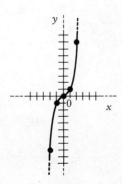

x	y
0	0
1	1
2	8
3	27

7. (a)

x	y
−1	−1
−2	−15
−3	−53

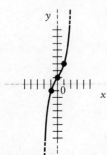

x	y
0	1
1	3
2	17
3	55

454 ANSWERS TO "(a)" EXERCISES

8. (a)

x	y
−1	3
−2	24
−3	81

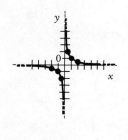

x	y
0	0
1	−3
2	−24
3	−81

9. (a)

x	y
$-\frac{1}{4}$	−4
$-\frac{1}{2}$	−2
−1	−1
−2	$-\frac{1}{2}$
−4	$-\frac{1}{4}$

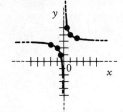

x	y
$\frac{1}{4}$	4
$\frac{1}{2}$	2
1	1
2	$\frac{1}{2}$
4	$\frac{1}{4}$

10. (a)

x	y
−.25	−1
−.5	1
−1	2
−2	2.5

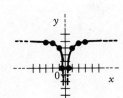

x	y
.25	7
.5	5
1	4
2	3.5
3	3.7

11. (a)

x	y
−.5	0
−1	3
−2	3.75
−3	3.9

x	y
.5	0
1	3
2	3.75
3	3.9

12. (a) $5
13. (a) When no additional time has passed, no interest is earned and the amount is equal to the starting amount.
14. (a) 35 mph, $5.25

Review Exercises Chapter 5

1. (a) $\{(-5y + \frac{3}{2}, y) \mid y \text{ any real number}\}$ $(-1, \frac{1}{2})$, $(4, -\frac{1}{2})$, $(\frac{3}{2}, 0)$
2. (a) $\{(3y^2 + 4, y) \mid y \text{ any real number}\}$ $(4, 0)$, $(7, 1)$, $(7, -1)$

ANSWERS TO "(a)" EXERCISES 455

3. (a) $\{(x, 5x + 12) \mid x \text{ any real number}\}$ (0, 12), (1, 17), (−2, 2)
4. (a) $\{(x, -2x^2 + 1) \mid x \text{ any real number}\}$ (0, 1), (1, −1), (−1, −1)
5. (a) $L = \dfrac{3G + 1}{20}$
6. (a)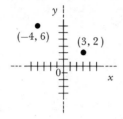
7. (a)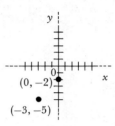
8. (a) (6, −3), (0, 6), (−1, −4)
9. (a)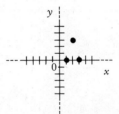
10. (a) $\{(-2, 1), (0, -5), (0, -1), (0, 3), (0, 7), (2, -3), (2, 5)\}$
11. (a) (1971, 15,000); in 1971 she sold $15,000 worth of goods.
    (1972, 20,000); in 1972 she sold $20,000 worth of goods.
    (1979, 60,000); in 1979 she sold $60,000 worth of goods.
12. (a)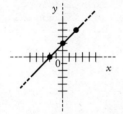

x	y
−2	0
0	2
2	4

13. (a)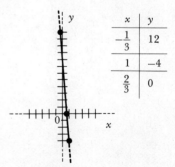

x	y
$-\frac{1}{3}$	12
1	−4
$\frac{2}{3}$	0

14. (a)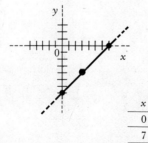

x	y
0	−7
7	0
3	−4

15. (a)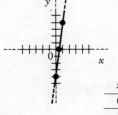

x	y
0	−4
1	4
$\frac{1}{2}$	0

456 ANSWERS TO "(a)" EXERCISES

16. (a)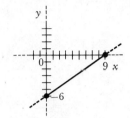

17. (a)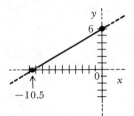

18. (a)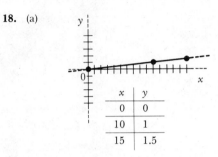

19. (a) $-\frac{2}{3}$ **20.** (a) 1 **21.** (a) 1 **22.** (a) 4 **23.** (a) $\frac{1}{3}$ **24.** (a) 1
25. (a) Slope = 7, y-intercept = 3 **26.** (a) Slope = $\frac{1}{2}$, y-intercept = -3

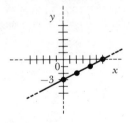

27. (a) $y = -5x + 2$ **28.** (a) $y = \frac{1}{4}x + 3$
29. (a) $y = 3x + 3$ **30.** (a) $y = \frac{2}{3}x + 7$
31. (a) $y = 0$ **32.** (a) $y = -3x + 2$
33. (a) $y = 2x - 11$ **34.** (a) $y = -x + 2$
35. (a) $y = -\frac{2}{25}x + 650$ **36.** (a)

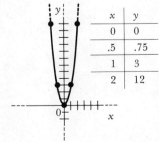

37. (a)

x	y
2	16
1	4
$\frac{1}{2}$	2
0	1
-1	$\frac{1}{4}$

38. (a)

x	y
2	25
1	4
$\frac{1}{2}$	$1\frac{3}{8}$
0	1
$-\frac{1}{2}$	$\frac{5}{8}$
-1	-2

39. (a) 8000 records

40. (a) 20 minutes

Calculator Activities Chapter 5 Review

1. (a)

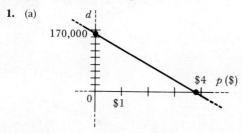

2. (a) $y = 1.119x - 529.7$

3. (a) Slope $= 1.613$, y-intercept $= -4.839$

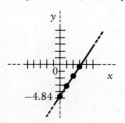

4. (a) $y = 935{,}394$

Exercises 6-1

1. (a) $y + 2x + 3 = 0$ and $x - y = 0$
$y = -2x - 3$ and $y = x$
Solution at $(-1, -1)$.

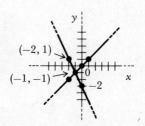

458 ANSWERS TO "(a)" EXERCISES

2. (a)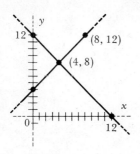

3. (a)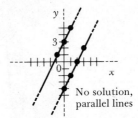
No solution, parallel lines

4. (a)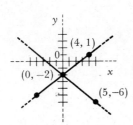

5. (a)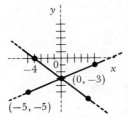

6. (a)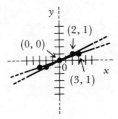

7. (a)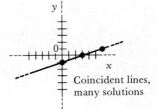
Coincident lines, many solutions

8. (a)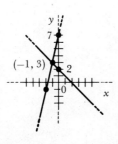

9. (a)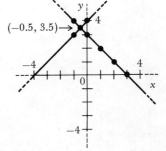

10. (a)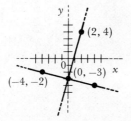

ANSWERS TO "(a)" EXERCISES

11. (a) Let $x =$ cost of one wallet, $y =$ cost of one dozen handkerchiefs
$3x + 5y = 54$ and $5x + 2y = 52$
$y = -\frac{3}{5}x + \frac{54}{5}$ and $y = -\frac{5}{2}x + 26$

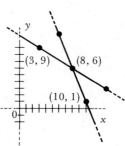

12. (a) Let $x =$ cost of one sweater, $y =$ cost of one animal
$4x + 3y = 63$ and $3x + 4y = 56$

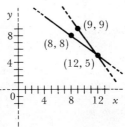

Exercises 6-2

1. (a) $x = 2, y = 0$
2. (a) $x = 4, y = -3$
3. (a) $3x - 4y = 6, 2x + y = 4$
   Take second equation and solve for y; substitute $y = -2x + 4$ in first equation:
   $3x - 4(-2x + 4) = 6$
   $3x + 8x - 16 = 6$
   $11x = 22$
   $x = 2, y = -2(2) + 4$
   $y = 0$

   $\checkmark\ 3(2) - 4(0) = 6$
   $6 = 6$
   $\checkmark\ 2(2) + 0 = 4$
   $4 = 4$

4. (a) $x = 3, y = -1$
5. (a) $x = 2, y = -1$
6. (a) $x = 2, y = 2$
7. (a) $x = -4, y = 2$
8. (a) $x = 2, y = -1$
9. (a) $x = -5, y = 2$
10. (a) $x = 1, y = -5$
11. (a) $x = \frac{3}{7}, y = \frac{13}{7}$
12. (a) $x = \frac{21}{19}, y = \frac{27}{19}$
13. (a) $x = 1, y = -\frac{2}{3}$
14. (a) $x = 3, y = 0.25$
15. (a) $x = -\frac{5}{3}, y = \frac{8}{3}$
16. (a) $x = 2, y = 3$

Calculator Activities 6-2

1. (a) $x = 1.117, y = -4.754$
2. (a) Many solutions
3. (a) $x = -0.006134, y = -2.0917$
4. (a) $x = 1.880, y = -1.520$

Exercises 6-3

1. (a) $\checkmark\ 3x - y = -5$
   $x - y = 7$

   $3x - y = -5$
   $\underline{-(x - y) = -(7)}$
   $2x\quad\ \ = -12$
   $x = -6$

   $3x - y = -5$
   $3(-6) - y = -5$
   $-18 - y = -5$
   $y = -13$

   $\checkmark\ 3(-6) - (-13) = -5$
   $-18 + 13 = -5$
   $-5 = -5$
   $\checkmark\ -6 - (-13) = 7$
   $7 = 7$

2. (a) $x = 2, y = -1$
3. (a) $x = 2, y = 8$
4. (a) $x = -1, y = 2$
5. (a) $x = -3, y = 2$
6. (a) $x = -1, y = 3$
7. (a) $x = \frac{1}{3}, y = \frac{1}{5}$
8. (a) $x = 1, y = 2$
9. (a) All reals such that $2x + 3y = 12$
10. (a) $x = 2, y = -1$
11. (a) $x = 2, y = 5$
12. (a) $x = 7, y = 3$
13. (a) $x = 24, y = 6$
14. (a) $x = 7, y = 0$
15. (a) $x = 22, y = 8$
16. (a) $x = 12, y = 6$
17. (a) No solution
18. (a) All solutions of $-x + 2y = 6$
19. (a) $m = 2, n = 3$
20. (a) $a = 0.2, b = 1$
21. (a) $x = -3.6, y = 22.4$
22. (a) $x = 0, y = 0$

Calculator Activities 6-3

1. (a) $x \approx 0.269, y \approx 0.111$
2. (a) $x \approx 1.698, y \approx -0.650$
3. (a) $x \approx 0.141, y \approx 0.141$
4. (a) $x \approx 189.648, y \approx 54.352$

Exercises 6-4

1. (a) Let x = number of men, y = number of women
Sum of the students is 31; therefore: $x + y = 31$
Difference between men and women is 5: $x - y = 5$
By the addition method: $2x\ \ \ = 36$
$x = 18$ men
Then $y = 13$ women

2. (a) $14 hours, $48 keypunch, $40 clerk
3. (a) Preys = 85, predators = 30
4. (a) Width = 6 centimeters, length = 21 centimeters (or, $w = 7$ and $L = 20$, etc.)
5. (a) Windspeed = 50 km/hr, plane in still air = 450 km/hr
6. (a) Head west for $1\frac{1}{4}$ hours, return trip $\frac{3}{4}$ hour.
7. (a) 52.5 gallons
8. (a) 9 qts
9. (a) 1600, 2040
10. (a) 6 units
11. (a) 20 calculators
12. (a) Fixed = $0.89, additional per pound $0.08

Nostalgia Problems 6-4

1. (a) Rye = $4, wheat = $6
2. (a) 25 pounds at 9¢, 75 pounds at 13¢

Calculator Activities 6-4

1. (a) Tomas = 14.23 hours, Cathy = 12.23 hours
2. (a) $2944.44

Review Exercises Chapter 6

1. (a)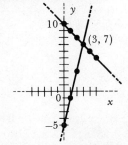

2. (a)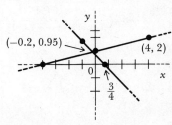

ANSWERS TO "(a)" EXERCISES   461

3. (a) $x + y = 12$, $x - y = 4$, ages are 8 and 4

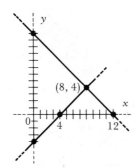

4. (a) $y = 5$, $x = 1$
6. (a) $x = -1$, $y = 0$
8. (a) $x = 24$, $y = 6$
10. (a) $x = 3$, $y = 2$
12. (a) $9.57 for the doubleknit; $5.14 for the orlon
13. (a) 6500, 500

5. (a) $x = 2$, $y = -7$
7. (a) $x = 7$, $y = 8$
9. (a) $x = -4$, $y = 3$
11. (a) 5 meters, 3 meters
14. (a) 30 and 12

Exercises 7-1

1. (a)
$$3x + y > 5$$
$$-3x + 3x + y > -3x + 5$$
$$y > -3x + 5$$
$$\{x, y \mid y > -3x + 5\}$$
If $x = 0$ then $y > 5$, so $(0, 6)$, $(0, 8)$, and $(0, 30)$ are solutions.

2. (a) $\{x, y \mid y \leq x - 1\}$; $(0, -3)$, $(5, 3)$, $(10, 8)$
3. (a) $\{x, y \mid y < -\frac{1}{3}x + \frac{4}{3}\}$; $(0, 1)$, $(3, 0)$, $(-6, 2)$
4. (a) $\{x, y \mid y > -2x + 15\}$; $(0, 15.5)$, $(4, 8)$, $(-2, 20)$
5. (a) $\{x, y \mid x > 0\}$; $(2, 12)$, $(1, -5)$, $(23, 14)$
6. (a) $\{x, y \mid y < -\frac{3}{2}\}$; $(6, -2)$, $(10, -6)$, $(-5, -9)$
7. (a) $\{x, y \mid y > \frac{7}{3}x + \frac{8}{3}\}$; $(0, 3)$, $(1, 6)$, $(-1, 1)$
8. (a) $\{x, y \mid y < \frac{11}{6}x - \frac{1}{2}\}$; $(0, -1)$, $(0, -5)$, $(2, 1)$
9. (a) $\{x, y \mid y < -x + \frac{1}{2}\}$; $(0, \frac{1}{4})$, $(-2, 2)$, $(6, -7)$
10. (a) $\{x, y \mid y \leq \frac{8}{11}x - \frac{10}{11}\}$; $(0, -1)$, $(0, 0)$, $(4, 2)$
11. (a) $\{(R, C) \mid R > C + 4090\}$; $(9000, 4000)$, $(11{,}000, 3000)$, $(130{,}000, 111{,}000)$
12. (a) Let $x =$ first city, $y =$ second city
$\{x, y \mid y \leq -x + 500{,}000\}$; $(200{,}000, 300{,}000)$, $(150{,}000, 200{,}000)$, $(90{,}000, 350{,}000)$
13. (a) $\{(I, P) \mid (I/P) \geq 0.12\}$; $(20, 100)$, $(10, 80)$, $(100, 600)$
14. (a) Let $x =$ first bacteria, $y =$ second bacteria
$\{(x, y) \mid -50 + y \leq x \leq 50 + y\}$; $(500, 450)$, $(600, 650)$, $(30, 40)$

Calculator Activities 7-1

1. (a) $\{x, y \mid y > 0.131x + 7.755\}$; $(0, 8)$, $(5, 10)$, $(-10, 7)$
2. (a) $\{x, y \mid y > -\frac{1}{660}x + 3.259\}$; $(0, 4)$, $(-700, 5)$, $(1800, 0.6)$
3. (a) $\{x, y \mid y \leq 2.518x\}$; $(0, 0)$, $(0, -3)$, $(3, 6)$
4. (a) $\{x, y \mid y \geq 0.25x - 0.1\}$; $(0, 0)$, $(0, 5)$, $(8, 2)$

462 ANSWERS TO "(a)" EXERCISES

Exercises 7-2

1. (a)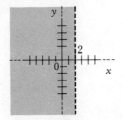

2. (a)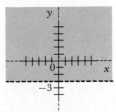

3. (a)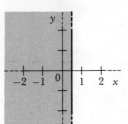

4. (a)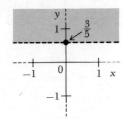

5. (a) $y \geq 3x + 1$; graph the straight line $y = 3x + 1$, which has a y-intercept of 1 and a slope of $\frac{3}{1}$. The inequality will include the line. Test with the point $(0, 0)$: $0 \geq 0 + 1$; $0 \geq 1$, not true! Therefore, the half-plane not including $(0, 0)$ is the solution set.

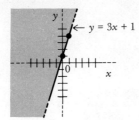

6. (a)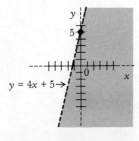

7. (a)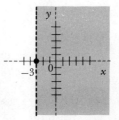

8. (a)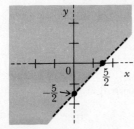

9. (a)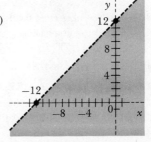

10. (a)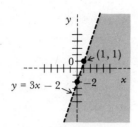

11. (a)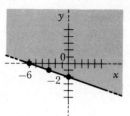

12. (a)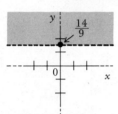

13. (a)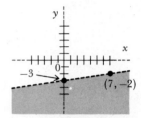

14. (a)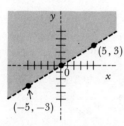

15. (a)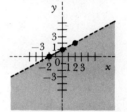

16. (a) Let y = profit, x = cost
$y > 4x$

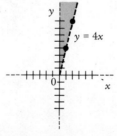

The line $y = 4x$ is graphed: it has a y-intercept of zero and a slope of $\frac{4}{1}$. The line itself is not included in the solution set. Since $(0, 0)$ is on the line, it is not a useful test point. Pick a point, $(5, 0)$: $0 > 5(4)$; $0 > 20$, which is not true. Therefore, the half-plane to the left of and above the line is the correct solution set.

17. (a) Let x = first account, y = second account amount
$0.08x + 0.07y > 700$

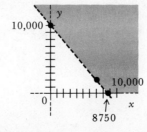

464 ANSWERS TO "(a)" EXERCISES

18. (a) Let $x =$ manufacturing cost, $y =$ profit
    $x + y \leq 14$

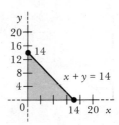

19. (a) Let $x =$ speed of plane, $y =$ speed of headwind
    $x - y \geq 160$

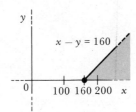

20. (a) $x > -2$
21. (a) The y-intercept is $y = 2$, the slope is $(2-0)/(0-6) = -\frac{1}{3}$. Using $y = mx + b$, the equation of the line is $y = -\frac{1}{3}x + 2$. Since the shaded portion is below the line, we need the "less than" symbol. The line itself is also included, since it is a solid line on the graph. $y \leq -\frac{1}{3}x + 2$
22. (a) $y \leq 0.25x$
23. (a) $y \leq 0.75x + 1.5$
24. (a) $y > \frac{2}{3}x + 2$
25. (a) $y < \frac{1}{9}x + \frac{14}{9}$

Calculator Activities 7-2

1. (a) 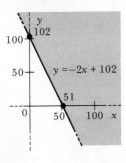   2. (a)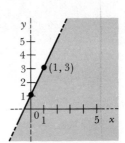

Exercises 7-3

1. (a) $y - x > 4$ and $y + 4x \leq 9$
    $y > x + 4$ and $y \leq -4x + 9$; the graph of the inequality $y > x + 4$ is the half-plane above the line $y = x + 4$; of the inequality $y \leq -4x + 9$ is the half-plane below and including the line $y = -4x + 9$. The intersection of these two half-planes is the graph of the given system.

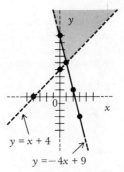

2. (a)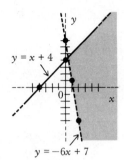

3. (a)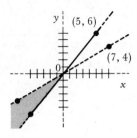

4. (a)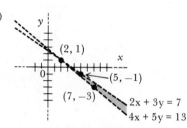

5. (a)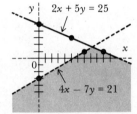

6. (a)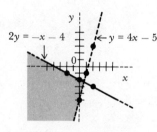

7. (a)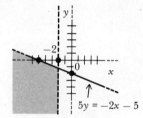

8. (a) $2x - 4y + 5 > -3$ or $-y \leq 5x - 7$; $y < \tfrac{1}{2}x + 2$ or $y \geq -5x + 7$; graph each of the associated equations, shading the half-plane below the first line, above the second, and taking all the shaded portions as the union.

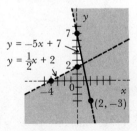

9. (a)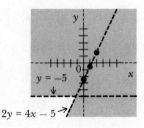

10. (a)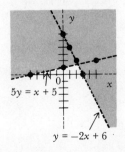

466 ANSWERS TO "(a)" EXERCISES

11. (a) $x \leq 3$
$y < \frac{2}{3}x$

12. (a) The lower line has a y-intercept of -5 and a slope of $-\frac{2}{3}$, giving the line $y = -\frac{2}{3}x - 5$; since the shading is above the line and not including it, we have the inequality $y > -\frac{2}{3}x - 5$. The other line has the same y-intercept of -5 and a slope of 3. The shading is to the left of (and above) the line and includes the line, giving the inequality $y \geq 3x - 5$. Thus, the system is
$$\begin{cases} y > -\frac{2}{3}x - 5 \\ y \geq 3x - 5 \end{cases}$$

13. (a) $\begin{cases} y > -\frac{1}{4}x + 3 \\ y < 7x - 5 \end{cases}$

14. (a) $\begin{cases} y \leq 5 \\ y < -\frac{3}{2}x + 5 \end{cases}$

15. (a) $x < -3$ or $y > -2x - 6$

16. (a) $y \leq 3$ or $y \leq -3x + 6$

17. (a) $x \geq 0, y \geq 0$
Let x = number of gifts costing \$1 each, y = number of gifts costing \$2 each. The total number of gifts is no more than 15; therefore, $x + y \leq 15$. The equals sign is included because it is "no more than." She can buy x gifts at \$1 and y gifts at \$2, spending no more than \$20, which gives the inequality $x + 2y \leq 20$. Then graphing the system $\begin{cases} x + y \leq 15 \\ x + 2y \leq 20 \end{cases}$
any point in the shaded area
gives a solution to the problem.

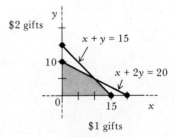

18. (a) Let x = number of combination pizzas, y = number of cheese pizzas,
$\begin{cases} x + y \leq 40 \\ 5x + 2y \leq 150, \text{ where } x \geq 0, y \geq 0 \end{cases}$

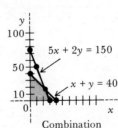

19. (a) Let x = amount in bank, y = amount in bonds
$\begin{cases} x + y \leq 30,000 \\ 0.0525x + 0.09y \geq 2000, \text{ where } x \geq 0, y \geq 0 \end{cases}$

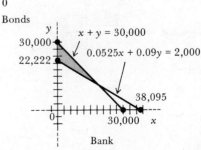

20. (a) Let x = number of fresh-water poles, y = number of surf-casting poles
$$\begin{cases} 4x + 6y \leq 180 \\ x + y \leq 35, \text{ where } x \geq 0, y \geq 0 \end{cases}$$

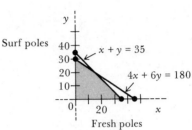

21. (a) Let x = number of predators, y = number of preys
$$\begin{cases} x + y < 500 \\ y \geq 200 + x, \text{ where } x \geq 0, y \geq 0 \end{cases}$$

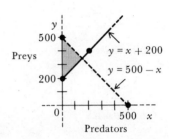

Calculator Activities 7-3

1. (a)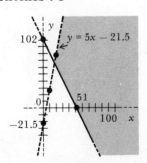

2. (a)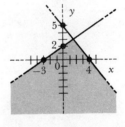

3. (a) $2.75x + 442.5 \geq 859.29$ says $x \geq 151.56$
$y \geq 2.75x + 442.5 + 1296.75$, where $x \geq 0, y \geq 0$

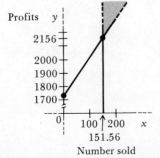

468 ANSWERS TO "(a)" EXERCISES

Review Exercises Chapter 7

1. (a) $\{(x,y) \mid y < \frac{5}{7}x + \frac{3}{7}\}$; $(0,0)$, $(1,1)$, $(-2,-4)$
2. (a) $\{(x,y) \mid y \geq \frac{3}{2}x - 4\}$; $(0,-3)$, $(2,5)$, $(4,19)$
3. (a) $\{(R,C) \mid R \geq C + 6012\}$; $(6012, 0)$, $(8000, 1000)$, $(7000, 500)$
4. (a) $\{(x,y) \mid y \leq -x + 2.75\}$; $(1,1)$, $(1, \frac{1}{2})$, $(\frac{1}{2}, 2)$
5. (a)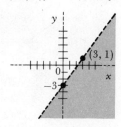
6. (a)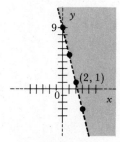
7. (a)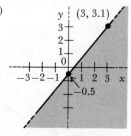
8. (a)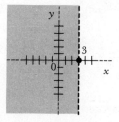
9. (a) Let C = cost, S = selling price
   $C \leq 0.75S$, where $C \geq 0$, $S \geq 0$
   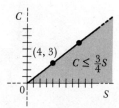
10. (a) Let l = length, w = width
    $2l + 2w \leq 14$, where $l \geq 0$, $w \geq 0$
    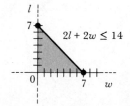
11. (a) $y > \frac{3}{2}x - 3$
12. (a) $y \geq -3$
13. (a) $y = 2x - 4$
    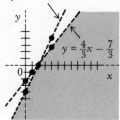
14. (a)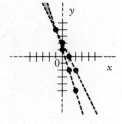

ANSWERS TO "(a)" EXERCISES **469**

15. (a)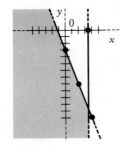

16. (a)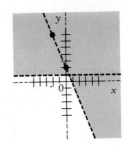

17. (a) $\begin{cases} y \leq 4x + 4 \\ y \leq -2x + 2 \end{cases}$

18. (a) $x \leq 4$
    $y < \frac{3}{4}x$

19. (a) Let $x =$ orange soda, $y =$ cherry cola
    $x + y \leq 25$
    $2x + 2.5y \leq 30$, where $x \geq 0$, $y \geq 0$

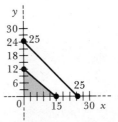

20. (a) Let $x =$ number of prime rib dinners, $y =$ number of chicken dinners
    $x + y \leq 84$
    $x \geq 2y$, where $x \geq 0$, $y \geq 0$

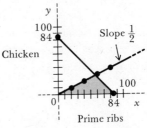

Calculator Activities Chapter 7 Review

1. (a) $\{(x, y) \mid y > 0.269x - 1.325\}$; $(0, -1)$, $(1, 0)$, $(5, 0.5)$
2. (a) $\{(x, y) \mid y > 20.05x - 23.37\}$; $(0, -20)$, $(2, 19)$, $(-1, -40)$
3. (a)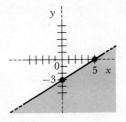
4. (a)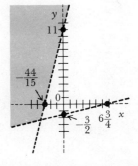

470 ANSWERS TO "(a)" EXERCISES

5. (a) Let x = number of $\frac{1}{4}$-pound size, y = number of $\frac{1}{2}$-pound size.
   $x + y \leq 144$
   $\frac{1}{4}x + \frac{1}{2}y \leq 51$, where $x \geq 0$, $y \geq 0$

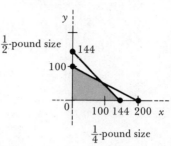

Exercises 8-1

1. (a) There are two terms, and there are no variables in the denominator; therefore, the expression is a binomial.
2. (a) Not a polynomial
3. (a) Monomial
4. (a) Trinomial
5. (a) Not a polynomial
6. (a) Not a polynomial
7. (a) Trinomial
8. (a) Binomial
9. (a) Not a polynomial
10. (a) Binomial
11. (a) The first term has the variable x and the coefficient 6, the second term, s, is the constant term.
12. (a) Coefficient 0.12, constant 0
13. (a) Coefficients 2 and 1, constant -7
14. (a) Coefficient 0.14, constant -0.5
15. (a) Coefficients 234 and -2, constant 445
16. (a) Coefficients 1 and 2, constant 1
17. (a) Coefficients 1, 1, and 5, constant 0
18. (a) Coefficients 26 and -14, constant 2
19. (a) Coefficients 2 and -1, constant 0
20. (a) Coefficient 9.04, constant -10.22
21. (a) Standard form requires the term to have descending exponents: $x^3 - 7x + 5$
22. (a) $2.3x + 5.7$
23. (a) $-7x^3 + 3x^2 + 2x$
24. (a) $\frac{2}{3}a^4 + \frac{1}{3}a^3 + \frac{1}{9}a + \frac{5}{9}$
25. (a) $y^3 - y^2 + 3y - 6$
26. (a) $-x^3 + 1.2x + 0.9$
27. (a) $17x$
28. (a) $3xy^2 - 2x^2y + 7xy^2$
    $3xy^2 + 7xy^2 - 2x^2y$
    $xy^2(3 + 7) - 2x^2y$
    $10xy^2 - 2x^2y$
29. (a) $3x^2 + 5x + 7$
30. (a) x^4y
31. (a) $7.7x^3 - 1.2x^2 - 5$
32. (a) $x^2y - \frac{1}{3}xy^2 + 8xy - 2$
33. (a) $x^2 - 3x + 2$, for $x = 3$
    $3^2 - 3(3) + 2$
    $9 - 9 + 2$
    2
34. (a) $\frac{1}{4}$
35. (a) -21.1
36. (a) 40 miles
37. (a) 897
38. (a) 0.28 cm/sec
39. (a) $50,000
40. (a) 300,000

Calculator Activities 8-1

1. (a) 7484
2. (a) -858.52
3. (a) 25 meters
4. (a) 100.18°F
5. (a) 4

Exercises 8-2

1. (a) $(6x^2 + 3x - 1) + (2x^2 + x + 4) = 6x^2 + 2x^2 + 3x + x - 1 + 4 = 8x^2 + 4x + 3$
2. (a) $-4x^2 + 5x + 2$
3. (a) $-\frac{1}{4}x^4 + \frac{1}{4}x^3 + x^2 + \frac{5}{2}x$
4. (a) $0.6a^3 - 6a^2 + 12.4a$
5. (a) $-y^2 + 5y - 1$
6. (a) $3w^2 + 2wz - 4z^2$
7. (a) $0.1x^3 + 0.3x^2 + 1.3x - 0.5$
8. (a) $12a^2 + 7a - 13$
9. (a) $-4t^3 + 4t^2 - 2t + 4$
10. (a) $-2x^3 - 2x + 10$
11. (a) $\begin{array}{r} x^2 - 4x + 1 \\ + 2x^2 - 7x + 9 \\ \hline 3x^2 - 11x + 10 \end{array}$
12. (a) $0.4y^2 - 1.3y + 0.3$
13. (a) $6m^3 - 4m^2 + 2m - 11$
14. (a) $\frac{13}{12}b^2 + \frac{1}{2}$
15. (a) $2y^3 + y^2 + y + 3$
16. (a) $-166y^4 - 728y^3 - 297y^2 + 177y - 963$
17. (a) $[(3x^2 + 5x - 1) + (5x^2 - 4x + 3)] - (9x^2 + x - 7)$
    $(3x^2 + 5x - 1 + 5x^2 - 4x + 3) - (9x^2 + x - 7)$
    $8x^2 + x + 2 - 9x^2 - x + 7$
    $-x^2 + 9$
18. (a) $10.6x^2 + 10.1xy - 2.1y^2$
19. (a) $11p^3 - 7p^2 + 3p - 6$
20. (a) $6a - b + 17c$
21. (a) $10x^2 + 2x + 6$
22. (a) $6a^2 - 5ab - 7b^2$
23. (a) $-8p^2 + p - 6$
24. (a) $-2\frac{1}{4}x + 1\frac{1}{2}$
25. (a) $12.31a^2 + 1.21ab + 3.13b^2$

Calculator Activities 8-2

1. (a) $1481x^4 - 1530x^3 - 2910x^2 - 477x - 5$
2. (a) $2.06a^2 - 1.45ab + 1.59b^2$
3. (a) $59x^5 - 1233.5x^4 + 406x^3 + 6.87x^2 - 7$

Exercises 8-3

1. (a) $3x^2 \cdot 15x^3 = 3 \cdot 15 \cdot x^2 \cdot x^3 = 45x^{2+3} = 45x^5$
2. (a) $-14x^3$
3. (a) $\frac{7}{8}x^5$
4. (a) $-0.12x^2y^2$
5. (a) $15a^3b^{10}$
6. (a) $315x^8y^5z^9$
7. (a) $\frac{2}{5}x^3y^4$
8. (a) $3.19a^5b^8c$
9. (a) $-\frac{1}{2}x^3y^8z^5$
10. (a) $1170w^2x^3y^{13}z^{16}$
11. (a) $(x^2y)^4 = x^{2 \cdot 4}y^{1 \cdot 4} = x^8y^4$
12. (a) $\frac{1}{8}p^6r^3$
13. (a) $1.331a^3b^3c^6$
14. (a) a^6b^4
15. (a) 1
16. (a) $-0.027p^6q^{12}r^3$
17. (a) $\frac{1}{256}a^8b^4c^{20}$
18. (a) $4.41x^6y^{12}z^4$
19. (a) a^8b^7
20. (a) $-m^{11}n^{20}$
21. (a) $8x^{12}y^6z^{18}$
22. (a) $-27x^{30}y^{27}$
23. (a) $64a^{29}b^{21}$
24. (a) $-864x^{36}y^{16}z^5$

Calculator Activities 8-3

1. (a) $7{,}320{,}500x^5y^4$
2. (a) $-13.3575m^4n^7$
3. (a) $-12.96x^3y^3z^4$
4. (a) $29{,}791x^6y^3$
5. (a) $-14.112a^3xy^5$

Exercises 8-4

1. (a) $13y(y + 2) = 13y \cdot y + 13y \cdot 2 = 13y^2 + 26y$
2. (a) $3z^2 - 4z$
3. (a) $4t^4 + 20t^3 + 8t^2$
4. (a) $32a^3b^2 + 41a^2b^3 - 23ab^4$

5. (a) $-30x^5y^2 + 20x^3y^2$
6. (a) $42a^8b + 105a^6b^4$
7. (a) $(2x + 3)(x - 5) = 2x \cdot x + 2x \cdot (-5) + 3 \cdot x + 3 \cdot (-5)$
$= 2x^2 - 10x + 3x - 15 = 2x^2 - 7x - 15$
8. (a) $4x^2 + 35x - 9$
9. (a) $8a^2 + 8ab - 6b^2$
10. (a) $x^2 - z^2$
11. (a) $2y^2 + 3y - 14$
12. (a) $14x^3 + 39x^2 + 4x - 15$
13. (a) $15y^3 - 32y^2 + 56y - 32$
14. (a) $-12m^3 + 4m^2 + 21m - 7$
15. (a) $x^3y^2 + 4x^2y + 3x$
16. (a) $6p^4 + 16p^3 - 49p^2 + 11p + 7$
17. (a)
$$\begin{array}{r} 3x + 4 \\ -4x + 7 \\ \hline 21x + 28 \\ -12x^2 - 16x \\ \hline -12x^2 + 5x + 28 \end{array}$$

18. (a) $6y^2 + 47y - 8$
19. (a) $-5m^2 + 8mn - 3n^2$
20. (a) $4x^2 - 9$
21. (a) $2a^3 + 7a^2 - 7a - 5$
22. (a) $20y^3 + 40y^2 + 5y + 4y^2z + 8yz + z$
23. (a) $-3x^4 - 2x^3 + 8x^2 + 3x - 4$
24. (a) $6x^3 - x^2 - 10x - 3$
25. (a) $-32a^3 + 84a^2b - 33ab^2 - 14b^3$
26. (a) $-2t^4 + 14t^3 - 37t^2 + 43t - 21$
27. (a) $29x - 27$
28. (a) $3x(4x + 5) + 2x(x - 5)$
$3x \cdot 4x + 3x \cdot 5 + 2x \cdot x + 2x \cdot (-5)$
$12x^2 + 15x + 2x^2 - 10x$
$14x^2 + 5x$
29. (a) $m^2 + 5m - 7$
30. (a) $\frac{5}{2}p^2 - 5p$
31. (a) $-13.3t^2 + 38.7t + 5.1$
32. (a) $660m^2 + 317m - 78$
33. (a) $17a^4 + 4a^3 - 32a^2 + 97a - 25$

Calculator Activities 8-4

1. (a) $11.36x^2 - 190.6x - 420$
2. (a) $13.448a^4 - 2.05a^3 - 10.824a^2b^2 + 2.9ab^2 - 11.6b^4$
3. (a) $2700y^3 - 1125y^2 + 60y - 25$
4. (a) $-6.36x^2 - 7.79x + 33.73$

Exercises 8-5

1. (a) $(x - 3)(2x + 1)$
$x \cdot 2x + x \cdot 1 - 3 \cdot 2x - 3 \cdot 1$
$2x^2 + x - 6x - 3$
$2x^2 - 5x - 3$
2. (a) $y^2 - 1$
3. (a) $3m^2 - m - 14$
4. (a) $4x^2 + 20x + 25$
5. (a) $9x^2 - 49$
6. (a) $16a^2 + 16ab + 4b^2$
7. (a) $81x^2 + 36x + 4$
8. (a) $m^2 + m - 2$
9. (a) $\frac{1}{4}y^2 - \frac{1}{16}$
10. (a) $25 - 80x + 64x^2$
11. (a) $36x^2 - y^2$
12. (a) $16x^2 - 49y^2$
13. (a) $4c^2 - 9d^2$
14. (a) $169a^2b^2 - 4y^2$
15. (a) $0.36x^2 - 1.08xy + 0.81y^2$
16. (a) $\frac{1}{25}x^2 - \frac{1}{25}x + \frac{1}{100}$
17. (a) $0.81m^2 - 1.44n^2$
18. (a) $28y^2 - 120y - 27$
19. (a) $16m^2n^4 - 40mn^3p^2 + 25n^2p^4$
20. (a) $4m^2 + 12mn + 9n^2 - p^2$

Calculator Activities 8-5

1. (a) $855x^2 + 435x - 240$
2. (a) $484t^2 - 9.24ty + 0.044y^2$
3. (a) $962,361x^2 - 213,444y^2$
4. (a) $0.087a^4b^2 + 1.436a^2bc + 1.76c^2$

Review Exercises Chapter 8

1. (a) Trinomial
2. (a) Monomial
3. (a) Not a polynomial
4. (a) Binomial
5. (a) Coefficients 1.5 and 2.8, constant -0.6
6. (a) Coefficients 8, -2, and $\frac{1}{2}$, constant $\frac{2}{3}$
7. (a) $-3x^3 + 5x + 2$
8. (a) $8.9m^3 + 2.3m^2 + 5.1m - 1.1$
9. (a) $4x + 4y + 1$
10. (a) $\frac{5}{2}x - \frac{1}{2}xy - 2y$
11. (a) $4\frac{1}{2}$
12. (a) 16.6
13. (a) $22
14. (a) 20 million barrels per day
15. (a) $6x^2 - 2x - 1$
16. (a) $-4.6y^2 - 1.1y + 4.1$
17. (a) $-\frac{3}{8}x^2 + \frac{5}{8}x + \frac{1}{2}$
18. (a) $-3y^2 + 5y - 7$
19. (a) $11x^2 + x - 10$
20. (a) $-7p^2 + 10p - 14$
21. (a) $-30x^7$
22. (a) $-7.13x^5y^7z^3$
23. (a) 1
24. (a) $-0.014x^{11}y^{25}$
25. (a) $9x^3 - 15x^2$
26. (a) $24x^2 - 8x - 2$
27. (a) $-6m^2 - 19m + 36$
28. (a) $-10x^3 + 26x^2 - 7x - 3$
29. (a) $6a^3 + 13a^2 - 41a + 12$
30. (a) $49x^2 - 9$
31. (a) $\frac{25}{64}x^2 + \frac{5}{8}x + 4$
32. (a) $16x^2 - 121$
33. (a) $2x^3 + 10x^2 + 10x - 4$
34. (a) $-x^3 + 6x^2 - 23x + 34$
35. (a) $\frac{18}{25}x^2 + x$

Calculator Activities Chapter 8 Review

1. (a) 1709
2. (a) 8.584, approx. 9 bacteria
3. (a) $8.24x^2 + 7.13x - 4.42$
4. (a) $967,572x^3y^5$
5. (a) $914,112a^5b^8$
6. (a) $3428b^2 - 990b - 606$

Exercises 9-1

1. (a) $\dfrac{x}{2}$
2. (a) $2x^3$
3. (a) $\dfrac{3.2x^3y}{1.6x^3y} = \dfrac{3.2}{1.6} \cdot \dfrac{x^3}{x^3} \cdot \dfrac{y}{y} = 2x^{3-3}y^{1-1} = 2x^0y^0 = 2$
4. (a) $\dfrac{4}{7}a^2b^2$
5. (a) $\dfrac{3n^7}{7m^4}$
6. (a) $\dfrac{1}{3r^3t}$
7. (a) $-\dfrac{2}{11x^3y}$
8. (a) $\dfrac{7x^3}{3z^2}$
9. (a) $-\dfrac{18p^5q}{r^4}$
10. (a) $\dfrac{6x^2}{y^5z^2}$
11. (a) $\dfrac{36x^5 + 12x^3}{6x^2} = \dfrac{12x^3(3x^2 + 1)}{6x^2} = 2x(3x^2 + 1)$
12. (a) $9m^2 + 7$
13. (a) $-7r^2 + 4r$
14. (a) $\frac{5}{2}x^3y^5 + \frac{3}{2}xy^4$
15. (a) $8x^2y^3 + 3x - 4$
16. (a) $-8abc^3 + 9bc^2 + \dfrac{1}{6a}$
17. (a) $8xyz - 2 - \dfrac{3}{xy}$
18. (a) $-\dfrac{2}{3}m^6n^5p^2 - \dfrac{5}{3}m^4n^3p + \dfrac{2}{p}$

Calculator Activities 9-1

1. (a) $\dfrac{38x^7}{y^2}$
2. (a) $\dfrac{1575}{y} - \dfrac{225y}{x}$

ANSWERS TO "(a)" EXERCISES

3. (a) $252a^4b^4$

5. (a) $-94,439,333$

4. (a) $-0.5xy^2 + 3 + \dfrac{3.5}{xy}$

Exercises 9-2

1. (a) $$x + 3 \overline{\smash{\big)}\, x^2 - 5x + 3} = x - 8 + \dfrac{27}{x+3}$$
    $$\begin{array}{r} x - 8 \\ \underline{x^2 + 3x } \\ -8x + 3 \\ \underline{-8x - 24} \\ 27 \end{array}$$

2. (a) $m - 6$
4. (a) $x - 6$
6. (a) $2x - y$
8. (a) $x + 3$

3. (a) $x + 11 + \dfrac{28}{x-3}$
5. (a) $3y + 4$
7. (a) $2x - 1 + \dfrac{1}{2x-1}$
9. (a) $x + 2$

10. (a) $$x - 4 \overline{\smash{\big)}\, x^3 - 2x^2 - 5x - 12} = x^2 + 2x + 3$$
    $$\begin{array}{r} x^2 + 2x + 3 \\ \underline{x^3 - 4x^2 } \\ 2x^2 - 5x \\ \underline{2x^2 - 8x } \\ 3x - 12 \\ \underline{3x - 12} \\ 0 \end{array}$$

11. (a) $y - 5 + \dfrac{2y + 17}{y^2 + 2}$

12. (a) $x^2 - 4x + 16$

13. (a) $5x^2 - 2xy + y^2 - \dfrac{y^3}{5x + 2y}$

14. (a) $2p^3 - 3p^2 - 5 + \dfrac{5}{2p+3}$

15. (a) $2y^2 - 3y + \dfrac{2}{3} + \dfrac{\tfrac{8}{3}y - \tfrac{16}{3}}{3y^2 + 2y - 4}$

16. (a) $3a^2 - 4ab - 11b^2$

Calculator Activities 9-2

1. (a) $10x - 30$

2. (a) $5.2x + 3.1$

Exercises 9-3

1. (a) $4x^2 - 6x$; both terms have a common x, and the largest common coefficient is a factor of 2; therefore, in factored form, $2x(2x - 3)$.
2. (a) $7y^3(3y^2 + 2)$
3. (a) $\tfrac{1}{4}p(p^2 - 3)$
4. (a) $a^4(17a + 19)$
5. (a) $-8a^3(4a^3 - 2a + 3)$
6. (a) $0.6x^2(2 - 6x^2 + x^3)$
7. (a) $9xy^5(5x^2y^2 - 2)$
8. (a) $2x(2x - 3y)$
9. (a) $x^2y^2(17xy^3 + 5y - 8)$
10. (a) $11x^2y(x^3y^4 - 2xy^2 + 5)$
11. (a) $\tfrac{1}{8}mn^2p(5m^5 + m - 7)$
12. (a) $5(a^2 + 6b^3 - 3a^2b^2)$
13. (a) $-3x^3(3x^4 - 27x^2 + 8)$
14. (a) $ac(6.2a^2b^5c - 21.7ac + 15.5b^2)$
15. (a) $12ab(2a^2b^8 - 3ab^4 + 12)$

Calculator Activities 9-3

1. (a) $12y^2z^2(15y - 19)$

2. (a) $0.13mn(48n - 40 + m)$

ANSWERS TO "(a)" EXERCISES

Exercises 9-4

1. (a) $81x^2 - 25$ is the difference of two squares; therefore, it factors into $(9x - 5)(9x + 5)$.
2. (a) $(4 - 3y)(4 + 3y)$
3. (a) $(8x - 11y)(8x + 11y)$
4. (a) Not factorable
5. (a) Not factorable
6. (a) Not factorable
7. (a) $(x^2 + y^2)(x - y)(x + y)$
8. (a) Not factorable
9. (a) $x^2 + 4x + 4$; x^2 and 4 are squares of monomials, and the middle term is $2AB = 4x = 2 \cdot x \cdot 2$; therefore, it factors into $(x + 2)^2$.
10. (a) $(a - 5)^2$
11. (a) Not factorable
12. (a) $(2x - 5)^2$
13. (a) $(7x - 3)^2$
14. (a) $(6y + 1)^2$
15. (a) $(6t - 11)^2$
16. (a) Not factorable
17. (a) $9x + 9y + ax + ay$
    $(9x + 9y) + (ax + ay)$
    $9(x + y) + a(x + y)$
    $(9 + a)(x + y)$
18. (a) $(m - 3)(m + 4)$
19. (a) $(4 - t)(r + 8)$
20. (a) $(2x + 1)(5x - 6)$
21. (a) $(2x - 3y)(2x + 3y)$
22. (a) $3x^3 + 12x^2 + 12x$
    $3x(x^2 + 4x + 4)$
    $3x(x + 2)^2$
23. (a) $2m(5m + 1)^2$
24. (a) Not factorable
25. (a) $3(8x - 7z)(8x + 7z)$
26. (a) $x^2(10y + 4x)(-7a + b)$
27. (a) $3x^3(6 - x)^2$
28. (a) $3(y^2 + 10)^2$

Calculator Activities 9-4

1. (a) $(19m - 11n)(19m + 11n)$
2. (a) $(1.01x + 0.92y)^2$
3. (a) $x^2(0.25x - 0.9y)^2$

Exercises 9-5

1. (a) $x^2 - x - 20 = (x + A)(x + B)$
    $= x^2 + (A + B)x + AB; A + B = -1, AB = -20$

A	B	(where larger factor is negative)
1	-20	$A + B = -19$
2	-10	$A + B = -10$
4	-5	$A + B = -1$

    Therefore, the factors are $(x - 5)(x + 4)$.
2. (a) $(x + 9)(x - 5)$
3. (a) $(p - 6)(p + 2)$
4. (a) $(m + 7)(m - 5)$
5. (a) $-(t + 12)(t - 1)$
6. (a) $(x + 11)(x - 2)$
7. (a) $2(x - 5)(x + 1)$
8. (a) $(2z - 1)(z + 10)$
9. (a) $(2x + 1)(x + 2)$
10. (a) $(8x - 1)(x + 6)$
11. (a) $(4p - 3q)(p - 6q)$
12. (a) $(2x - 5)(3x + 8)$
13. (a) $(3x - 5)(4x - 1)$
14. (a) $(2m - n)(m + 3n)$
15. (a) $(2x + 1)(x + 1)$
16. (a) $(2y - 3)^2$
17. (a) $(4x - 1)(x + 1)$
18. (a) $18x^3 - 21x^2 - 30x$
    $3x(6x^2 - 7x - 10)$
    $3x(6x + 5)(x - 2)$
19. (a) $(5x + 9)^2$

476 ANSWERS TO "(a)" EXERCISES

20. (a) $(8m^2 + 9n^2)(8m^2 - 9n^2)$
22. (a) $-(6y + 5)(y - 4)$
24. (a) $3x(2a^2 + 4a - 7)$
26. (a) $(x^2 + 1)(x - 1)(x + 1)$
28. (a) $(3a + 4b)^2$
30. (a) $3x(3x - 5)(4x - 1)$
32. (a) $y^2(y^2 + 9)(y - 2)(y + 2)$
34. (a) $(-2m + n)(4m + n)$

21. (a) $y^2(3x^2 + 10x - 2)$
23. (a) $2xy(6x^2 + 17x - 14y)$
25. (a) $3(2 - m)(2 + m)$
27. (a) $3(9x^3 - 1)$
29. (a) $x^2(x^2 + 4)$
31. (a) $2(x^2z^2 + 4)(xz - 2)(xz + 2)$
33. (a) $(a + 2b)(1 - x)$

Calculator Activities 9-5

1. (a) $(29x - 12)(14x + 3)$
2. (a) $18(9m - 16)(5m - 25)$

Exercises 9-6

1. (a) $x^2 - 7x + 12 = 0$ $x = 3$ $x = 4$
$(x - 3)(x - 4) = 0$ √ $3^2 - 7(3) + 12 = 0$ √ $4^2 - 7(4) + 12 = 0$
$x - 3 = 0$ or $x - 4 = 0$ $9 - 21 + 12 = 0$ $16 - 28 + 12 = 0$
$x = 3$ or $x = 4$ $0 = 0$ $0 = 0$

2. (a) $-1, -5$
4. (a) $0, -4$
6. (a) $-\frac{1}{2}$
8. (a) $4, -1$
10. (a) $-\frac{2}{5}$
12. (a) $5, 3$
14. (a) $\frac{2}{5}, -3$
16. (a) 0

3. (a) $\frac{1}{4}, -\frac{2}{3}$
5. (a) -4
7. (a) $6, -1$
9. (a) $0, 5$
11. (a) $0, 11$
13. (a) $0, 3$
15. (a) $3, -2$
17. (a) Let n = north-south boundary
$n + 4$ = east-west boundary
$n(n + 4) = 252$
$n^2 + 4n = 252$
$n^2 + 4n - 252 = 0$
$(n + 18)(n - 14) = 0$
$n + 18 = 0$ or $n - 14 = 0$
$n = -18$ or $n = 14$
Since n is a distance, it cannot be negative; therefore, $n = 14$ kilometers for the north-south boundary, and $n + 4 = 18$ kilometers for the east-west boundary.

18. (a) $5m$ by $17m$
20. (a) 80 seconds
22. (a) 1 centimeter by 1 centimeter

19. (a) $-3, -1, 1, 3$
21. (a) 4 skateboards

Nostalgia Problems 9-6

1. (a) 18 years old

Calculator Activities 9-6

1. (a) $-3.889, 0.136$
3. (a) 148 workers, 393 workers

2. (a) $0.933, 0.88$
4. (a) 65 kilometers

Review Exercises Chapter 9

1. (a) $4x$
2. (a) $3x^3y$
3. (a) $\dfrac{n^6}{3m}$
4. (a) $3r^2s - 2r$
5. (a) $-6mn^4p + 5.5n^2p - 2.5$
6. (a) $x - 3$
7. (a) $3m - 4$
8. (a) $25y^4 + 20y^2 + 16$
9. (a) $6x^2(x^3 + 2x^2 - 3x + 7)$
10. (a) $5ab^2c(5a^2 - 3a^2bc + 4)$
11. (a) $(2x - 5)^2$
12. (a) $8y(3x - 2y)(3x + 2y)$
13. (a) $(5a + 4)(3a - 1)$
14. (a) $(5x + 3)(x + 3)$
15. (a) $(5y + 3)(y - 2)$
16. (a) Not factorable
17. (a) $-(8x - 1)(x + 6)$
18. (a) $x(x^2 + 1)(x + 1)(x - 1)$
19. (a) $y^2(y - 10)(y - 16)$
20. (a) $(11 - 14a)(11 + 14a)$
21. (a) $-1, -2$
22. (a) $2, -\tfrac{3}{4}$
23. (a) $8, -1$
24. (a) $3, -1$
25. (a) $12, 14, 16$ or $-2, 0, 2$
26. (a) 1 second, $3\tfrac{1}{2}$ seconds

Calculator Activities Chapter 9 Review

1. (a) $41x^{10}y^{18}$
2. (a) $12mn - \dfrac{5p}{m}$
3. (a) $9x + 24$
4. (a) $43(3x - 5y)(3x + 5y)$
5. (a) $(0.21m - 0.03)(9m + 11)$
6. (a) $0, 3$
7. (a) 2 meters

Exercises 10-1

1. (a) $3x^2$
2. (a) $\dfrac{-400a^3b^4}{-80ab} = 5a^{3-1}b^{4-1} = 5a^2b^3$
3. (a) $\dfrac{m - 4}{m}$
4. (a) $2y - 3$
5. (a) $\dfrac{x^2 - x}{x^2 + 2x} = \dfrac{x(x - 1)}{x(x + 2)} = \dfrac{x - 1}{x + 2}$
6. (a) $\dfrac{x}{y}$
7. (a) $\dfrac{x - 1}{y}$
8. (a) $\dfrac{2x + 3}{x - 4}$
9. (a) $\dfrac{y + 2}{y - 2}$
10. (a) $\dfrac{x + 4y}{4y}$
11. (a) $a - 5$
12. (a) $\dfrac{a^2(x^2 - y^2)}{x^2}$
13. (a) $\dfrac{1}{18xy}$
14. (a) $\dfrac{a + b}{a - b}$
15. (a) $\dfrac{-1}{x + 7}$

Calculator Activities 10-1

1. (a) $\dfrac{4a - 22}{3b - 11}$
2. (a) $\dfrac{16y - 44}{13y + 17}$

Exercises 10-2

1. (a) $\dfrac{5}{3a^2b} \cdot \dfrac{21b^2}{7a} = \dfrac{5 \cdot 21}{3 \cdot 7} \cdot \dfrac{b^2}{a^2b \cdot a} = \dfrac{5b}{a^3}$
2. (a) $\dfrac{3}{4n}$

478 ANSWERS TO "(a)" EXERCISES

3. (a) $\frac{56}{25}$

4. (a) $\frac{ay}{x}$

5. (a) $\frac{x^3}{y^3}$

6. (a) 4

7. (a) $\frac{(2m-3)(m+4)}{(m+3)(5m-2)}$

8. (a) $\frac{2a-1}{6a-3} \cdot \frac{6a+18}{2a-6} = \frac{2a-1}{3 \cdot (2a-1)} \cdot \frac{6(a+3)}{2(a-3)} = \frac{a+3}{a-3}$

9. (a) $\frac{3s(t-2)}{(t-3)(t+2)}$

10. (a) $\frac{3x-6}{x-2}$

11. (a) $\frac{x-y}{x+y}$

12. (a) $\frac{t+2}{t+3}$

13. (a) $\frac{a}{a+5}$

14. (a) $\frac{4a-5}{3a-2}$

15. (a) $\frac{x-1}{x+1}$

16. (a) $\frac{1}{2t}$

17. (a) $\frac{x^3(x+1)}{-20}$

Calculator Activities 10-2

1. (a) $\frac{48y(21x-5)}{13x+11}$

2. (a) $\frac{(14u-13v)z^3}{14(14u+13v)}$

Exercises 10-3

1. (a) $\frac{2x+3y+5}{9y}$

2. (a) $\frac{3x^2+42y^2}{7xy}$

3. (a) $\frac{6x^2+13x+4}{6xy}$

4. (a) $\frac{9}{m-n}$

5. (a) $\frac{5}{6t-12r} - \frac{4}{3t+6r} = \frac{5}{6(t-2r)} - \frac{4}{3(t+2r)}$ LCD $= 6(t-2r)(t+2r)$

$\frac{5(t+2r) - 4[2(t-2r)]}{6(t-2r)(t+2r)} = \frac{5t+10r-8t+16r}{\text{LCD}} = \frac{-3t+26r}{6(t-2r)(t+2r)}$

6. (a) $\frac{5a+6}{6b}$

7. (a) $\frac{x}{x^2-y^2}$

8. (a) $\frac{-7x+6}{3x(x+3)}$

9. (a) $\frac{x^2+x+2}{x^2-1}$

10. (a) $\frac{3x-5}{2(x^2-4)}$

11. (a) $\frac{-5}{(y+2)(y+3)}$

12. (a) $\frac{10a-7b}{(a-b)(2a-b)}$

13. (a) $\frac{2x+6y}{x^2-y^2}$

14. (a) $\frac{3m^2-1}{m^2(m+1)}$

15. (a) $\frac{30x^2+15x-2}{18x^3}$

16. (a) $\frac{9b^3+22b^2-4b-30}{b^2(b-5)(b+3)}$

17. (a) $\frac{c^2+4cd+d^2}{c^2d^2}$

18. (a) $\frac{3m^2+19m-20}{(m-2)^2(m+3)}$

Calculator Activities 10-3

1. (a) $\dfrac{215x^2 - 47x - 87}{15x(5x - 3)(x - 14)}$

Exercises 10-4

1. (a) $\dfrac{\frac{3}{8} - \frac{1}{4}}{1 + \frac{1}{2}} = \dfrac{\frac{3}{8} - \frac{1}{4}}{1 + \frac{1}{2}} \cdot \dfrac{8}{8} = \dfrac{3 - 2}{8 + 4} = \dfrac{1}{12}$

2. (a) $\dfrac{85}{28}$

3. (a) $x + y$

4. (a) $\dfrac{1}{a - 3}$

5. (a) $\dfrac{\dfrac{m}{m+n} - \dfrac{n}{n-m}}{\dfrac{n}{m+n} + \dfrac{m}{n-m}} \cdot \dfrac{(m+n)(m-n)}{(m+n)(m-n)} = \dfrac{mn - m^2 - mn - n^2}{n^2 - mn + m^2 + mn} = \dfrac{-m^2 - n^2}{n^2 + m^2} = -1$

6. (a) $\dfrac{24}{y}$

7. (a) $\dfrac{x^2 + 1}{-2x}$

8. (a) $\dfrac{3y + 4}{4y - y^2}$

Calculator Activities 10-4

1. (a) $\dfrac{771}{43{,}120} \approx .0179$

2. (a) $\dfrac{x(768x^2 + 591x - 260)}{(16x - 5)(x^2 - 493x - 533)}$

Exercises 10-5

1. (a) $\dfrac{4x^2}{9y^4}$

2. (a) $\left(\dfrac{4a^2b}{5ab^3}\right)^3 = \left(\dfrac{4a}{5b^2}\right)^3 = \dfrac{64a^3}{125b^6}$

3. (a) $\dfrac{x^2}{9y^6}$

4. (a) $\dfrac{(x - 1)^2}{(x + 1)^2}$

5. (a) $\dfrac{25y^4}{9x^6}$

6. (a) $\dfrac{-64x^{15}}{27y^9z^9}$

7. (a) 1

8. (a) $\left(\dfrac{x^{-3}(-4x + 1)^2}{y^{-2}(2y - 3)^3}\right)^{-2} = \left(\dfrac{y^2(-4x + 1)^2}{x^3(2y - 3)^3}\right)^{-2} = \left(\dfrac{x^3(2y - 3)^3}{y^2(-4x + 1)^2}\right)^2 = \dfrac{x^6(2y - 3)^6}{y^4(-4x + 1)^4}$

9. (a) $\dfrac{4x^2(z + 3)^6}{z^6(x + 1)^4}$

10. (a) $\dfrac{3x}{2}$

11. (a) $\dfrac{125b^6}{a^9(a + b)^6}$

12. (a) $\dfrac{(4x + 3)^2}{(4x - 3)^6}$

Calculator Activities 10-5

1. (a) $\dfrac{1331x^6}{1728y^6}$

2. (a) $\dfrac{x^2}{196y^{10}}$

Exercises 10-6

1. (a) $\dfrac{3x - 2}{x + 4}$ for $x = 3$: $\dfrac{3(3) - 2}{3 + 4} = \dfrac{9 - 2}{7} = \dfrac{7}{7} = 1$

2. (a) $-\dfrac{2}{5}$

3. (a) $\dfrac{17}{21}$

4. (a) 2

5. (a) $\dfrac{(5x + 1)(3x - 2)}{2x - 3}$ for $x = 0$: $\dfrac{(0 + 1)(0 - 2)}{0 - 3} = \dfrac{(1)(-2)}{-3} = \dfrac{2}{3}$

6. (a) $-\frac{4}{5}$
7. (a) -8
8. (a) $\frac{8.75}{6}$ or 1.4583
9. (a) $-\frac{5}{2}$
10. (a) Set the denominator $a^2 + a - 2$ to zero and solve; those solutions are the excluded values.
$a^2 + a - 2 = 0$
$a + 2 = 0$ or $a - 1 = 0$
$a = -2$ or $a = 1$
Therefore, the expression has no value when $a = -2$ or $a = 1$.
11. (a) $-4, 3$
12. (a) $5, -4$
13. (a) $-\frac{1}{2}, -5$
14. (a) $s = 0, r = 2s$
15. (a) 4
16. (a) $\frac{1}{2}, 7$

Calculator Activities 10-6

1. (a) 3.915
2. (a) 934.376
3. (a) 20.083
4. (a) 1.117
5. (a) -0.529
6. (a) $\frac{5}{14}, -\frac{13}{15}$

Exercises 10-7

1. (a) x cannot have the value -4 or 6

$\checkmark \quad \dfrac{3}{26+4} = \dfrac{2}{26-6}$

$\dfrac{3}{x+4} \cdot (x+4)(x-6) = \dfrac{2}{x-6} \cdot (x+4)(x-6)$

$\qquad \dfrac{3}{30} = \dfrac{2}{20}$

$\qquad 3(x-6) = 2(x+4)$

$\qquad \dfrac{1}{10} = \dfrac{1}{10}$

$\qquad 3x - 18 = 2x + 8$

$\qquad x = 26$

2. (a) $-\frac{3}{2}$ 3. (a) 2 4. (a) 4 5. (a) -4 6. (a) 6 7. (a) 2
8. (a) 6 9. (a) 8 10. (a) -2 11. (a) 1 12. (a) $-\frac{1}{2}$
13. (a) All real numbers except -1
14. (a) All real numbers except 1 and -1

15. (a) $k = \dfrac{1}{2}\left(\dfrac{e_1}{b_1} + \dfrac{e_2}{b_2}\right)$

16. (a) $g = \dfrac{2T}{a^2(d_1 - d_2)}$

$2k = \dfrac{e_1}{b_1} + \dfrac{e_2}{b_2}$

$2k - \dfrac{e_2}{b_2} = \dfrac{e_1}{b_1}$

$b_1\left(2k - \dfrac{e_2}{b_2}\right) = e_1$

17. (a) $R = \dfrac{E}{I} - \dfrac{r}{2}$

18. (a) $r_1 = \dfrac{r_2 F(n-1)}{r_2 - F(n-1)}$

19. (a) $u = \dfrac{n_1 rv}{v(n_2 - n_1) - rn_2}$

20. (a) $v = \dfrac{R}{(n+a)^2} - \dfrac{R}{(m+b)^2}$

Calculator Activities 10-7

1. (a) $\dfrac{403}{69}$ or 5.841

ANSWERS TO "(a)" EXERCISES

Exercises 10-8

1. (a) Let $h =$ hours it takes Julia, then

	Total hours	Amount of job in one hour
Jaime	7	$\frac{1}{7}$
Julia	h	$\frac{1}{h}$
Both	4	$\frac{1}{4}$

$$\left(\frac{1}{7}+\frac{1}{h}\right)\cdot 28h = \frac{1}{4}\cdot 28h$$
$$4h + 28 = 7h$$
$$28 = 3h$$
$$9\frac{1}{3} = h \text{ (hours for Julia alone)}$$

2. (a) -11

3. (a) Let $s =$ speed of riverboat in still water, then

	r	$\cdot\ t$	$= d$
Up river	$s - 3$	$\frac{26}{s-3}$	26
Down river	$s + 3$	$\frac{38}{s+3}$	38

Since the times are the same:
$$\frac{26}{s-3} = \frac{38}{s+3}$$
$$26(s + 3) = 38(s - 3)$$
$$26s + 78 = 38s - 114$$
$$192 = 12s$$
$$16 = s \text{ (speed in km/hr)}$$

4. (a) 8 km/hr **5.** (a) 15 km/hr
6. (a) $4\frac{4}{5}$ days **7.** (a) $2\frac{2}{5}$ days
8. (a) 120 cookies

Nostalgia Problems 10-8

1. (a) $200 and $160 **2.** (a) $4800

Review Exercises Chapter 10

1. (a) $\frac{1}{b}$ **2.** (a) $m - 5$

3. (a) $\frac{1}{b(a + 4)}$ **4.** (a) $\frac{m}{m+5}$

5. (a) $\frac{5x - 1}{(x+1)(x-2)}$ **6.** (a) $\frac{5a + b + 4b^2}{3a - b}$

7. (a) $\frac{2t^2 - 16t - 1}{(t+3)^2(t-2)}$ **8.** (a) $\frac{y^2 + 2y - 3}{y(y+1)(y-1)}$

9. (a) $\frac{2a}{2a + b}$ **10.** (a) $\frac{5x^2 + 2x}{3 - x^2}$

11. (a) $\frac{2y^3 + 20}{10y^2 - 15y}$ **12.** (a) $\frac{x^4 y^2}{4}$

13. (a) $\frac{125a^9}{216b^{12}}$ **14.** (a) $\frac{4r^8(s-t)^8}{9}$

15. (a) 4 **16.** (a) -1 **17.** (a) $-\frac{1}{3}$ **18.** (a) $-\frac{3}{2}, 5$ **19.** (a) $\frac{53}{11}$

ANSWERS TO "(a)" EXERCISES

20. (a) 4

22. (a) $v = \dfrac{tw}{1 - Tw}$

21. (a) $k = \dfrac{s}{a + b + c}$

23. (a) $3\dfrac{1}{13}$ hours

Calculator Activities Chapter 10 Review

1. (a) $\dfrac{17a - 13b}{12ab}$
2. (a) 1
3. (a) $\tfrac{7}{3}(102x - 7)$
4. (a) $\dfrac{65x^2 - 137x + 14}{270x^3 - 540x^2 + 170x}$
5. (a) -0.209
6. (a) Has no value at $y = 0$, $x = \tfrac{7}{9}, -\tfrac{19}{15}$
7. (a) 3.219 hours

Exercises 11-1

1. (a) 0.48
2. (a) $\dfrac{7}{9} = 9 \overline{)7.000}^{\,0.777} = 0.777\ldots$
3. (a) $-0.5454\ldots$
4. (a) $0.0714285\ldots$
5. (a) $-0.23076923\ldots$
6. (a) $\dfrac{127}{100}$
7. (a) $0.1777\ldots$ $\quad 100x - 10x = 17.7\ldots - 1.7\ldots$
   Let $x = 0.1777\ldots$ $\quad 90x = 16$
   $$x = \dfrac{16}{90}, \text{ or } \dfrac{8}{45}$$
8. (a) $\dfrac{3293}{999}$
9. (a) $\dfrac{24}{45}$
10. (a) $\dfrac{-11}{6}$
11. (a) $\sqrt{\dfrac{1}{4}} = \dfrac{\sqrt{1}}{\sqrt{4}} = \dfrac{1}{2}$
12. (a) 5
13. (a) 0.6
14. (a) 0.2
15. (a) -2
16. (a) -7
17. (a) -2
18. (a) $\tfrac{11}{12}$
19. (a) 5
20. (a) 0.2

Calculator Activities 11-1

1. (a) 4.47

Exercises 11-2

1. (a) $\sqrt{32} = \sqrt{16 \cdot 2} = \sqrt{16}\sqrt{2} = 4\sqrt{2}$
2. (a) $3\sqrt[3]{2}$
3. (a) $6\sqrt{0.2} = \tfrac{6}{5}\sqrt{5}$
4. (a) $7\sqrt{2}$
5. (a) $\dfrac{2\sqrt{13}}{3\sqrt{3}}$
6. (a) $10\sqrt[3]{2}$
7. (a) $3\sqrt[4]{2}$
8. (a) 1
9. (a) $\dfrac{5\sqrt{3}}{2\sqrt{5}}$
10. (a) $0.2\sqrt[3]{2}$
11. (a) $\dfrac{2\sqrt[3]{2}}{3}$
12. (a) $-4\sqrt{5}$
13. (a) $\sqrt{a^2 bc} = \sqrt{a^2} \cdot \sqrt{bc} = a\sqrt{bc}$
14. (a) $2xy\sqrt{x}$
15. (a) $6p\sqrt{2pq}$
16. (a) $\dfrac{3}{m^2\sqrt{7}}$

ANSWERS TO "(a)" EXERCISES

17. (a) $9m^2\sqrt{m}$
18. (a) $\dfrac{7\sqrt{2tv}}{z^2\sqrt{11z}}$
19. (a) $\dfrac{xy^2}{z^3}$
20. (a) $\dfrac{2a\sqrt[4]{a}}{b^2}$
21. (a) $6a\sqrt{3}$
22. (a) $-3ab\sqrt[3]{ab}$

Exercises 11-3

1. (a) $\sqrt{2}+\sqrt{50}=\sqrt{2}+5\sqrt{2}=6\sqrt{2}$
2. (a) $\sqrt{5}$
3. (a) $9\sqrt{2}$
4. (a) $20\sqrt{2}$
5. (a) $-2\sqrt{3}$
6. (a) $-13\sqrt{2}$
7. (a) $12\sqrt{6}+6\sqrt{5}$
8. (a) $92\sqrt{2}-12\sqrt{3}$
9. (a) $\sqrt{0.5}\cdot\sqrt{200}=\sqrt{0.5\cdot 200}=\sqrt{100}=10$
10. (a) 0.6
11. (a) 12
12. (a) $-30\sqrt{2}$
13. (a) $84\sqrt{2}$
14. (a) $6\sqrt{2}$
15. (a) $\sqrt{10}(\sqrt{5}+2\sqrt{10})=\sqrt{50}+2\sqrt{100}=5\sqrt{2}+20$
16. (a) $3+\sqrt{0.6}$
17. (a) $24-6\sqrt{2}$
18. (a) $11.8+3.4\sqrt{7}$
19. (a) $-11-8\sqrt{8}$
20. (a) $25+8\sqrt{10}$
21. (a) $176-10\sqrt{7}$
22. (a) 99
23. (a) $\dfrac{1}{\sqrt{2}-1}=\dfrac{1}{(\sqrt{2}-1)}\dfrac{(\sqrt{2}+1)}{(\sqrt{2}+1)}=\dfrac{\sqrt{2}+1}{2-1}=\sqrt{2}+1$
24. (a) $6\sqrt{5}+12$
25. (a) $-\dfrac{10\sqrt{3}+20}{4}$
26. (a) $\dfrac{4+\sqrt{7}}{6}$
27. (a) $\dfrac{3\sqrt{6}-12\sqrt{3}+5\sqrt{2}-20}{2}$
28. (a) $-\dfrac{\sqrt{30}+2\sqrt{10}+4\sqrt{42}+8\sqrt{14}}{107}$

Exercises 11-4

1. (a) $(\sqrt{y}+3)(\sqrt{y}-5)=\sqrt{y\cdot y}+3\sqrt{y}-5\sqrt{y}-15=y-2\sqrt{y}-15$
2. (a) $10a-33\sqrt{a}-7$
3. (a) $28x+17\sqrt{xy}-3y$
4. (a) $3y-2\sqrt{3y}+1$
5. (a) $10x-32\sqrt{x}-32$
6. (a) $81a-9$
7. (a) $60x-37\sqrt{10x}-20$
8. (a) $6xy-2\sqrt{3xy}-12$
9. (a) $\dfrac{1}{\sqrt{x}-1}=\dfrac{1}{(\sqrt{x}-1)}\dfrac{(\sqrt{x}+1)}{(\sqrt{x}+1)}=\dfrac{\sqrt{x}+1}{x-1}$
10. (a) $\dfrac{12\sqrt{a}-28}{9a-49}$
11. (a) $\dfrac{2x+5\sqrt{xy}}{x}$
12. (a) $\dfrac{-5\sqrt{a}+10\sqrt{5}}{2a-40}$
13. (a) $\dfrac{36\sqrt{2x}+60}{18x-25}$
14. (a) $\dfrac{2x+2\sqrt{2x}+1}{2x-1}$
15. (a) $\dfrac{3y+10\sqrt{yz}+7z}{y-z}$
16. (a) $\dfrac{a-2\sqrt{ab}+b}{a-b}$

ANSWERS TO "(a)" EXERCISES

Calculator Activities 11-4

1. (a) 0.317
2. (a) -5.907

Exercises 11-5

1. (a) $\frac{4}{5}$
2. (a) $4 + \sqrt{2x} = 12$ $\checkmark$ $4 + \sqrt{2 \cdot 32} = 12$
   $(\sqrt{2x})^2 = (8)^2$ $4 + \sqrt{64} = 12$
   $2x = 64$ $4 + 8 = 12$
   $x = 32$ $12 = 12$
3. (a) 13
4. (a) $-\frac{34}{3}$
5. (a) $\frac{3}{2}$
6. (a) $\frac{49}{20}$
7. (a) $\frac{2}{9}$
8. (a) 1
9. (a) Extraneous solution
10. (a) 1
11. (a) $\frac{1}{2}$
12. (a) Solution extraneous
13. (a) $a = 6$ meters, $b = 8$ meters
    $c = \sqrt{a^2 + b^2}$
    $c = \sqrt{6^2 + 8^2}$
    $c = \sqrt{36 + 64}$
    $c = \sqrt{100}$
    $c = 10$ meters
14. (a) 4 centimeters
15. (a) 4.8 meters approx.
16. (a) 12.2 centimeters approx.
17. (a) 455.4 meters approx.
18. (a) 12.8 meters
19. (a) 4π or 12.57 seconds
20. (a) 14.14 kilograms
21. (a) 90 kilograms

Review Exercises Chapter 11

1. (a) $0.\overline{54}$
2. (a) $-0.\overline{307692}$
3. (a) $\frac{121}{40}$
4. (a) $\frac{24}{45}$
5. (a) $\frac{7}{25}$
6. (a) 0.12
7. (a) 2
8. (a) $9\sqrt{3}$
9. (a) $10\sqrt[3]{5}$
10. (a) $\frac{5\sqrt{3}}{6}$
11. (a) $17x\sqrt{x}$
12. (a) $3yz\sqrt[3]{yz^2}$
13. (a) $\frac{4m^2n^2}{pr^3}\sqrt{\frac{3m}{p}}$
14. (a) $-5\sqrt{3}$
15. (a) $22\sqrt{7}$
16. (a) 120
17. (a) $-189.28\sqrt{3}$
18. (a) $6 - 6\sqrt{3}$
19. (a) $42 - 6\sqrt{10}$
20. (a) $\sqrt{7} + 2$
21. (a) $-11 - 8\sqrt{2}$
22. (a) $6x - 28\sqrt{x} + 30$
23. (a) $5x - 4\sqrt{10xy} + 6y$
24. (a) $\frac{10\sqrt{x} + 5}{4x - 1}$
25. (a) $\frac{10a - 29\sqrt{a} + 21}{25a - 49}$
26. (a) $\frac{9}{7}$
27. (a) Extraneous solution
28. (a) $\frac{9}{4}$
29. (a) Extraneous solution
30. (a) Extraneous solution
31. (a) 1.32 centimeters approx.
32. (a) 60.6 centimeters approx.
33. (a) 22.2 centimeters approx.

Calculator Activities Chapter 11 Review

1. (a) 0.714285 ...
2. (a) 3.9
3. (a) -0.081

Exercises 12-1

1. (a)
$$x(x+3) = 4$$
$$x^2 + 3x = 4$$
$$x^2 + 3x - 4 = 0$$
$$(x+4)(x-1) = 0$$
$$x+4 = 0 \text{ or } x-1 = 0$$
$$x = -4 \text{ or } x = 1$$

$x = -4$
✓ $-4(-4+3) = 4$
$-4(-1) = 4$
$4 = 4$

$x = 1$
✓ $1(1+3) = 4$
$1(4) = 4$
$4 = 4$

2. (a) $4, -7$
3. (a) $\frac{3}{2}, -\frac{3}{2}$
4. (a) 2
5. (a) $\frac{1}{2}, -\frac{2}{3}$
6. (a) $2, 1$
7. (a) $2, \frac{1}{3}$
8. (a) $5, -4$
9. (a) $3, 2$
10. (a) $-3, -4$
11. (a) Let w = width, then the length $= w + 2$, area $= w(w+2)$
New width $= w + 7$, new length $= w + 2 + 2 = w + 4$, area $= (w+7)(w+4)$
$$(w+7)(w+4) = w(w+2) + 91$$
$$w^2 + 11w + 28 = w^2 + 2w + 91$$
$$9w = 63$$
$w = 7$ meters originally in width
$w + 2 = 9$ meters originally in length
12. (a) 6 and 10, or -9 and -5
13. (a) 6 feet

Nostalgia Problems 12-1

1. (a) 12 horses

Calculator Activities 12-1

1. (a) $3750

Exercises 12-2

1. (a) $x^2 = 49$
$\sqrt{x^2} = \pm\sqrt{49}$
$x = \pm 7$

✓ $(\pm 7)^2 = 49$
$49 = 49$

2. (a) ± 11
3. (a) $\pm\frac{\sqrt{3}}{2}$
4. (a) $\pm\frac{13}{6}$
5. (a) ± 2
6. (a) $\frac{1 \pm 3\sqrt{2}}{3}$
7. (a) $\pm\sqrt{6}$
8. (a) $2 \pm 4\sqrt{3}$
9. (a) ± 5
10. (a) $\pm\frac{\sqrt{10}}{10}$

Calculator Activities 12-2

1. (a) ± 1.72 approx.
2. (a) $0.825, -1.225$ approx.
3. (a) $850, 576$

Exercises 12-3

1. (a) $x^2 + 18x$
$\frac{18}{2} = 9, 9^2 = 81$
$x^2 + 18x + 81$

2. (a) $a^2 - 10a + 25$

ANSWERS TO "(a)" EXERCISES

3. (a) $x^2 - x + \frac{1}{4}$
5. (a) $b^2 + 9b + \frac{81}{4}$
7. (a) $y^2 - 17y + \frac{289}{4}$
9. (a) $a^2 + 19a + \frac{361}{4}$
11. (a) $m^2 - \frac{3}{2}m + \frac{9}{16}$
13. (a) $x^2 - 1.2x + 0.36$

4. (a) $y^2 + 3y + \frac{9}{4}$
6. (a) $x^2 - 3x + \frac{9}{4}$
8. (a) $y^2 - 7y + \frac{49}{4}$
10. (a) $x^2 - 98x + 2401$
12. (a) $b^2 + \frac{1}{3}b + \frac{1}{36}$
14. (a) $m^2 + 0.6m + 0.09$

Exercises 12-4

1. (a)
$$x^2 + 10x + 9 = 0$$
$$x^2 + 10x + 25 = -9 + 25$$
$$(x+5)^2 = 16$$
$$x + 5 = \pm 4$$
$$x = -5 \pm 4$$
$$x = -9, -1$$

$x = -9$
✓ $(-9)^2 + 10(-9) + 9 = 0$
$81 - 90 + 9 = 0$
$0 = 0$

$x = -1$
✓ $(-1)^2 + 10(-1) + 9 = 0$
$1 - 10 + 9 = 0$
$0 = 0$

2. (a) $9, -5$
4. (a) $\dfrac{5 \pm \sqrt{133}}{6}$
6. (a) $7 \pm \sqrt{77}$
8. (a) $\dfrac{3 \pm \sqrt{41}}{4}$
10. (a) $\dfrac{3 \pm \sqrt{5}}{4}$
12. (a) $-8 \pm 4\sqrt{2}$
14. (a) $8, \frac{2}{3}$

3. (a) $12, -14$
5. (a) $3, -\frac{1}{3}$
7. (a) $\dfrac{4 \pm \sqrt{46}}{6}$
9. (a) $\dfrac{-3 \pm \sqrt{17}}{4}$
11. (a) $10 \pm 2\sqrt{10}$
13. (a) $\dfrac{5 \pm \sqrt{7}}{6}$

Exercises 12-5

1. (a) $x^2 + 2x - 2 = 0$, $a = 1$, $b = 2$, $c = -2$
$$x = \frac{-2 \pm \sqrt{2^2 - 4(1)(-2)}}{2(1)}$$
$$x = \frac{-2 \pm \sqrt{4+8}}{2} = \frac{-2 \pm \sqrt{12}}{2} = \frac{-2 \pm 2\sqrt{3}}{2}$$
$$x = -1 \pm \sqrt{3}$$

2. (a) $3 \pm 2\sqrt{3}$
4. (a) $-\frac{1}{2}, -\frac{5}{3}$
6. (a) $\dfrac{3 \pm \sqrt{33}}{4}$
8. (a) $\frac{3}{2}, 1$
10. (a) No real solution
12. (a) $4 \pm \sqrt{15}$

3. (a) $\dfrac{2 \pm \sqrt{2}}{2}$
5. (a) No real solution
7. (a) No real solution
9. (a) $\frac{5}{2}, -1$
11. (a) $\dfrac{2 \pm \sqrt{14}}{5}$

Calculator Activities 12-5

1. (a) 0.171, −1.671
2. (a) No real solution
3. (a) 6.9 seconds, 2.7 seconds

Exercises 12-6

1. (a)

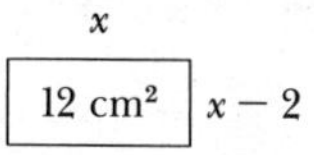

$A = lw$, where $l = x$, $w = x - 2$

$x(x-2) = 12$

$x^2 - 2x - 12 = 0$, $a = 1$, $b = -2$, $c = -12$

$$x = \frac{-(-2) \pm \sqrt{(-2)^2 - 4(1)(-12)}}{2(1)}$$

$$x = \frac{2 \pm \sqrt{4 + 48}}{2} = \frac{2 \pm \sqrt{52}}{2}$$

$x = 4.61$, or -2.61 (approx.)

We cannot have a negative length, therefore, length = 4.61 centimeters, width = 2.61 centimeters

2. (a) $2\frac{1}{3}$ or -3
3. (a) 2.68 centimeters
4. (a) 1.7 km/hr
5. (a) 12 boxes
6. (a) 14 shares total (some partial shares)
7. (a) 4 or −5
8. (a) 4m by 2m

Nostalgia Problems 12-6

1. (a) 7 persons
2. (a) 8 and 6
3. (a) 256 sq. yds.
4. (a) A relieved 120 persons, B relieved 80 persons

Calculator Activities 12-6

1. (a) 0.81 second
2. (a) 6.156 thousands or 0.244 thousands
3. (a) 13.6 kilometers

Exercises 12-7

1. (a)

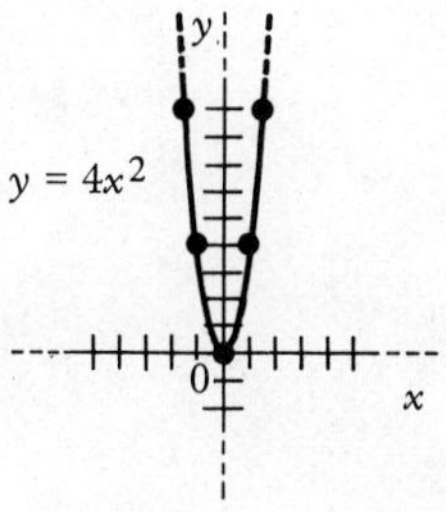

x	y
0	0
±1	4
±2	1
±2	16
±.5	1
±1.5	9

2. (a)

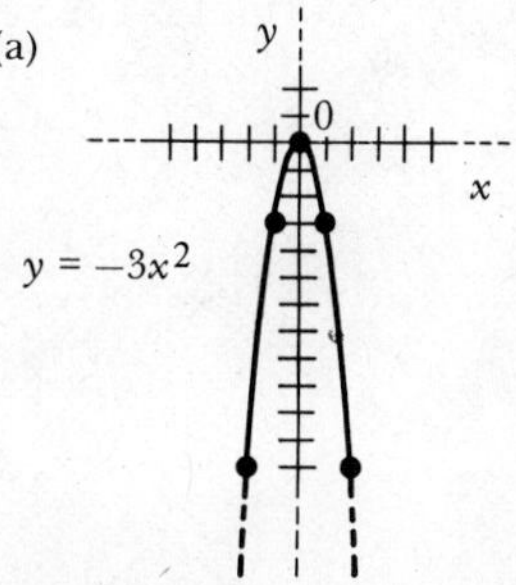

x	y
0	0
±.5	−1.5
±1	−3
±1.5	6.75
±2	−12

488 ANSWERS TO "(a)" EXERCISES

3. (a)

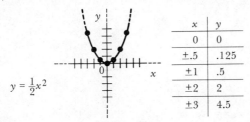

$y = \frac{1}{2}x^2$

x	y
0	0
±.5	.125
±1	.5
±2	2
±3	4.5

4. (a)

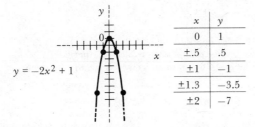

$y = -2x^2 + 1$

x	y
0	1
±.5	.5
±1	−1
±1.3	−3.5
±2	−7

5. (a)

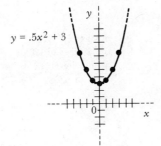

$y = .5x^2 + 3$

x	y
0	0
±1	3.5
±2	5
±3	7.5

6. (a)

$y = 2(x - 1)^2 - 1$

x	y
0	1
.5	−.5
1	−1
1.5	−.5
2	1
−1	7
3	7

7. (a)

$y = -3(x + 2)^2 + 2$

x	y
−3	1
−2.5	1.25
−2	2
−1.5	1.25
−1	−1

8. (a)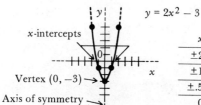

x	y
-2	5
-1	0
$-.5$	-1
0	-1
$.5$	0
1	2
2	9

9. (a)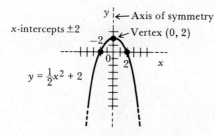

x	y
± 2	5
± 1	-1
$\pm .5$	-2.5
0	-3

10. (a)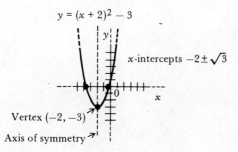

11. (a)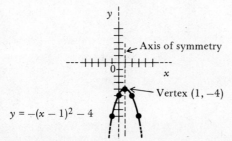

12. (a)

490 ANSWERS TO "(a)" EXERCISES

13. (a)

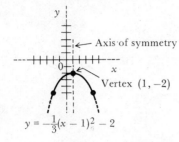

14. (a)

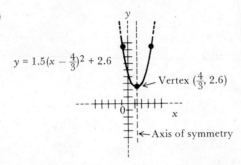

15. (a)

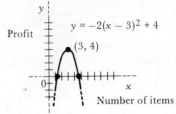

(b) x-intercepts $3 \pm \sqrt{2}$ are break-even points

16. (a)

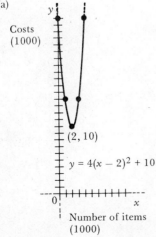

17. (a)

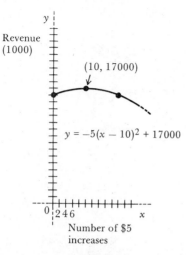

Review Exercises Chapter 12

1. (a) $6, -2$

2. (a) $-\frac{1}{2}, 3$

3. (a) $2, -8$

4. (a) $\pm \dfrac{\sqrt{35}}{7}$

5. (a) $x^2 + 7x + \dfrac{49}{4}$

6. (a) $m^2 - \dfrac{15}{2}m + \dfrac{225}{16}$

7. (a) $-4 \pm \sqrt{17}$

8. (a) $\dfrac{-1 \pm \sqrt{37}}{4}$

9. (a) $\dfrac{1 \pm \sqrt{101}}{10}$

10. (a) $\dfrac{5 \pm \sqrt{13}}{6}$

11. (a) 30.2 km/hr

12. (a)

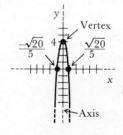

13. (a)

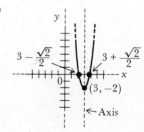

14. (a)

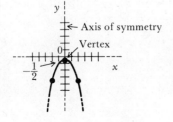

15. (a)

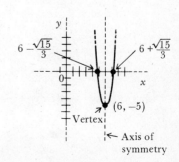

16. (a)

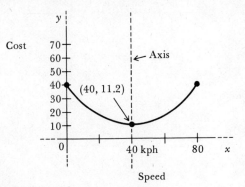

Nostalgia Problems Chapter 12 Review

1. (a) 9 mph and 6 mph

Calculator Activities Chapter 12 Review

1. (a) −0.260 thousand, or down $260
2. (a) ±1.767
3. (a) 2.951, −0.604

Chapter-Opening Photo Problems

1. 252 m^2; 3.9 or 4 cans
2. $1750.33
3. 4.065 hours; or 4 hours, 4 minutes
4. Number of cartons ≤ 300
6. 0.75 liter
7. Let x = number of magazine ads, y = number of TV commercials $\{x \geq 3 \cap y \geq 5 \cap 15x + 45y \leq 300\}$
9. 138.$\overline{8}$ meters
10. 20.6 hours
11. 15.8 kilometers

Index

A

Absolute value, 51
Addition:
   method of solving equations, 199
   property of equations, 83
   property of inequalities, 121
   of real numbers, 50
Additive:
   property of equality, 23
   property of order, 48
Additive inverse, 25
Algebraic expression, 381
Area:
   of a circle, 17
   of a rectangle, 17
   of a trapezoid, 18
   of a triangle, 19
Associative property:
   additive, 24
   multiplicative, 24
Axis:
   of symmetry of a parabola, 418
   x-axis, 143
   y-axis, 143

B

Base of exponents, 62
Binomial, 248

C

Celsius, 17
Circle:
   area of, 17
   circumference of, 17
Closure property, 24
   additive, 24
   multiplicative, 24
Coefficient, 247
Coincident lines, 165, 190
Commutative property, 24
   additive, 24
   multiplicative, 24
Completing the square, 402
Complex fraction, 332
Composite number, 32
Constant term of a polynomial, 246
Coordinate(s), 144
   plane, 143
   of a point, 144
   x-coordinate, 144
   y-coordinate, 144
Coordinate system, 143
Cube, 62

D

Decimal:
   infinite nonrepeating, 362
   infinite repeating, 361
Degree of a polynomial, 248
Difference of two squares, 271, 290
Discriminant, 408
Distributive property:
   additive, 25
   multiplicative, 25

E

Edge of a half-plane, 224
Empty set, 5
Equation(s), 13
   graphing of, 169
   of a line, 165
   nonlinear, 176
   in one variable, 78
   systems, 188
   in two variables, 137

Equivalent, 82
Exponent, 62
   laws of, 258, 260, 279, 336
   zero, 64
Exponential curves, 178
Extraneous solution, 385

F

Factor, 27
   of integer, 32
   of monomial, 287
Factoring:
   by grouping, 294
   monomial, 287
Fahrenheit, 17
Finite decimal, 361
Flowchart, 88
Foil method, 268
Formulas, 14, 110
   from geometry, 36

G

Graph, 145
   of an equation, 169
   of an inequality, 223
   of a (pair of) number(s), 44
Greater than, 44, 48
Greatest common factor (GCF), 288
Greatest monomial factor, 288
Grid, 142
   plot, 142
   points, 142

H

Half-plane, 224
Horizontal method:
   of adding two polynomials, 254
   of multiplying two polynomials, 264

I

I_r, I, 7
Identity element:
   additive, 25
   multiplicative, 25
Identity property, 25
Index, 364

Inequality, 117
   graph of, 223
Infinite decimals, 361
Integers, 2
Intercepts, 156
   x- and y-, 422
Intersection, 5
   of two lines, 189
Inverse element:
   additive, 25
   multiplicative, 25
Inversive property, 25
Irrational numbers, 3

L

Laws of exponents:
   first, 258
   second, 260
   third, 260
   fourth, 279
   fifth, 336
Laws of radicals:
   first, 367
   second, 367
Least common denominator (LCD), 325, 328
Least common multiple (LCM), 33
Left side of equation, 13
Left side of inequality, 117
Less than, 44, 48
Like monomials, 248
Linear equations in two variables, 151
Linear inequalities in two variables, 219
Loop of a flowchart, 88

M

Mathematical expression, 9
Monomial:
   factors, 287
   like, 248
   product, 258
Multiple, 32, 287
Multiplication of real numbers, 50
Multiplication property:
   for equations, 83
   for inequalities, 124

Multiplicative:
  identity, 25
  inverse, 25
  property of equality, 23
  property of order, 48

N

Natural number, 49
Negative, 25
  exponent, 64
  number, 49
  slope, 164
Nonlinear equations, 176
Null set, 5
Number line, 44
Numerical expression, 7

O

Open ray, 128
Operations, 3
Order, 43
  additive property, 48
  multiplicative property, 48
  transitive property, 48
  trichotomy property, 48
Ordered pairs, 138
Origin, 45, 143

P

Parabola, 177
  axis of, 418
  vertex of, 418
Parallel lines, 65, 190
Perfect square, 292
  integers, 69
Plot on a grid, 142
Point of intersection, 189
Polynomial(s), 246
  difference, 254
  product, 262
  quadratic, 304
  standard form, 248
  sum, 253
  term, 246
Positive, 49
  slope, 164

Power, 62
Prime number, 32
Products:
  of monomials, 258
  of polynomials, 262
Properties of addition and
        multiplication, 24, 25
Properties of equality, 23
Pythagorean theorem, 68, 387

Q

Q, 7
Quadrants, 144
Quadratic:
  equation, 305, 395
  equation in two variables, 417
  formula, 407
  polynomial, 304

R

R, 7
Radical:
  equation, 385
  sign, 67, 363
Radicand, 364
Rational:
  equation, 345
  expression, 282, 315, 324
  number, 2
Rationalizing the denominator, 378
Ray, 128
  open, 128
Real numbers, 3
  operations, 50
Reciprocal, 25
Rectangle, 17
Reflexive property of equality, 23
Right side of equation, 13
Right side of inequality, 117

S

Scientific notation, 63
Set(s), 4
  element of, 4
  empty, 5
  intersection, 5

Set(s) (cont.):
  null, 5
  solution, 119
  subset, 6
  union, 5
Simplest form, 316
Simplified form, 368, 369
Slope:
  of a line, 161, 162
  of a horizontal line, 166
  of a vertical line, 166
Slope-intercept form, 169
Solution:
  of equation, 78
  set, 119
Solving an equation, 82
  addition method, 199
  substitution method, 193
Sphere:
  surface area, 18
  volume, 18
Square, 62
  root, 67
Standard form of a polynomial, 248
Subset, 6
Substitution method of solving
    equations, 193
Subtraction, 52
  of rational expressions, 324
  of real numbers, 50
Sum of polynomials, 253
Symmetric property, 23
System of equations, 188

T

Term(s), 27
  of a polynomial, 246

Transitive property:
  of equality, 23
  of order, 48
Trapezoid, area, 18
Trichotomy property, 48
Trinomial, 248

U

Union, 5

V

Value:
  absolute, 51
  of a variable, 7
Variable, 7
Vertex of a parabola, 418
Vertical method:
  of adding two polynomials, 254
  of multiplying two polynomials, 264

X

x-axis, 143
x-coordinates, 144

Y

y-axis, 143
y-coordinate, 144

Z

Zero exponent, 64